NATURAL HAZARDS

NATURAL HAZARDS

Editor

Theophil Brodbeck

scitus
academics

Natural Hazards

Edited by **Theophil Brodbeck**

This Book is Printed on Acid-Free Paper. ♾

ISBN: 9781681179018

Library of Congress Control Number: 2017948929

SCITUS Academics LLC,

616, Corporate Way, Suite 2, 4766,

Valley Cottage, NY 10989

www.scitusacademics.com

Notice

Printed in United States of America.

Preface

A natural hazard is an extreme event that occurs naturally and causes harm to humans – or to other things that we care about, though usually the focus is on humans. Natural disasters happen all over the world and they can be utterly devastating for people's lives and the environments in which they live. Although natural disasters are caused by nature and there is nothing that we can do to prevent them happening, there are many different natural causes that lead to natural disasters, and being aware of these causes enables us to be better prepared when such disasters do arrive. Because hazards are threats of harm mainly to human systems, human activities play a large role in how severe a hazard is. For example, when large numbers of people crowd into floodplains and low-lying areas, they are putting themselves in harm's way, increasing the severity of potential floods. Earthquakes are another common natural disaster that can cause many fatalities. The movements of the plates in the earth's crust cause them. Tsunamis can also be caused by underwater volcanic eruptions. Volcanic eruptions are another natural disaster, and they are caused by magma escaping from inside the earth. Although hazards can originate in different sources and systems, such as atmospheric, hydrologic, oceanographic, volcanologic, seismic, neotectonic, the environmental impacts are equally catastrophic. The current phases of physical exposure of human beings and economic assets have been comparatively designed by historical patterns of settlements. Beneficial climatic and soil conditions that have spurred economic activities are in many cases also associated to hazard-prone landscapes. Both volcanic slopes and flood plains areas have historically attracted human habitation. Where settlement patterns have contributed to configure risk scenarios, new forces, such as population growth and increased rural/urban migration, act as dynamic pressures contributing to changing patterns in increasing people's exposure to hazards.

This volume' Natural Hazards' addresses the needs of various users, interested in a better understanding of hazards and their more efficient management. It deals with approaches to managing disaster - assessing hazards, emergencies and disaster impacts demonstrates the array of information that is critical for improving disaster management. The book reflects major management components of the disaster continuum (the nature of risk, hazard, vulnerability, planning, response and adaptation) in the context of threats that derive from nature. It covers updated information on a variety of natural hazards potentially deriving into disasters, i.e. tropical storms, avalanches, coastal floods. Hopefully, the book will be valuable to those who invest their efforts in building communities robust to natural disasters.

Table of Contents

CHAPTER 1

Lightning Occurrence and Social Vulnerability

Ronald L. Holle[1] and Mary Ann Cooper[2]

ABSTRACT

The occurrence of lightning in time and space around the world is well known. Lightning fatalities and injuries are well delineated in the United States; however, there is much less information about lightning impacts on people in the developing world. It is estimated that between 6000 and 24,000 people are killed globally per year, and 10 times as many are injured. The fatality rate per capita has become very low in the developed countries during the past century due to the availability of lightning-safe structures and vehicles, less labor-intensive agriculture, and other factors, but this reduction has not occurred where people continue to work and live in lightning-unsafe situations. Lightning safety advice often mistakenly expects that the direct strike is most common, but ground current, direct contact, side flash, and upward streamers are much more frequent mechanisms. In developed countries, the injury:death ratio is approximately 10:1, meaning that 90% survive but may have permanent disabling injuries. The proximate cause of death is cardiac arrest and anoxic brain injury at the time of the lightning strike, and, at this time, the damage from a lightning strike cannot be reversed or decreased in survivors. Lightning vulnerability in many developing countries continues to be a major issue due to widespread exposure during labor-intensive agriculture during the day when thunderstorms are the most frequent and while occupying lightning-unsafe dwellings at night.

Keywords: lightning, lightning strikes, lightning fatalities, lightning safety, lightning occurrence

1. INTRODUCTION

More lightning occurs in clouds than that strikes the ground; and as the cloud moves above the earth, opposite charges are induced on the surface of the earth and on objects on the ground under the cloud. Upward streamers, not usually visible from these objects, will reach up and attempt to connect with the downward-moving lightning channel. A cloud-to-ground flash has one or more return strokes. When you see lightning flickering, those are return strokes within the same flash. There are about four strokes per flash, when averaged over a large sample size. The first stroke comes faintly to ground and contacts the surface of the earth. The stroke may attach to a tree, open land, the ocean, or other objects. Then the light fills the channel going upward. The next stroke in the flash will likely come down from the cloud to the ground in the same channel as the first stroke, and so on. Occasionally, one of the subsequent strokes will stray from the pre-existing channel and come to ground one or two kilometers away.

One of the important features of this well-known mechanism of lightning strikes is the search radius. The cloud-to-ground channel makes its way downward in step leaders of about 50-m lengths (**Figure 1**). At the lower tip of each step, the channel searches for a feature that makes a convenient connection to ground. Higher up in the cloud, there is nothing to strike, so the step leader keeps coming toward the ground as each step attempts to reach out and connect to the ground. Branching occurs that is generally downward. Only when the lowest tip of the channel is 30–50 m from the surface of the earth does it 'decide' what to strike.

This last step leader, not usually visible from a distance, is nearly vertical and connects one of the upward streamers emanating from objects such as trees, poles, and sometimes open water. The most likely connections will be made to objects that are tall, isolated, and pointed. But if that object is more than 30–50 m away from the lowest tip of the downward leader, the lightning channel may come all the way to ground, close to a tall building or tower.

Figure 1. Cloud-to-ground lightning photographed from Oro Valley, Arizona, USA (©R. Holle).

2. LIGHTNING OCCURRENCE

The location, time of year and day, and number of lightning occurrences throughout the world are known quite well. **Figure 2** shows the latest 10-year map of the United States cloud-to-ground flash density per square kilometer. With respect to time of day, about two-thirds of lightning occurs during the afternoon from noon to 1800 local time. Cloud-to-ground lightning is the most frequent along the coasts of Florida and the Gulf of Mexico, and generally decreases to the north and west as the number of days with substantial low-level moisture decreases in frequency. Large variations in lightning frequency also occur in the western states where large elevation changes affect thunderstorm formation.

Figure 3 shows the global occurrence of lightning strokes. The highest densities are over tropical and subtropical coastlines, near large elevation changes, and east coasts at middle latitudes. The lowest lightning frequency is over oceans, the polar regions, and west coasts at middle latitudes. The dominance of land lightning is due to daytime heating that can produce updrafts rising to altitudes where temperatures are colder than freezing, the altitudes where lightning initiates. Since much of outdoor human activity occurs during daytime, the juxtaposition of lightning with people can be expected to result in fatalities and injuries.

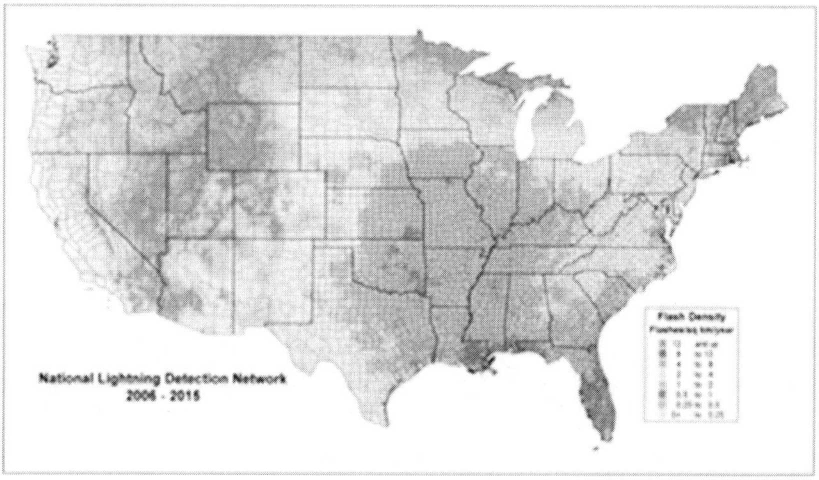

Figure 2. Cloud-to-ground flash density per square kilometer per year over the contiguous United States from the National Lightning Detection Network from 2006 through 2015 (Courtesy Vaisala, Inc.).

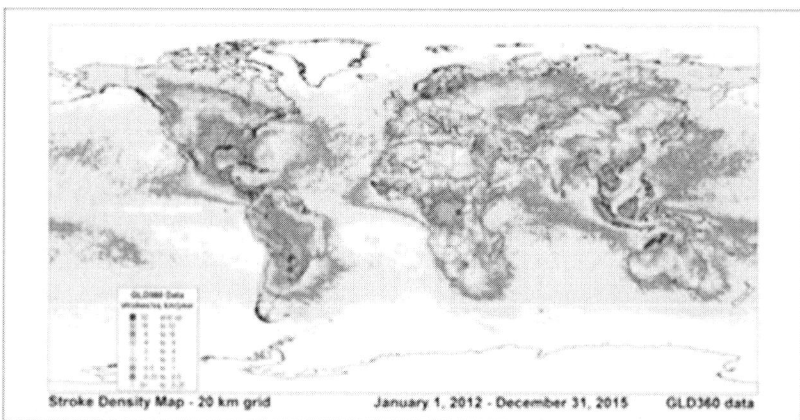

Figure 3. Lightning stroke density per square kilometer per year over the world from the Global Lightning Dataset GLD360 from 2012 through 2015 (Courtesy Vaisala, Inc.).

Nearly everywhere in the world has a season with more lightning than others. In the middle latitudes about two-thirds of lightning occurs during meteorological summer (June, July, and August in the Northern Hemisphere and December, January, and February in the Southern Hemisphere). In the Tropics, the passage of the equatorial trough or the summer monsoon strongly affects the frequency of lightning. The equatorial trough (also known as the Intertropical Convergence Zone, or ITCZ) is a somewhat continuous east-west area of rain and thunderstorms that stays within about 20 degrees latitude of the equator, shifting northward (southward) during the Northern (Southern) Hemisphere summer. For locations near the equator, the equatorial trough crosses locations twice a year which can result in two rainy seasons whereas more subtropical locations have only a single rainy season. The Asian Monsoon can be considered to be a variation of the equatorial trough that is strongly affected by the large land mass of Asia and the Himalayas. The equatorial trough generally moves northward in the Northern Hemisphere summer and brings tropical moisture from the south, often resulting in higher thunderstorm and lightning frequencies, and then reverses to send dry air flowing from the north during the Northern Hemisphere winter when lightning occurrence is much less likely.

3. LIGHTNING FATALITIES AND INJURIES

3.1. United states fatalities and injuries

According to the U.S. National Weather Service (NWS), lightning fatalities have averaged 32 per year over the past decade. **Figure 4** shows the latest decade of available fatality data with the top panel showing the number of fatalities by state while the lower panel indicates the fatality rate per million people. The fatality rate turns out to be rather different from the actual number of fatalities, and indicates how the number of lightning flashes and the number of fatalities are related, but not as directly as may be expected. States with the highest fatality rates in the western United States appear to indicate a region with drier air and lower rainfall rates that give the perception that lightning is not as much of a threat as when it is raining heavily in other locations in the United States.

The necessary ingredients of lightning frequency and population density in the United States have been combined in a recent study that is able to replicate the primary locations of observed lightning fatalities [1]. This approach shows the concentration of fatalities in urban areas that have moderate to high lightning frequencies. Such a study has not yet been attempted in other countries where the lightning risk is very different due to such factors as the availability of lightning-safe buildings and vehicles, agricultural participation, and related societal differences.

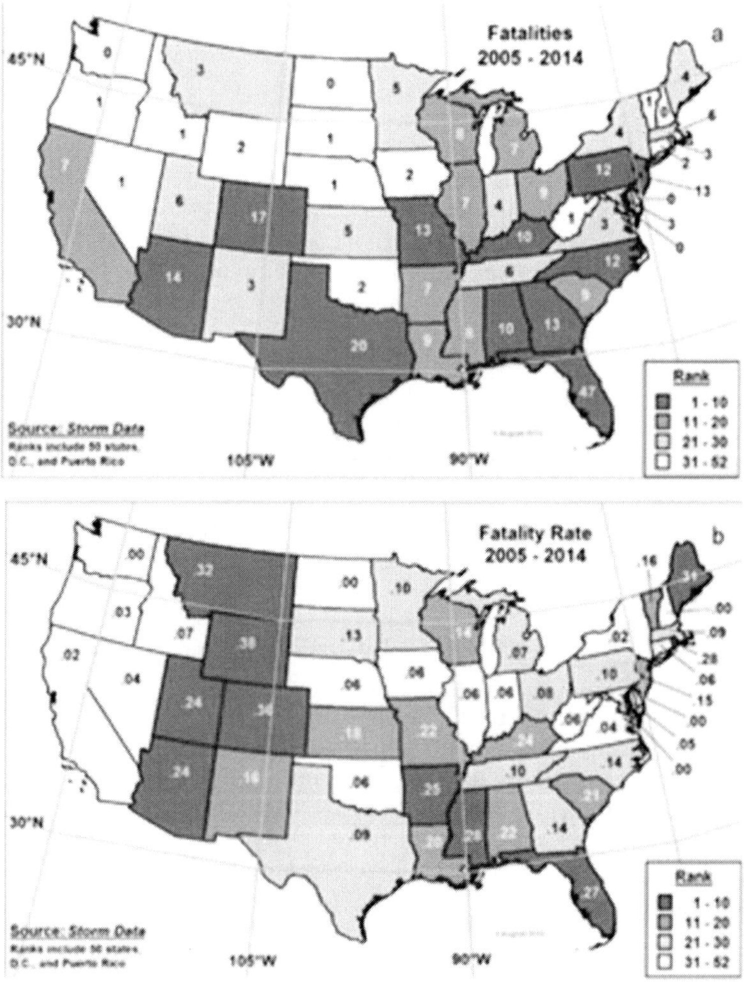

Figure 4. Ranks of lightning fatalities per state (a) and fatality rate per million people (b) from 2005 through 2014 based on the United States National Weather Service website [2].

Underreporting of lightning casualties has been a problem in past years in the United States, but much less in the past couple of decades. A primary reason for underreporting is that about 90% of all lightning fatalities and injuries are to one person at a time, which leads to the tendency for such events to be reported less often in the media than multiple-casualty events. With the inception of the Lightning Safety Awareness Team, a multidisciplinary group sponsored by the National Weather Service, it is probable that every fatality is documented [2], but an estimate is that only about 70% of the injuries reach the reporting system that is maintained by all NWS offices across the country. An additional issue is that underreporting may seem to be occurring when there is actually a definition issue. The National Weather Service in its *Storm Data* publication does not include secondary casualties due to lightning. For example, a house caught on fire at night due to lightning that results in a fatality or injury is not counted as a lightning impact, since the primary cause is coded as fire not lightning. A useful method for estimating injury underreporting is the ratio of injuries to deaths. An intensive study over three full years across the state of Colorado showed that about ten injuries occur per fatality [3]. When the ratio is less than ten, an indication exists that not enough injuries have been reported and documented. The 10:1 injury:death ratio is assumed to apply to the United States and more developed countries with their similar socio-economic infrastructures, providing widespread availability of lightning-safe buildings and vehicles.

The lightning fatality rate per million people in the United States has dropped by more than two orders of magnitude since 1900 (**Figure 5**). Similar trends have been observed in many more developed countries of the world with published lightning fatality rates during the same period [4]. Also shown is the rural percentage of the population, which decreased from 60% in 1900 to fewer than 20% at present. The United States population not only transitioned out of a mainly labor-intensive agricultural society a century ago, but also moved into more substantial home and workplace buildings with grounded wiring and plumbing, together with the ready availability of fully enclosed metal-topped vehicles, better medical care, and greatly improved meteorological information about thunderstorms. All of these factors are present in more developed countries that have lower lightning death rates.

Figure 6 graphs United States lightning fatality types one hundred years apart, comparing the 1890s (top) with the 2005–2014 period (bottom). During the late nineteenth century, being indoors was the most common situation for lightning fatalities, while agriculture and outdoors were also a large component of the events. During the decade starting in 2005, indoor fatalities have become quite rare, agriculture events have greatly reduced from a century ago as farming became mechanized, and lightning-safe buildings and vehicles became common. Recreation and sports situations are relatively more frequent now than earlier. However, the overstated scenario of golf fatalities has been exceeded substantially in recent years by hiking, climbing, boating, and other water-related activities [2].

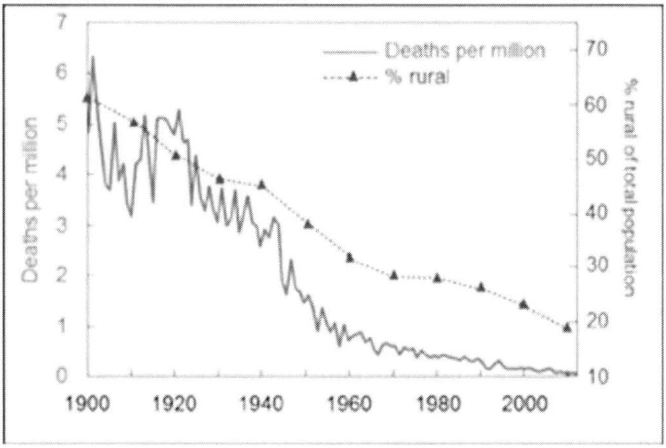

Figure 5. Solid red line: United States lightning deaths per million people from 1900 to 2013. Dashed blue line: percent rural population.

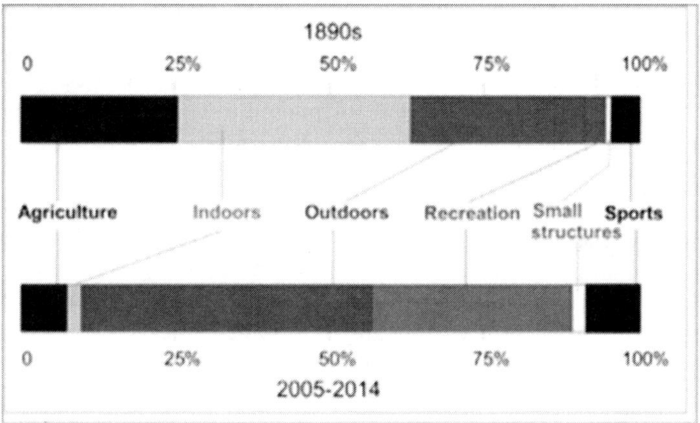

Figure 6. Comparison of the percentage of types of the United States lightning fatalities in the 1890s versus 2005 through 2014 [20].

3.2. Global fatalities and injuries

Worldwide, **Figure 7** shows information on lightning fatality rates that has been collected in a number of countries during the last quarter century; nevertheless, significant gaps exist in our knowledge of the absolute numbers [4]. Studies have estimated the number of deaths per year attributed to lightning globally anywhere from a few thousand to 24,000

[5–7]; however, much uncertainty exists due to the limited sample size (**Figure 7**). One country of particular interest is Malawi with a very high rate of 84 lightning deaths per million people per year that far exceeds the rate in any other country [4]. The 1008 annual fatalities for this small but populous country may represent a very complete data collection method that is what adjacent countries in the region should report, so very large numbers of lightning fatalities may be actually occurring but we do not know them. If the "10:1 (injury:death) rule" is appropriate for developing countries, then lightning is causing a large number of deaths and injuries worldwide, regardless of our inability to state the actual numbers.

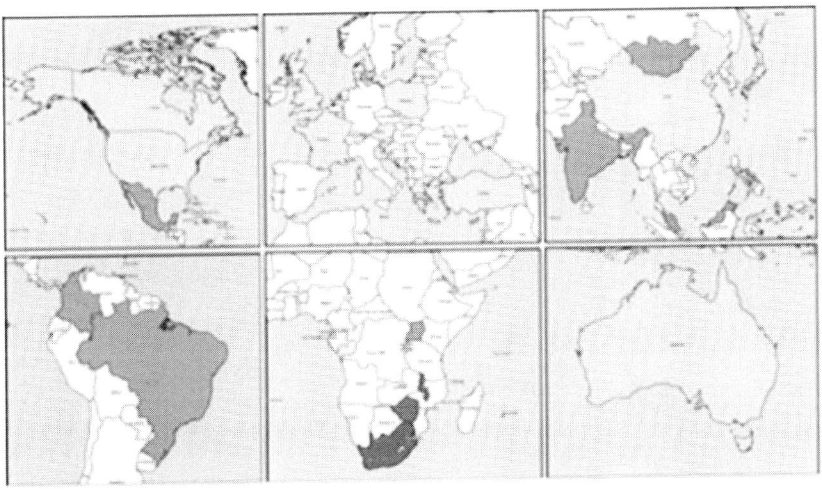

Figure 7. Lightning fatality rate per million people per year by continent. Red shading indicates rate >5.0 fatalities per million per year, orange is 0.6 to 5.0, and yellow is 0.5 or less. White indicates no national summaries have been published for datasets ending in 1979 or later (Updated with permission from Holle [4]).

In contrast with the United States, many populous less-developed countries have as much as 90% of the population living and working in lightning-unsafe locations and situations. During the day, many people are involved in labor-intensive agriculture, fishing in open boats, walking to market, or inside schools without recourse to safety in an appropriate building or vehicle [8] (**Figures 8** and **9**). At night, people live in lightning-unsafe dwellings without adequate wiring, plumbing, or metal structural components that can carry lightning into the ground without affecting individuals inside [9] (**Figure 10**).

Figure 8. Many forms of labor-intensive work are far from lightning safety such as (a) fishermen in open boats on Lake Victoria, the second largest lake in the world, and (b) fields in Nepal (courtesy ACLENet and M.A. Cooper).

In the developing world, the injury:death ratio is expected to be lower since fewer lightning-safe locations are available, resulting in a higher proportion of deaths. In addition, more people die per lightning event in the developing world than in the United States, particularly in agriculture and school events. For developing countries, counting both "primary" and "secondary" causes may be preferable until fatality rates decrease and the availability of safe locations increases.

Figure 9. Most forms of transportation or going to market in developing countries are not lightning safe such as (a) taxi in India, (b) ox cart in India, and (c) bota-bota taxi in Nepal (©M.A. Cooper).

Figure 10. An estimated 90% of sub-Saharan buildings and housing are not lightning-safe, such as (a) farming settlement with thatch and sheet metal roofs on mud brick walls in Zambia, (b,c) combination shops and homes in India (©M.A. Cooper).

In more developed countries, the dominant profile of lightning casualties is the young male. Risk-taking in recreation, workplaces, and organized sports tends to be dominated by this group of males between about 15 and 30 years old. However, in lesser developed countries, the distribution is substantially more equal between female and male, and the ages are much more disperse, both because of more widespread exposure. A recent study of labor-intensive agriculture in mainly India and Bangladesh shows that 47% of the fatalities and injuries were females as they work during the daytime when thunderstorms are most frequent [8]. The lack of lightning-safe dwellings, schools, and workplaces means that all ages and both genders are equally vulnerable at all times.

3.3. Damage to property and indirect impacts

In developed countries, precautions against the effects of lightning to property, electronics, and utility lines are usually routine and subject to building codes using well-accepted practices for public buildings such as churches, hospitals, and schools. In developing countries, the effects of lightning damage to property can have not only direct effects on the structure but also indirect economic effects. The impacts include food spoilage from lack of refrigeration after electrical failure, electrical parts and repairs are unaffordable or not available for days, hospitals are without power, and databases and expensive, irreplaceable electronics are damaged. These adverse effects can occur in countries that are already struggling with other pressing issues such as drought, HIV, underemployment, or civil strife. Individual families can suffer not only when one or more of their members is injured or killed by lightning but also when their livestock, often the major measure of wealth, are killed en masse by lightning (**Figure 11**).

Figure 11. Cattle killed by a cloud-to-ground lightning strike in South Africa (©I. Jandrell).

4. LIGHTNING INJURY MECHANISMS AND SAFETY

A widespread misconception exists that the direct strike is the most common lightning injury mechanism. A direct result of this misconception is the mistaken development of the most lightning safety avoidance rules that address the stroke coming straight down and striking a person in the head. This is the least common mechanism. Instead, there are five primary mechanisms of lightning injuries (**Figure 12**).

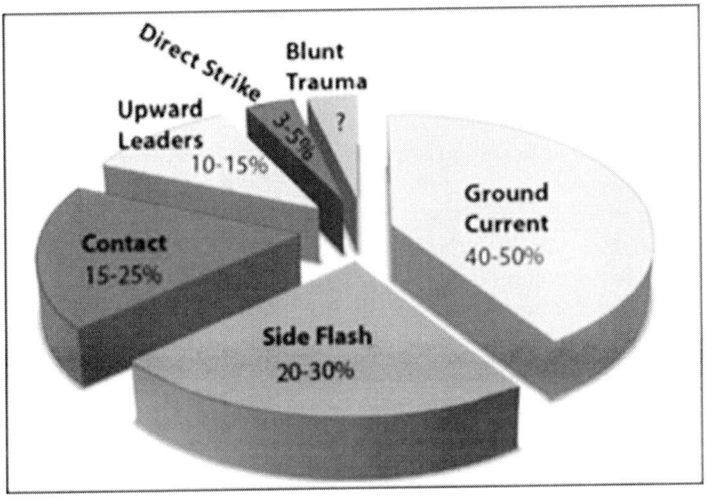

Figure 12. Mechanisms of lightning injury and death.

The five mechanisms lead to a very different conclusion regarding lightning avoidance than the direct strike—everywhere outside is unsafe from lightning. Lowering oneself in height is not sufficient. In the order of highest frequency:

- Ground current: The most common mechanism appears to be ground current, where lightning strikes the surface of the earth and spreads to nearby people and can affect a large number of people.
- Side flash: This occurs when trees, poles, towers, and many other objects that are not necessarily tall are struck and a portion of the lightning's effects jumps to a nearby person.
- Contact: Being in contact with conducting paths such as plumbing or wiring either outdoors or inside structures can be dangerous when they are struck at a distance by a cloud-to-ground lightning flash.
- Upward leader: Occurs when an upward leader is induced from a person and rises to meet the downward-traveling stepped leader from cloud to the ground. The upward streamer is strong enough to cause

injury even when the lightning channel is not completed through the person.

- Direct strike: The least common mechanism.
- Blunt trauma: Occurs with or separately from all of these mechanisms as a person is thrown or when too near where the lightning strikes [10].

Lightning safety myths abound. One cannot anticipate with certainty where lightning will strike the ground or what it will strike. At best, lightning can only be described statistically as more likely to hit certain types of objects: tall, isolated, and/or pointed. Safety messages that stress avoiding standing in a certain way, not holding specific objects, being a certain distance from tall objects, and variations on these concepts are not reliable. Safety messages that emphasize what is on the feet are irrelevant; after lightning rips apart several kilometers of air on its journey from cloud to ground, a thin rubber shoe is overwhelmed and immaterial. Part of the perpetuation of many false myths is due to the fact that around 90% of lightning casualties survive. What may seem to be a factor in an individual case, and go on to assume mythological quality, usually does not generalize to a population.

It is estimated that around 10% of lightning casualties are related to trees [11]. Perhaps, a third of all cloud-to-ground lightning flashes around the globe attach to trees. Once lightning strikes a tree, the current comes down the trunk and spreads horizontally (ground current). It also produces side flashes to people or animals who are close by, such as those seeking to stay dry under the canopy (**Figure 11**) and are also close to the trunk. Some people may suffer contact injury if they are touching the tree at the moment lightning strikes it. Finally, blunt and penetrating trauma can occur when bark and tree limbs explode outward at a high speed up to tens of meters away.

Only two reliable safe locations exist from lightning. One reliably safe location is inside a substantial well-constructed building with wiring, plumbing, and perhaps metal structural members. Such buildings where people work or live are able to provide a path for a lightning strike into the ground without causing harm. The other reliably safe location is inside a fully enclosed metal-topped vehicle. The effects of such buildings and vehicles are similar to that of a Faraday cage where the current flows outside of people within the structure or vehicle. Direct strikes to such buildings and vehicles can be frightening and sometimes have disconcerting impacts. However, such property damage is massively preferred to people being outside of such locations.

Unsafe structures include anywhere with the word shelter attached—beach, sun, shade, rain, or bus shelter. While they can be made safe, most people cannot tell if that is the case (**Figure 13**). One should always assume they are not lightning-safe because they will likely not surround a person inside with a certain path for lightning to follow. Similarly, any other structure is unsafe when made of mud, brick, or thatch without specifically designed metal-conducting paths for the current to follow.

Figure 13. Lightning-unsafe small structure. Note sign recommending a safe place elsewhere rather than staying at this location (©R. Holle).

Unsafe vehicles include motorcycles, convertibles, golf carts, tuk-tuks, bota-botas, four-wheelers, and similar vehicles. A summary of motorcycle lightning events, often resulting in deaths, is in [12]. A common misconception refers to rubber tires being of relevance. When lightning strikes a fully enclosed metal-topped vehicle, the current flows through the metal structure around the people inside, then exits through the ground. The tires are the shortest path to ground, so they may explode or flatten. Tires are damaged as an effect of the lightning strike, but they did not protect people inside, instead safety is provided by the metal structure surrounding the person inside.

5. LIGHTNING INJURY

5.1. Effects of lightning on people

A myriad of injuries from lightning have been reported including damage to the ears, eyes, skin, heart, and brain [13]. The proximate cause of death is cardiac arrest and anoxic brain injury at the time of the strike, even if resuscitation delays the legal pronouncement for a few days.

Most people assume that lightning causes a significant burn injury but, in developed countries, burns tend to be superficial and insignificant and lightning causes more neurological injury and blunt trauma. At the time of the strike, injured persons often suffer keraunoparalysis, a paralytic state lasting minutes to hours with loss of sensation affecting the lower limbs more than the upper limbs. In developed countries, keraunoparalysis usually resolves without treatment, although some may have permanent weakness. In developing countries, where mud brick, thatched roofs, and other insubstantial buildings are the norm, keraunoparalysis may prevent even the most robust person from escaping as burning thatch falls on them, resulting in reports by journalists of "charred bodies" [14].

Lightning-injured persons may suffer temporary or permanent neurological problems including chronic pain syndromes and cognitive damage similar to those reported in post-concussive syndrome with inability to multitask, attention-deficit, memory problems, learning difficulty, irritability, and inability to return to their previous level of employment [15, 16]. Disability may significantly affect a family's socioeconomic status if the survivor is unable to return to work or needs chronic care. A further setback to the victim's family, particularly in developing countries such as in Africa, is a common belief that a family affected by lightning injury has been "cursed". This may force the family to leave their community, home, and employment to start over in a new community where their tragedy is unknown [17].

5.2. Treatment of lightning injury

While the effects of lightning injury can be treated, currently there is no way to reverse or decrease the damage that is set in motion when the strike occurs. Lightning Strike and Electric Shock Survivors International (LSESSI) is a support group that has helped hundreds of survivors and their families [16]. Treatment is standard for pain syndromes, anoxic brain injury, and cognitive disability. Unfortunately, this type of care is expensive and seldom available in developing countries.

As in most injuries and illnesses, prevention is far better than caring for those injured by lightning and, in developed countries, lightning injury prevention is simple and cost-effective [18]. However, decreasing lightning injuries in developing countries is a much more complex task than in countries where lightning-safe structures and vehicles are common and close by. In cases where lightning affected pupils and staff at unsubstantial schools in developing nations, a review of over 100 events in the past decade found 200 deaths and 700 injuries. These events occurred most often in primary and high schools with many situations involving dozens of children per event. Partially in response to such events, the African Centre for Lightning and Electromagnetics Network (www.ACLENet.org) was established in 2013. The confluence of lightning frequency, personal vulnerability at work, school, and home in less developed countries in Africa, as well as in Southeast Asia [19], makes this a timely endeavor.

6. CONCLUSIONS

The location, time, and frequency of cloud-to-ground lightning around the globe have become quite well known. About two-thirds of cloud-to-ground lightning occurs between noon and 1800 local time, and about two-thirds occur in summer in the middle latitudes. Lightning is more frequent along coastlines and near large topographic features in the Tropics and subtropics. However, due to socio-economic factors, the number of fatalities and injuries is not as closely related to lightning frequency as might be expected. In the United States and other developed countries, the fatality rate per capita is more than two orders of magnitude lower than a century ago. This rate decrease is attributable to a shift from rural to urban settings where fewer people are involved in labor-intensive agriculture. Many lightning injuries and deaths in the developed world are from leisure activities, often related to various types of activities in the vicinity of water bodies [2]. At least as important has been improvements to the quality of buildings that are usually safe from lightning due to grounding according to codes that provide safety to people inside. In addition, the widespread availability of fully enclosed metal-topped vehicles provides mobile lightning-safe locations almost everywhere. Additional factors include better medical treatment and improved understanding of thunderstorms and their associated lightning threat.

In contrast, many developing countries continue to have very high per-capita fatality and injury rates. A large portion of their population is involved in labor-intensive agriculture during the daytime when thunderstorms are more common. No lightning-safe locations are typically available to these people, often nearly equally male and female, while working in the fields. In addition, dwellings occupied outside of working hours often are not lightning-safe due to their construction of mud brick and thatch or sheet metal roofing. As a result, an estimate is that as many as 24,000 deaths and 240,000 injuries occur from lightning globally every year [7], almost entirely in the less-developed countries of the world. These figures remain elusive, due to sporadic data gathering in the most lightning-vulnerable locations where data collection will be slow to improve in the near future such as equatorial countries of Africa and Asia. Injury prevention is far better than taking care of people after they are injured; both scenarios are relatively easier in developed countries. Unfortunately, the infrastructure and housing in developing countries precludes easy answers like "When Thunder Roars, Go Indoors", and lightning remains a substantial threat to entire villages, schools, and populations.

REFERENCE

1. Roeder WP, Cummins BH, Cummins KL, Holle RL, Ashley WS. Lightning fatality risk map of the contiguous United States. Natural Hazards, 2015;79:1681-1692. DOI 10.1007/s11069-014-1254-9.
2. National Oceanic and Atmospheric Administration, National Weather Service, www.lightningsafety.noaa.gov, Victims/Medical, Analysis of Recent Lighting Fatalities.
3. Cherington M, Walker J, Boyson M, Glancy R, Hedegaard H, Clark S. Closing the gap on the actual numbers of lightning casualties and deaths. In: Preprints of the 11th Conference on Applied Climatology, 10–14 January 1999; Dallas, Texas, American Meteorological Society; 1999. pp. 379-380.
4. Holle RL. The number of documented global lightning fatalities. In: Preprints of the 6th International Lightning Meteorology Conference, 18-21 April 2016; San Diego, California, Vaisala; 2016, 4 pp.
5. Gomes R, Ab Kadir MZA. A theoretical approach to estimate the annual lightning hazards on human beings. Atmospheric Research, 2011;101:719-725. doi:10.1016/j. atmosres. 2011.04.020.
6. Cardoso I, Pinto Jr. O, Pinto IRCA, Holle RL. A new approach to estimate the annual number of global lightning fatalities. In: Preprints of the 14th International Conference on Atmospheric Electricity (ICAE), 8–11 August 2011; Rio de Janeiro, Brazil, 2011. 4 pp.
7. Holle RL, López RE. A comparison of current lightning death rates in the U.S. with other locations and times. In: Preprints of the International Conference on Lightning and Static Electricity, 16–19 September 2003; Blackpool, England, Royal Aeronautical Society, paper 103-34 KMS; 2003. 7 pp.
8. Holle RL. Lightning-caused deaths and injuries related to agriculture. In: Preprints of the 6th International Lightning Meteorology Conference, 18–21 April 2016; San Diego, California, Vaisala; 2016.
9. Holle RL. Lightning-caused casualties in and near dwellings and other buildings. In: Preprints of the International Lightning Meteorology Conference, 21–22 April 2010; Orlando, FL, Vaisala; 2010. 19 pp.
10. Blumenthal R, West NJ. Investigating the risk of lightning's pressure blast wave. South African Journal of Science, 2015;111(3/4): 5; article 2014-0187.
11. Holle RL. Lightning-caused deaths and injuries In the vicinity of trees. In: Preprints of the 6th Conference on the Meteorological Applications of Lightning Data, 6–10 January 2013; Austin, Texas, American Meteorological Society; 2013. 8 pp.

12. Cooper MA, Holle RL. Casualties from lightning involving motorcycles. In: Preprints of the International Conference on Lightning and Static Electricity, 28–30 August 2007; Paris, France; 2007. paper Ic07/PPRKM02.
13. Cooper MA, Holle RL, Andrews CJ, Blumenthal R. Lightning injuries. In: Auerbach P, editor. Wilderness Medicine, 6th ed., Elsevier Mosby; Philadelphia, PA. 2012. pp. 60-101.
14. Cooper MA. Whether the medical aspects of lightning injury are different in developing countries. In: Preprints of the 32nd International Conference on Lightning Protection, 2–7 September 2012; Vienna, Austria, DOI:978-1-4673-1897-6/12/.
15. Cooper MA. Disability, not death, is the main problem with lightning injury. National Weather Digest, 2001;25:43-47.
16. Cooper MA, Marshburn S. Lightning strike and electric shock survivors, International. NeuroRehabilitation, 2005;20:43-37.
17. Mulder MB, Msalu L, Caro T, Salerno J. Remarkable rates of lightning strike mortality in Malawi. PLoS One, 2012;7(1). DOI:10.1371/journal.pone.0029281.
18. Roeder WP, Holle RL, Cooper MA, Hodanish S. Communicating lightning safety effectively. In: Preprints of the 5th Conference on the Meteorological Applications of Lightning Data, 23–27 January 2011; Seattle, Washington, American Meteorological Society; 2011. 18 pp.
19. Gomes R, Kithil R, Ahmed M. Developing a lightning awareness program model for third world based on American-South Asian experience. In: Preprints of the 28th International Conference on Lightning Protection, 17–21 September 2006; Kanazawa, Japan; pp. 1240-1243.
20. López RE, Holle RL. Changes in the number of lightning deaths in the United States during the twentieth century. Journal of Climate, 1998;11:2070-2077.

CHAPTER 2

Rangeland management and climate hazards in drylands: Dust storms, desertification and the overgrazing debate

Nick Middleton

[1] *St Anne's College, University of Oxford, Oxford, UK*

ABSTRACT

This paper examines the theory and supporting evidence for links between desertification, drought and dust storms with a particular focus on studies undertaken in and around the Gobi Desert. Overgrazing of rangeland by pastoralists has been the most commonly cited cause of desertification in global drylands for more than 30 years, but the evidence supporting this link is not always convincing. Nonetheless, overgrazing, desertification and dust storms are frequently connected, regardless. Drought is another well-known and important driver of vegetation cover change. Distinguishing between vegetation cover adversely affected by drought and that reduced by grazing is imperative for policy makers because identifying the incorrect driver of vegetation change risks the development of inappropriate policy.

Keywords: Rangeland, Pastoralism, Desertification, Dust storms, Drought, Gobi Desert

1. INTRODUCTION

Deserts and semideserts, commonly referred to as drylands, are defined by their aridity: their water resources are limited because precipitation is characteristically scarce and unreliable and evaporation is typically high. These areas are important global environments because they cover more than 40 % of the world's land surface and are home to over 2 billion people (Middleton and Thomas 1997; Safriel et al. 2005). Drylands are found on every continent, but most drylands are located in developing countries and 90 % of dryland inhabitants live in these countries (Safriel et al. 2005). Globally, about half of all dryland inhabitants are poor, about a billion people in total, dubbed the 'forgotten billion' because they have habitually been neglected in development processes (Middleton et al. 2011).

These people are involved in many occupations, but herding livestock is the most extensive form of land use. Indeed, desert-marginal rangelands support about 50 % of the world's livestock (Allen-Diaz et al. 1996) and pastoralism remains the only viable livelihood for many rural dryland populations. Where comprehensive estimates have been made, pastoralism has been shown to contribute significantly to the GDP of dryland developing country economies: for example, about 10 % of GDP in Mali, 20 % in Kyrgyzstan and 30 % in Mongolia (WISP 2008). In these and many other countries, no other drylands production system is more productive or supports as large a human population.

Drylands in general are also important to the Earth system for a range of key resources and other ecosystem services. Some dryland vegetation communities, such as the temperate grasslands, rival tropical rainforests for their species richness (Dengler et al. 2014), while their regulating services include important stores of carbon and effects on regional and even global climate (Yatagai and Yasunari 1995).

Pastoralists in drylands inevitably depend directly upon a highly variable natural resource base for their livelihoods, and this variability is driven primarily by climatic factors that present a range of natural hazards (Middleton and Sternberg 2013). Principal among these is the variability of precipitation. Rainfall totals often vary greatly from year to year and over short distances, typically producing 'resource mosaics' of water, nutrients, biodiversity and productivity. Long experience of coping with the vagaries of their environment has enabled traditional pastoralists to develop numerous strategies for managing risk and variability. Patterns of pastoral mobility underpin local adaptive capacities, although herders remain vulnerable to natural perturbations, particularly those driven by the highly variable climate regimes.

Rangeland management sometimes produces unintended outcomes, and one frequently cited result of mismanagement by pastoralists is overgrazing, leading to a loss of vegetation cover and accelerated soil erosion, two common forms of land degradation, or desertification as it is known in drylands (Middleton and Thomas 1997). However, both vegetation cover change and accelerated erosion—by wind and/or water—can also result from solely climatic drivers, without the influence of any rangeland mismanagement, and the issue of overgrazing is controversial (Vetter 2005; Reid et al. 2014). Different views on the extent of overgrazing in drylands have in turn divided opinions about the sustainability of pastoralism in general, with consequent implications for dryland development policies.

The aim of this paper is to examine the theory and empirical evidence for links between desertification by overgrazing, drought and dust storms with a particular focus on the Gobi Desert, a major world dryland area where mobile pastoralism has a very long history stretching back well over a 1000 years (Lattimore 1938; Di Cosmo 1994; Endicott 2012). This review is followed by assessments of how contemporary governments in China and Mongolia view pastoralism and rangeland use and their policies towards them.

2. DUST STORMS, DESERTIFICATION AND DROUGHT

The entrainment and transport of soil dust by wind is a common phenomenon in many drylands, and dust storms are recognised as of increasing environmental and human significance at local and global levels. As a hazard, desert dust has serious implications for human society via impacts on health, agriculture, transport and communications (Goudie and Middleton 2006). In and around the Gobi Desert, much attention has been devoted to challenges associated with dust transported over long distances to major regional cities, including Beijing (Liu et al. 2014), Seoul and Tokyo (Kashima et al. 2016). The long-term beneficial aspects of dust deposition are also evident in China on the Loess Plateau, an important farming region for many centuries (Huang et al. 2002).

Dust storms occur when strong, turbulent winds blow over dry, unconsolidated, fine-grained surface materials where vegetation cover is sparse if not altogether absent. Many dust sources are naturally devoid of vegetation thanks to a dryland climate. For instance, the world's largest source of desert dust is in the Bodélé Depression, a very sparsely populated area of Chad with plentiful supplies of fine-grained sediments provided by an ancient lake bed (Bristow et al. 2009). Elsewhere, dust is entrained from environments that have become susceptible to wind action through human influence. These situations include lake beds that have been desiccated due to society's use of water, as in the Aral Sea in Central Asia (Issanova et al. 2015), and agricultural fields left bare after harvests and/or ploughing, as during the notorious Dust Bowl years on the Great Plains of the USA (McLeman et al. 2014). The relative contribution of sources significantly influenced by human action to global dust emissions is a subject for debate with estimates varying from 50 % (Tegen et al. 1996) to insignificant (Prospero et al. 2002).

Vegetation protects soil surfaces from wind erosion. One outcome of the US Dust Bowl was a considerable increase in research into how wind erosion can be controlled in agricultural areas, and the importance of vegetation was quickly realised. Indeed, the establishment and maintenance of sufficient vegetative cover is often referred to as the 'cardinal rule' for controlling wind erosion (Skidmore 1986).

A number of other human activities—besides cultivation—can result in the degradation and reduction of vegetation cover, with an increase in wind erosion a possible outcome. These include logging, the use of fire, clearance for urban development and roads, and the effects of intensive grazing by domestic livestock.

The overuse of rangelands, caused by allowing too many animals or inappropriate types of animals to graze, is frequently cited as the most widespread cause of desertification, defined in the UN Convention to Combat Desertification (UNCCD) as 'land degradation in arid, semiarid, and dry sub-humid areas resulting from various factors, including climatic variations and human activities'. At the global scale, it is estimated that up to 73 % of the world's rangelands may be degraded (Lund 2007). Local estimates suggest that 90 % of the natural grasslands in China (Yang et al. 2005) and 70 % in Mongolia (Regdel et al. 2012) are degraded.

Rangeland degradation can take many forms, including impacts on flora, fauna and soil properties. Variables that have been used as indicators to assess degradation on rangelands include vegetation diversity, total vegetation cover, shrub cover, grass cover, forb cover, total vegetation biomass, belowground biomass, seedling survival, rodent species richness, rodent species diversity, cryptogamic crust cover, litter cover, litter biomass, soil bulk density, infiltration rate, soil organic matter, level of soil carbon and soil erosion rate. All of these indicators are interrelated in numerous ways, but some variables have a more immediate and obvious effect on susceptibility to wind erosion (total vegetation cover, cryptogamic crust cover) than others. Of course, as the UNCCD definition of desertification allows, rangeland degradation, and hence all these indicators used to assess degradation, can also be affected by natural, abiotic factors, notably variable rainfall.

Drought is the principal natural hazard faced by communities living in most drylands, and it can wreak significant, widespread and long-lasting damage on herding communities via declines in forage and water availability. Droughts frequently result in huge losses of livestock for pastoralists, most notably in Africa (Huho et al. 2011). The negative impact drought has on vegetation cover also has effects on soil erodibility, often leading to enhanced dust storm activity (Middleton 1985; McTainsh et al. 2005). It is not always easy to determine the relative importance of natural and human-induced drivers of vegetation loss and any consequent soil loss. Indeed, different interpretations of vegetation change drivers have created considerable controversy and confusion in the desertification literature, particularly on the margins of the Sahara (Tucker et al. 1991; Nicholson et al. 1998) but also more recently in the Gobi (Piao et al. 2005; Sternberg et al. 2015).

3. THE OVERGRAZING DEBATE

The impact of grazing by domestic livestock in rangelands has also generated considerable debate in recent decades. Much of the discussion has revolved around different ideas of rangeland dynamics in general and two paradigms in the ecology and management of rangelands in particular. On the one hand is a model based on the largely static 'climax community' of Clementsian succession (Clements 1916). This is an equilibrium model that stresses the importance of biotic feedbacks on vegetation composition, cover and productivity. Principal among these feedbacks are those driven by livestock populations and their density in any particular area. Range management under this model therefore focuses on stocking rates, carrying capacity and range condition assessment.

By contrast, the more dynamic non-equilibrium view of rangelands has it that vegetation condition is driven primarily by stochastic abiotic factors, notably variable rainfall. This results in highly variable and unpredictable primary production. Under this model, livestock populations are thought to have negligible feedback on the vegetation as their numbers rarely reach equilibrium with a resource base that is constantly fluctuating.

There are studies reported in the scientific literature to support both approaches to rangeland dynamics. Further, as Jones (2000) points out, the results of studies conducted in different plant communities, or at different sites representing the same community, frequently contradict one another. He cites surveys of grazing impact in an arid shrub/bunchgrass community in two adjacent valleys in the US southwest, one survey concluding that vegetative cover was greater in ungrazed areas (Johansen and St. Clair 1986) and the other finding that total vegetation cover was greater in grazed areas (Brotherson and Brotherson 1981). The lack of consensus regarding potentially detrimental effects of livestock grazing on rangelands is epitomised in two papers from Sahelian Africa. Hein (2006) concludes that a 10-year enclosure experiment in Senegal shows high grazing pressure negatively affects rangeland productivity, but Retzer (2006) reanalyses the same data to show that the vegetation on the 'high grazing' site has great resilience and produced greater forage for 2 years after a major drought than the less intensively grazed site.

Discussion about the relative merits of the two approaches to understanding rangeland dynamics has been intensified when the debate progresses to encompass views on how best to manage grazing in rangelands. In this context, both natural and social sciences are represented since the focus is on the sustainability of pastoralist land use. Views on the sustainability of grazing have also diverged, largely reflecting the different ideas on rangeland ecology. One approach adopts the equilibrium model and follows the arguments put forward in Hardin's (1968) Tragedy of the Commons idea. This view argues that traditional pastoralists frequently graze their livestock in unsustainable numbers because they pursue their short-term survival at the expense of the long-term conservation of commonly owned natural resources. The opposing view adopts a non-equilibrium understanding of rangelands, in which grazing has only a marginal influence on vegetation dynamics. Under this philosophy, a traditional 'opportunistic' grazing strategy matches the stocking rate to the available forage in any particular year.

In the former approach, traditional pastoralism is seen as unsustainable and in need of modernisation (Lamprey 1983). The latter view leads to the opposite conclusion: pastoralist land use practices are an effective and sustainable response to the highly variable natural environment (Behnke et al. 1993).

Although views on the sustainability of grazing in rangelands have often become polarised, matching the opposing paradigms of rangeland ecology, recent studies indicate that it is not an either/or issue. Most arid and semiarid rangelands encompass elements of both equilibrium and non-equilibrium approaches and both grazing and variable rainfall are important drivers of vegetation dynamics on different temporal and spatial scales (Vetter 2005; Gillson and Hoffman 2007). Hence, any sustainable approach to management must take account of temporal variability and spatial heterogeneity (Joly et al. 2012).

Of course there are also many nuances to a debate that is frequently presented in simple dualistic terms. The notion that a certain level of grazing has the effect of enhancing plant productivity has a considerable history, although plants respond to a whole suite of environmental factors, and the variety of plant responses is subject to the constraints of plant genetics, developmental stage and the plant tissues affected (McNaughton 1983). Further, different plants respond to being eaten by animals in different ways (Matches 1992). Similarly, different species of domestic livestock have different forage demands and foraging behaviours and can therefore reasonably be expected to have different impacts on vegetation communities. One form of vegetation change that is commonly associated with heavy grazing in some grasslands is an increase in biomass and abundance of woody plant species that are frequently thorny or unpalatable, with the suppression of herbaceous plant cover. This bush encroachment is another widely recognised form of rangeland degradation (D'Odorico et al. 2012).

The direct defoliation effects of grazing or browsing are just one set of impacts associated with livestock. Soil surfaces and soil profiles are materially affected by compaction caused by animal trampling, and excretion by grazing animals has impacts on vegetation, soils and surface waters (Bilotta et al. 2007). The time and space scales of observations on which we base our understanding also have consequences. One study in the Eastern Cape, South Africa concluded that historical land use change was critical: particularly intense erosion was associated with cultivated land that had been abandoned and reverted to grazing (Rowntree et al. 2004). In a global review of ecosystem sensitivity to grazing by large herbivores, Milchunas and Lauenroth (1993) concluded that an evolutionary history involving grazing animals and the local environment was a critical factor in determining negative impacts of grazing on productivity.

4. DESERTIFICATION IN THE GOBI DESERT

The Gobi is one of the world's largest deserts although, like many deserts, its boundaries are dynamic and ill-defined. Establishing the extent of any particular desert is complicated by the use of localised names for areas within large deserts (Middleton 2009), but this paper follows Sternberg et al. (2015) by including within the Gobi a large part of Mongolia and contiguous desert areas in northern China known locally as the Tengger, Ordos, Mu Us, Badain Jaran, Junggar, Gurbantunggut, Qubqi and Otindag.

There are many reports in the literature indicating that desertification is a serious issue in the Gobi (Fullen and Mitchell 1994; Yang et al. 2005;

Johnson et al. 2006; Regdel et al. 2012; Qi et al. 2013). Among the causes, an increase in the number of domestic livestock is frequently cited as a major driver of degradation in Mongolia (Sasaki et al. 2008; Hilker et al. 2014) and overgrazing also features as a prominent reason for desertification in China (Zhao et al. 2005; Han et al. 2008).

However, not all agree about the extent or severity of desertification in the Gobi or about the importance of the human impact. In a review, Wang et al. (2008) conclude that there is surprisingly little unassailable evidence to support the claim that desertification in China is primarily due to human impacts. Similarly, Khishigbayar et al. (2015) suggest that many claims about rangeland degradation in Mongolia have weak empirical support. There are also numerous studies that emphasise abiotic drivers of vegetation change, particularly the significance of the precipitation regime (Fernández-Giménez and Allen-Diaz 2001; Stumpp et al. 2005; Jeong et al. 2011). Assessments of rangeland condition and desertification also vary widely. In Mongolia, the most commonly cited proportion of rangelands degraded is 70 %, as cited above, but Addison et al. (2012) point out that a variety of scales, indicators and sampling regimes have thrown up estimates ranging from 9 to 90 % of the entire country.

Of course the UNCCD definition of desertification cited above means that rangeland degradation can be the result of both climatic variability and human activities (e.g. Liu et al. 2013), perhaps acting in tandem and not necessarily at the same time. However, some of the disagreements in the literature may also be the result, at least partly, of the methodology used. Khishigbayar et al. (2015) point out that there are plenty of remote sensing studies reporting widespread degradation of Mongolian rangelands, as indeed there are in China, but very few long-term field studies of vegetation responses to shifts in climate and stocking densities, a general observation that applies to most other rangelands. Yang et al. (2005) also criticise the use of some remotely sensed data sources in assessments of desertification in China, as well as highlighting the uncertainty of baseline assessments. Interestingly, Zemmrich (2007) assesses the impact of grazing in Mongolia's arid desert steppe and finds that grazing influence increases with decreasing spatial scale of analysis.

5. POLICIES TO COMBAT DESERTIFICATION

Despite the difficulties and disagreements surrounding our understanding of change in dryland landscapes, regions that are inherently variable in biophysical terms and prone to stochastic disturbances, there remains a prevalent view of desertification as being widespread and of overgrazing as a core cause in both Mongolia and China, a perspective that has informed policies. The design of policy responses, particularly many of those applied to Chinese grasslands, incorporates a specific aim of reducing the generation of dust and sand storms.

Recent large-scale projects in China that target rangelands, at least in part, include the Pasture to Grassland programme implemented in all of China's pastoral areas from 2003 and the Beijing–Tianjin Sand Source Control programme in Inner Mongolia from 2001 (Dalintai et al. 2011). Under Pasture to Grassland, grazing on existing pastures is banned (either permanently, for a few years, or seasonally) in specific zones; Beijing–Tianjin Sand Source Control involves a mix of measures, including grazing prohibition or rotation or rest, and the conversion of cropland to forest or grassland (though not for grazing). One of China's most ambitious projects to combat desertification and control dust storms is an afforestation project, the Three Norths Forest Shelterbelt programme (also known as the Great Green Wall or GGW), which was launched in 1978 and is not scheduled to be complete until 2050 (Wang et al. 2010a). Additional 'ecological restoration projects' are focused on other land uses associated with soil erosion problems, such as the Grain-for-Green programme designed to convert cropland to forest and grassland (Lei et al. 2012).

Assessments of the efficacy of these projects in remediating desertification in general, and specifically in preventing dust storms, do not all agree. Tan and Li (2015) show that the GGW programme greatly improved vegetation cover, as measured by the normalised difference vegetation index (NDVI), and effectively reduced dust storm intensity in northern China. A similar conclusion about vegetation is reached by Zhang et al. (2016) in their assessment of several afforestation programmes, although Wu et al. (2013) found that vegetation response to the first 10 years of the Beijing–Tianjin Sand Source Control programme was distinctly patchy. After a decade-long fencing experiment to assess the impacts of a grazing ban on sandy rangelands in Inner Mongolia, Su et al. (2015) warn that one unexpected outcome is likely to be bush encroachment, often shown in NDVI studies as a form of greening. They conclude that some degree of grazing by livestock may provide critical ecosystem services by regulating woody plant encroachment.

Other researchers point to warmer temperatures and greater rainfall as being more important drivers of greening trends in northern China (Piao et al. 2007; Sternberg et al. 2015), and the value of large-scale afforestation, a primary tool for controlling desertification and soil erosion in China, has been questioned due to low tree survival rates. Cao (2008) argues that inappropriate project design—planting trees in rangelands—has resulted in declining soil moisture content and a worsening of wind erosion in China's dryland regions. Qu et al. (2011) emphasise the significance of government policies aimed at controlling natural resource degradation in rural China for the success of economic policies introduced since 2004 to stimulate rural incomes. However, despite the importance of realising the 'new socialist countryside', these authors also conclude that degradation of natural grassland and wind erosion have become much more severe in recent decades. That said, Huang et al. (2013) report that rangeland restoration programmes have resulted in a systematic increase in the area of grassland in most northern and western regions and that restored, ungrazed grasslands have contributed to intensified primary production in these dryland areas.

Tree planting initiatives have also been promoted in Mongolia, where state-sponsored planting programmes date back at least to the 1970s in many parts of the country, including forestry offices in the Gobi region (Dorjsuren and Sainbayar 2004). A major new national initiative was launched in the Gobi in 2005, when the government pledged considerable investment for a large-scale planting scheme designed to reduce dust storms and mitigate the impacts of desertification. The scheme, variously known as the Great Green project or the Korea-Mongolia Greenbelt Project (since planting sites are co-funded with Korean bilateral aid money), is intended to extend east–west across the country for 2500 km with a width of at least 600 m. High initial success rates have been reported, but historical survival rates of tree planting in the Gobi over the longer term have been minimal (Mühlenberg et al. 2006).

A number of dust storm studies in the Gobi region point to rainfall and other climatic drivers of change as being of greatest importance. Lee and Sohn (2011) found that most dust source areas in China experienced a continuous decrease over a 34-year data period (1974–2007) which they attribute to enhanced precipitation and its effects on soil moisture and surface vegetation. Fan et al. (2014) emphasise the effect of earlier spring vegetation green-up in suppressing dust storm activity. Zhu et al. (2008) refer to a weakening of the westerly jet stream in northern China and Mongolia to explain the declining trend. Lee and Sohn also conclude that an increase in dust occurrence after the mid-1990s over eastern Mongolia and Chinese Inner Mongolia may be largely influenced by drought conditions and associated soil moisture change, while Kurosaki et al. (2011) emphasise the importance of vegetation residue from the previous year and its effect on erodibility in explaining the recent increase in dust outbreaks over East Asia. Forty years of dust event data from Seoul in South Korea show an increasing trend from the early 1980s (Kim 2008) which is attributed to an eastward shift in the regional source of dust (from Mongolia to Manchuria) but also to changes in upper-level synoptic transport patterns. Intensified atmospheric circulation and an associated increase in wind speed as well as changes in snow cover, soil moisture and potential evaporation rates throughout Inner Mongolia and Manchuria are considered to be the main causes of the increasing number of dusty days in Korea. No mention is made of policies to control desertification or dust storms in any of these dust storm papers.

6. DISCUSSION

The deserts and semideserts of China and Mongolia are well established as significant sources of desert dust, although on the global scale the most important regional source is not the Gobi but the Taklamakan Desert (Goudie and Middleton 2006), located further west in the Tarim Basin. The regional preeminence of the Taklamakan is confirmed by many studies (e.g. Shao and Wang 2003; Zhang et al. 2003), but its standing as a source of long-distance dust transport is debated (see Bory et al. 2003; Zhang et al. 2008). Several specific sources within the Gobi are also identified. Prominent among these are the Badain Jaran (Zhang et al. 2003), the numerous alluvial fans along the Hexi Corridor in Gansu province (Derbyshire et al. 1998) and the Inner Mongolia Plateau with adjacent areas of south-eastern Mongolia (Wang et al. 2004).

The standing of rangelands in this picture remains unclear. There are many uncertainties and gaps in our knowledge about how grasslands may act as dust sources (Shinoda et al. 2011). In a rare field experiment, Hoffmann et al. (2008) found that grazing intensity was a significant factor affecting wind erosion on plots in the Xilingele grassland, Inner Mongolia, but that these local sources provided just 5 % of the total horizontal dust flux and some of this was material recently deposited from distant sources and re-emitted. However, areas with significant livestock populations (the lowlands of the eastern Mongolian Plateau, including Dornogovi province in Mongolia, and the Otindag and Horqin sandy lands of Inner Mongolia) were identified by Zhang et al. (2008) as the greatest source of regional dust emissions in their assessment based on satellite imagery during the period 2000–2006. Zhang et al. (2003) also highlight Otindag and Horqin, along with the Mu Us Desert, as areas with increased dust emissions over the previous 20 years, but dismiss them as minor sources, concluding that efforts to reduce desertification in these areas have had little effect on Asian dust emissions.

These uncertainties are of course compounded by the doubts surrounding the overall extent and severity of desertification in the Gobi, the debate over the relative importance of biotic vs abiotic drivers of vegetation dynamics in rangelands, and the sustainability of pastoralism in general. All these factors are to some extent facets of different paradigms of rangeland ecology.

Contemporary policies towards pastoralism in rural China have multiple objectives (Addison and Greiner 2015). Generally, they aim is to improve both the environment and the livelihoods of pastoralists. They are designed to address the former by restoring grassland ecology, combatting desertification and reducing dust storms, and the latter by supplementing the incomes of herders and encouraging them away from traditional forms of extensive pastoralism. Central facets of these mandatory schemes for pastoralists include a shift towards exclusive user rights enforced by fencing, prescribed carrying capacities per hectare where grazing is allowed, and grazing bans elsewhere, with cash payments to compensate pastoral households, or to incentivise herd sizes being kept below set stocking rates.

There is no doubt that these schemes are significantly altering the livelihoods of a great many pastoralists living in China and that a critical aspect of these changes is a trend towards less mobility. Some have criticised the way in which the privatisation of grassland user rights has undermined the pastoral system's capacity to respond to drought and other adverse weather events through herd movement, increasing vulnerability to environmental change and resulting, as Li and Huntsinger (2011) suggest, in greater impoverishment and environmental deterioration. In some cases, these schemes have resulted in sedentary lifestyles in towns and cities through an associated resettlement policy frequently called 'ecological migration' (Yeh 2005). Ecological migration programmes have also affected pastoralists in other drylands in China, notably in Tibet where the goal is to move 100,000 herders off pastures to begin new lives in towns and cities (Wang et al. 2010b). Both Yeh and Wang et al. also note that the herders targeted by these ecological migration programmes are predominantly from ethnic minority groups.

Across the border in Mongolia, policies are also designed to address the issue of rangeland degradation (including dust storm prevention, although it is less of an issue than in China) and to improve the welfare of herders. Government in Mongolia is generally less interventionist since the collapse of socialism in the early 1990s, which resulted in the disappearance of local collectives—influenced by central government and preexisting customary institutions—that had previously allocated pastures and supported pastoralism through the provision of fodder, livestock, transport and veterinary services (Sneath 2003). Livestock were privatised with the transition to a market economy and nationally their numbers increased sharply, along with a general shift from sheep to goats, responding to growing international demand for cashmere. The increase in livestock numbers, particularly goats, is frequently cited as a major cause of desertification in Mongolia (Dorj et al. 2013), although many herders have suffered devastating losses during droughts and extreme winters (Sternberg 2010; Fernández-Giménez et al. 2012). One marked response to this situation has been a trend towards supporting localised forms of common property management by organising so-called community-based rangeland management (CBRM) groups with the goals of improving rangeland conditions and herder well-being. Many of these initiatives have received support from overseas development agencies and NGOs. However, the success of these CBRM groups in improving rangeland condition in the Mongolian Gobi is a matter for debate (Addison et al. 2013) and their social benefits are by no means clear-cut (Ulambayar et al. 2016).

The Mongolian government's Land Law, introduced in 1994 to fill the institutional land tenure vacuum resulting from the end of socialism, authorised contracts for pastoral resources such as pastures and campsites. However, the law's lack of clarity and its implementation at the local level have had a range of not always intended effects. Herders' user rights have often been sidelined by an expansion of minerals licensing across the countryside, leaving pastoralists seemingly powerless in the face of mining activities (Upton 2012). The Land Law has also affected herder mobility. Fernández-Giménez and Batbuyan (2004) concluded that the poor had become more mobile and the wealthy more sedentary due to changes in access to pastoral resources. Further evidence of declining mobility among herders in Mongolia for economic reasons has been presented by Lkhagvadorj et al. (2013), but it is not a central ambition of government to encourage herders to settle in the same way it is in China. Lkhagvadorj et al. (2013) suggest that reduced seasonal migrations have left herders more vulnerable to a shortage of fodder, a finding supported by Middleton et al. (2015) who highlight the consequences of such vulnerability during severe climatic events: loss of livestock on a sometimes large scale. Wesche and Retzer (2005) state that a nomadic way of life appears to be a 'crucial precondition for sustainable land use' in the Gobi Desert steppes.

7. CONCLUSIONS

This paper investigates the links between dust storms, desertification, climate variability and pastoralism as an entry point for comparing policies for rangeland management in the Gobi Desert of Mongolia and China. It assesses the received wisdom that overgrazing of rangeland by pastoralists has been the most common cause of desertification on both sides of the border and concludes that the evidence supporting this link is not always convincing. Indeed, many of the issues pertaining to the desertification debate are surrounded by controversy and contradictory findings in this part of the world, as in others. Underlying many of these controversies is a fundamental tension between two alternative views of rangeland dynamics.

Policies towards using (or not using) the grassland resource in the Gobi Desert are substantially different either side of the political frontier between China and Mongolia. Whether deliberate or not, these contrasting policies appear to be driven in part by an adherence to, or at least an influence from, contrasting paradigms in our understanding of rangeland ecology.

China's rangeland policies represent an unambivalent example of management on the basis of carrying capacities and Clementsian succession theory. In order to tackle what are considered to be the unwanted outcomes of overstocking stemming from the classic Tragedy of the Commons, individual household tenure of grasslands has been introduced alongside degrees of restriction on grazing in certain areas deemed highly susceptible to desertification and dust storm generation. A notable effect of these policies is a marked move away from traditional mobile pastoralism.

In Mongolia, by contrast, although the same conceptual framework of stable-state rangelands and carrying capacity also holds considerable sway, the policies towards rangelands are markedly different. Prominent among these differences is a willingness to encourage the communal management of rangelands through pasture user groups. Government policies do not actively promote the sedentarisation of mobile herders, although there is some evidence to suggest that Mongolian pastoralists are becoming less mobile anyway due to land tenure and economic factors.

There are many nuances to the debate on the sustainability of pastoralism in drylands, and current thinking in rangeland ecology has it that most arid and semiarid rangelands encompass elements of both equilibrium and non-equilibrium approaches and that grazing and variable rainfall are both important drivers of vegetation dynamics on different temporal and spatial scales. However, the state of pastoralism either side of the political boundary between Mongolia and China is very different. In Mongolia, pastoralism continues to be backed by government policies because it is viewed as an enduring land use practice that is an essentially sustainable response to the highly variable natural environment, albeit that livestock dynamics follow the typically 'boom and bust' cycle in line with non-equilibrium theory.

In China, by contrast, traditional pastoralism is clearly viewed as unsustainable and in need of modernisation and is therefore unlikely to feature prominently in the 'new socialist countryside'.

ACKNOWLEDGMENTS

The author acknowledges the 'Gobi Hazards' project funded by the Leverhulme Trust as a central support for significant parts of this research. Fieldwork in China was supported financially by the International Science and Technology Cooperation Program of China (No. 2013DFA91700). Two anonymous reviewers also made some valuable comments.

REFERENCES

1. Addison J, Greiner R (2015) Applying the social–ecological systems framework to the evaluation and design of payment for ecosystem service schemes in the Eurasian steppe. Biodivers Conserv. doi:10.1007/s10531-015-1016-3
2. Addison J, Friedel M, Brown C, Davies J, Waldron S (2012) The critical review of degradation assumptions applied to Mongolia's Gobi Desert. Rangeland J 34(2):125–137
3. Addison J, Davies J, Friedel M, Brown C (2013) Do pasture user groups lead to improved rangeland condition in the Mongolian Gobi Desert? J Arid Environ 94:37–46
4. Allen-Diaz B, Chapin FS, Diaz S, Howden M, Puigdefabregas J, Stafford Smith M (1996) Rangelands in a changing climate: impacts, adaptations and mitigation. In: Watson WT, Zinyowera MC, Moss RH, Dokken DJ (eds) Climate change 1995—impacts, adaptation and mitigation. Cambridge University Press, Cambridge, pp 131–158
5. Behnke RH, Scoones I, Kerven C (1993) Range ecology at disequilibrium: new models of natural variability and pastoral adaptation in African Savannas. Overseas Development Institute, London
6. Bilotta GS, Brazier RE, Haygarth PM (2007) The impacts of grazing animals on the quality of soils, vegetation, and surface waters in intensively managed grasslands. Adv Agron 94:237–280
7. Bory AJM, Biscaye PE, Grousset FE (2003) Two distinct seasonal Asian source regions for mineral dust deposited in Greenland (NorthGRIP). Geophys Res Lett 30(4):1167
8. Bristow CS, Drake N, Armitage S (2009) Deflation in the dustiest place on Earth: the Bodélé Depression, Chad. Geomorphology 105(1):50–58
9. Brotherson JD, Brotherson WT (1981) Grazing impacts on the sagebrush communities of central Utah. Great Basin Nat 41:335–340
10. Cao S (2008) Why large-scale afforestation efforts in China have failed to solve the desertification problem. Environ Sci Technol 42:1826–1831
11. Clements FE (1916) Plant succession: an analysis of the development of vegetation. Carnegie Institution of Washington, Washington DC
12. Dalintai B, Yanbo L, Jianjun C (2011) The Eurasian Steppe: history of utilization and policies on the rangeland. In: Fernández-Giménez ME, Wang X, Baival B, Klein J, Reid R (eds) Restoring community connections to the land. CABI, Wallingford, pp 51–68
13. Dengler J, Janišová M, Török P, Wellstein C (2014) Biodiversity of Palaearctic grasslands: a synthesis. Agric Ecosyst Environ 182:1–14
14. Derbyshire E, Meng X, Kemp RA (1998) Provenance, transport and characteristics of modern aeolian dust in western Gansu Province, China, and interpretation of the Quaternary loess record. J Arid Environ 39:497–516
15. Di Cosmo N (1994) Ancient Inner Asian nomads: their economic basis and its significance in Chinese history. J Asian Stud 53:1092–1126
16. D'Odorico P, Okin GS, Bestelmeyer BT (2012) A synthetic review of feedbacks and drivers of shrub encroachment in arid grasslands. Ecohydrology 5(5):520–530

17. Dorj O, Enkhbold M, Lkhamyanjin S, Mijiddorj K, Nosmoo A, Puntsagnamil M, Sainjargal U (2013) In: Heshmati GA, Squires VA (eds) Mongolia: Country features, the main causes of desertification and remediation efforts. In combating desertification in Asia, Africa and the Middle East. Springer, Dordrecht, pp 217–229
18. Dorjsuren C, Sainbayar D (2004) Review of the a forestation and desertification control strategies in Mongolia. FAO, Ulaanbaatar
19. Endicott E (2012) A history of land use in Mongolia: the thirteenth century to the present. Springer, Berlin
20. Fan B, Guo L, Li N, Chen J, Lin H, Zhang X, Shen M, Rao Y, Wang C, Ma L (2014) Earlier vegetation green-up has reduced spring dust storms. Scientific reports 4:6749
21. Fernández-Giménez M, Allen-Diaz B (2001) Vegetation change along gradients from water sources in three grazed Mongolian ecosystems. Plant Ecol 157:101–118
22. Fernández-Giménez ME, Batbuyan B (2004) Law and disorder: local implementation of Mongolia's Land Law. Dev Change 35:141–166
23. Fernández-Giménez ME, Batkhishig B, Batbuyan B (2012) Cross-boundary and cross-level dynamics increase vulnerability to severe winter disasters (dzud) in Mongolia. Glob Environ Change 22:836–851
24. Fullen MA, Mitchell DJ (1994) Desertification and reclamation in north-central China. Ambio 23(2):131–135
25. Gillson L, Hoffman MT (2007) Rangeland ecology in a changing world. Science 1136577(53):315
26. Goudie A, Middleton N (2006) Desert dust in the global system. Springer, Berlin
27. Han JG, Zhang YJ, Wang CJ, Bai WM, Wang YR, Han GD, Li LH (2008) Rangeland degradation and restoration management in China. Rangeland J 30:233–239
28. Hein L (2006) The impacts of grazing and rainfall variability on the dynamics of a Sahelian rangeland. J Arid Environ 64:488–504
29. Hilker T, Natsagdorj E, Waring RH, Lyapustin A, Wang Y (2014) Satellite observed widespread decline in Mongolian grasslands largely due to overgrazing. Glob Change Biol 20(2):418–428
30. Hoffmann C, Funk R, Wieland R, Li Y, Sommer M (2008) Effects of grazing and topography on dust flux and deposition in the Xilingele grassland, Inner Mongolia. J Arid Environ 72:792–807
31. Huang CC, Pang J, Huang P, Hou C, Han Y (2002) High-resolution studies of the oldest cultivated soils in the southern Loess Plateau of China. Catena 47:29–42
32. Huang L, Xiao T, Zhao Z, Sun C, Liu J, Shao Q, Wang J (2013) Effects of grassland restoration programs on ecosystems in arid and semiarid China. J Environ Manag 117:268–275
33. Huho JM, Ngaira JK, Ogindo HO (2011) Living with drought: the case of the Maasai pastoralists of northern Kenya. Educ Res 2:779–789
34. Issanova G, Abuduwaili J, Galayeva O, Semenov O, Bazarbayeva T (2015) Aeolian transportation of sand and dust in the Aral Sea region. Int J Environ Sci Technol 12(10):3213–3224
35. Jeong S, Ho C, Brown M, Kug J, Piao S (2011) Browning in desert boundaries in Asia in recent decades. J Geophys Res. doi:10.1029/2010JD014633
36. Johansen JR, St. Clair LL (1986) Cryptogamic soil crusts: recovery from grazing near Camp Floyd State Park, Utah, USA. Great Basin Nat 46:632–640
37. Johnson DA, Sheehy DP, Miller D, Damiran D (2006) Mongolian rangelands in transition. Sécheresse 17:133–141
38. Joly F, Saïdi S, Begz T, Feh C (2012) Key resource areas of an arid grazing system of the Mongolian Gobi. Mong J Biol Sci 10(1–2):13–24
39. Jones A (2000) Effects of cattle grazing on North American arid ecosystems: a quantitative review. West N Am Nat 60(2):155–164
40. Kashima S, Yorifuji T, Bae S, Honda Y, Lim YH, Hong YC (2016) Asian dust effect on cause-specific mortality in five cities across South Korea and Japan. Atmos Environ 128:20–27
41. Khishigbayar J, Fernández-Giménez ME, Angerer JP, Reid RS, Chantsallkham J, Baasandorj Y, Zumberelmaa D (2015) Mongolian rangelands at a tipping point? Biomass and cover

are stable but composition shifts and richness declines after 20 years of grazing and increasing temperatures. J Arid Environ 115:100–112
42. Kim J (2008) Transport routes and source regions of Asian dust observed in Korea during the past 40 years (1965–2004). Atmos Environ 42(19):4778–4789
43. Kurosaki Y, Shinoda M, Mikami M (2011) What caused a recent increase in dust outbreaks over East Asia? Geophys Res Lett 38:L11702
44. Lamprey HF (1983) Pastoralism yesterday and today: the overgrazing problem. In: Bourlière F (ed) Tropical savannas: ecosystems of the world. Elsevier, Amsterdam, pp 643–666
45. Lattimore O (1938) The geographical factor in Mongol history. Geogr J 91:1–16
46. Lee EH, Sohn BJ (2011) Recent increasing trend in dust frequency over Mongolia and Inner Mongolia regions and its association with climate and surface condition change. Atmos Environ 45:4611–4616
47. Lei DE, Shangguan ZP, Rui LI (2012) Effects of the grain-for-green program on soil erosion in China. Int J Sediment Res 27:120–127
48. Li W, Huntsinger L (2011) China's grassland contract policy and its impacts on herder ability to benefit in Inner Mongolia: tragic feedbacks. Ecol Soc 16(2):1
49. Liu YY, Evans JP, McCabe MF, de Jeu RAM, van Dijk A, Dolman AJ, Saizen I (2013) Changing climate and overgrazing are decimating Mongolian steppes. Plos One 8(2):e57599
50. Liu Q, LiuY Yin J, Zhang M, Zhang T (2014) Chemical characteristics and source apportionment of PM 10 during Asian dust storm and non-dust storm days in Beijing. Atmos Environ 91:85–94
51. Lkhagvadorj D, Hauck M, Dulamsuren Ch, Tsogtbaatar J (2013) Pastoral nomadism in the forest-steppe of the Mongolian Altai under a changing economy and a warming climate. J Arid Environ 88:82–89
52. Lund HG (2007) Accounting for the world's rangelands. Rangelands 29:3–10
53. Matches AG (1992) Plant response to grazing: a review. J Prod Agric 5:1–7
54. McLeman RA, Dupre J, Ford LB, Ford J, Gajewski K, Marchildon G (2014) What we learned from the Dust Bowl: lessons in science, policy, and adaptation. Popul Environ 35(4):417–440
55. McNaughton SJ (1983) Compensatory plant growth as a response to herbivory. Oikos 40:329–336
56. McTainsh G, Chan YC, McGowan H, Leys J, Tews K (2005) The 23rd October 2002 dust storm in eastern Australia: characteristics and meteorological conditions. Atmos Environ 39:1227–1236
57. Middleton N (1985) Effect of drought on dust production in the Sahel. Nature 316:431–434
58. Middleton N (2009) Deserts: a very short introduction. Oxford University Press, Oxford
59. Middleton N, Sternberg T (2013) Climate hazards in drylands: a review. Earth-Sci Rev 126:48–57
60. Middleton N, Thomas D (1997) World atlas of desertification, 2nd edn. Arnold, London
61. Middleton N, Stringer L, Goudie A, Thomas D (2011) The forgotten billion: MDG achievement in the drylands. UNDP-UNCCD, New York
62. Middleton N, Rueff H, Sternberg T, Batbuyan B, Thomas D (2015) Explaining spatial variations in climate hazard impacts in western Mongolia. Landsc Ecol 30(1):91–107
63. Milchunas DG, Lauenroth WK (1993) Quantitative effects of grazing on vegetation and soils over a global range of environments. Ecol Monogr 63(4):327–366
64. Mühlenberg M, Batkhishig T, Dashzeveg T, Drößler L, Neusel B, Tsogtbaatar J (2006) Lessons from tree planting initatives in Mongolia. Mongolia Discussion Papers, East Asia and Pacific Environment and Social Development Department. World Bank, Washington DC
65. Nicholson SE, Tucker CJ, Ba MB (1998) Desertification, drought, and surface vegetation: an example from the West African Sahel. Bull Am Meteorol Soc 79:815–829
66. Piao S, Fang J, Liu H, Zhu B (2005) NDVI-indicated decline in desertification in China in the past two decades. Geophys Res Lett. doi:10.1029/2004GL021764

67. Piao S, Fang J, Friedlingstein P, Ciais P, Viovy N, Demarty J (2007) Growing season extension and its impact on terrestrial carbon cycle in the northern hemisphere over the past 2 decades. Glob Biogeochem Cycles 21:GB3018. doi:10.1029/2006GB002888
68. Prospero JM, Ginoux P, Torres O, Nicholson SE, Gill TE (2002) Environmental characterization of global sources of atmospheric soil dust identified with the NIMBUS-7 TOMS absorbing aerosol product. Rev Geophys 40(1):1002
69. Qi F, Wei L, Haiyang X (2013) Comprehensive evaluation and indicator system of land desertification in the Heihe River Basin. Nat Hazards 65:1573–1588
70. Qu F, Kuyvenhoven A, Shi X, Heerink N (2011) Sustainable natural resource use in rural China: recent trends and policies. China Econ Rev 22(4):444–460
71. Regdel D, Dugarzhav C, Gunin P (2012) Ecological demands on socioeconomic development of Mongolia under climate aridization. Arid Ecosyst 2:1–10
72. Reid RS, Fernández-Giménez ME, Galvin KA (2014) Dynamics and resilience of rangelands and pastoral peoples around the globe. Annu Rev Environ Resour 39:217–242
73. Retzer V (2006) Impacts of grazing and rainfall variability on the dynamics of a Sahelian rangeland revisited—new insights from old data. J Arid Environ 67:157–164
74. Rowntree K, Duma M, Kakembo V, Thornes J (2004) Debunking the myth of overgrazing and soil erosion. Land Degrad Dev 15(3):203–214
75. Safriel U, Adeel Z, Niemeijer D, Puigdefabregas J, White R, Lal R, Winslow M, Ziedler J, Prince S, Archer E, King C (2005) Chapter 22: dryland systems. Millennium Ecosystem Assessment. Ecosystems and human well-being. World Resources Institute, Washington DC, pp 623–662
76. Sasaki T, Okayasu T, Jamsran U, Takeuchi K (2008) Threshold changes in vegetation along a grazing gradient in Mongolian rangelands. J Ecol 96:145–154
77. Shao Y, Wang J (2003) A climatology of Northeast Asian dust events. Meteorol Z 12(4):187–196
78. Shinoda M, Gillies JA, Mikami M, Shao Y (2011) Temperate grasslands as a dust source: knowledge, uncertainties, and challenges. Aeolian Res 3(3):271–293
79. Skidmore EL (1986) Wind erosion control. Clim Chang 9:209–218
80. Sneath D (2003) Land use, the environment and development in post-socialist Mongolia. Oxford Dev Stud 31(4):441–459
81. Sternberg T (2010) Unravelling Mongolia's extreme winter disaster of 2010. Nomad People 14:72–86
82. Sternberg T, Rueff H, Middleton N (2015) Contraction of the Gobi Desert, 2000–2012. Remote Sens 7:1346–1358
83. Stumpp M, Wesche K, Retzer V, Miehe G (2005) Impact of grazing livestock and distance from water points on soil fertility in southern Mongolia. Mt Res Dev 25:244–251
84. Su H, Liu W, Xu H, Wang Z, Zhang H, Hu H, Li Y (2015) Long-term livestock exclusion facilitates native woody plant encroachment in a sandy semiarid rangeland. Ecol Evol 5(12):2445–2456
85. Tan M, Li X (2015) Does the Green Great Wall effectively decrease dust storm intensity in China? A study based on NOAA NDVI and weather station data. Land Use Policy 43:42–47
86. Tegen I, Lacis A, Fung I (1996) The influence of mineral aerosols from disturbed soils on the global radiation budget. Nature 380:419–422
87. Tucker C, Dregne H, Newcomb W (1991) Expansion and contraction of the Sahara Desert from 1980 to 1990. Science 253:299–301
88. Ulambayar T, Fernández-Giménez ME, Baival B, Batjav B (2016) Social outcomes of community-based rangeland management in Mongolian steppe ecosystems. Conserv Lett. doi:10.1111/conl.12267
89. Upton C (2012) Mining, resistance and pastoral livelihoods in contemporary Mongolia. In: Dierkes J (ed) Change in Democratic Mongolia: social relations, health, mobile pastoralism, and mining (vol 25). Brill, Leiden, pp 223–248
90. Vetter S (2005) Rangelands at equilibrium and non-equilibrium: recent developments in the debate. J Arid Environ 62:321–341
91. Wang X, Dong Z, Zhang J, Liu L (2004) Modern dust storms in China: an overview. J Arid Environ 58:559–574

92. Wang X, Chen F, Hasi E, Li J (2008) Desertification in China: an assessment. Earth-Sci Rev 88:188–206
93. Wang X, Zhang C, Hasi E, Dong Z (2010a) Has the Three Norths Forest Shelterbelt Program solved the desertification and dust storm problems in arid and semiarid China? J Arid Environ 74:13–22
94. Wang Z, Song K, Hu L (2010b) China's largest scale ecological migration in the three-river headwater region. Ambio 39:443–446
95. Wesche K, Retzer V (2005) Is degradation a major problem in semi-desert environments of the Gobi region in southern Mongolia. Erforschung Biologischer Ressourcen der Mongolei 9:133–146
96. WISP (World Initiative for Sustainable Pastoralism) (2008) Forgotten Services, Diminished Goods: understanding the agroecosystem of pastoralism. WISP Policy Note 8
97. Wu Z, Wu J, Liu J, He B, Lei T, Wang Q (2013) Increasing terrestrial vegetation activity of ecological restoration program in the Beijing–Tianjin sand source region of China. Ecol Eng 52:37–50
98. Yang X, Zhang K, Jia B, Ci L (2005) Desertification assessment in China: an overview. J Arid Environ 63:517–531
99. Yatagai A, Yasunari T (1995) Interannual variations of summer precipitation in the arid/semi-arid regions in China and Mongolia: their regionality and relation to the Asian summer monsoon. J Meteorol Soc Jpn Ser 2(73):909–923
100. Yeh ET (2005) Green governmentality and pastoralism in western China: 'converting pastures to grasslands'. Nomad People 9:9–30
101. Zemmrich A (2007) Equilibrium or non-equilibrium ecosystems? Scaling effects in rangeland ecosystems of western Mongolia. Erdkunde 61:335–343
102. Zhang XY, Gong SL, Zhao TL, Arimoto R, Wang YQ, Zhou ZJ (2003) Sources of Asian dust and role of climate change versus desertification in Asian dust emission. Geophys Res Lett 30(24):2272
103. Zhang B, Tsunekawa A, Tsubo M (2008) Contributions of sandy lands and stony deserts to long-distance dust emission in China and Mongolia during 2000–2006. Glob Planet Change 60(3):487–504
104. Zhang Y, Peng C, Li W, Tian L, Zhu Q, Chen H, Fang X, Zhang G, Li G, Mu X, Li Z (2016) Multiple afforestation programs accelerate the greenness in the 'Three North' region of China from 1982 to 2013. Ecol Ind 61:404–412
105. Zhao H-L, Zhao X-Y, Zhou R-L, Zhang T-H, Drake S (2005) Desertification processes due to heavy grazing in sandy rangeland, Inner Mongolia. J Arid Environ 62:309–319
106. Zhu C, Wang B, Qian W (2008) Why do dust storms decrease in northern China concurrently with the recent global warming? Geophys Res Lett 35(18):L18702

CHAPTER 3

Estimation of river flood damages in Jakarta, Indonesi

Pini Wijayanti[1,2], Xueqin Zhu[1], Petra Hellegers[1], Yus Budiyono[3,4] Ekko C. van Ierland[1]

1 Environmental Economics and Natural Resources Group, Wageningen University, Wageningen, The Netherland

2 Bogor Agricultural University, Bogor, Indonesi

3 Institute for Environmental Studies, VU University, Amsterdam, The Netherland

4 Agency for the Assessment and Application of Technology, Jakarta, Indonesi

ABSTRAC

Flooding is a serious problem in Jakarta, and detailed estimation of flood damage is necessary to design optimal flood management strategies. This study aims to estimate flood damage in a densely populated area in Jakarta by means of a survey, to develop the relationship between flood characteristics and flood damage, and to compare the damage estimates from the survey with the damage estimates obtained by a flood damage model for Jakarta, i.e. the damage scanner model. We collected data on economic losses of the January 2013 flood in a survey of flood-affected households and business units in *Pesanggrahan* River. The actual flood damage in the survey area is US$ 0.5 million for the residential sector and US$ 0.7 million for the business sector. The flood damage for a similar event in the same area based on the damage scanner model is estimated to be US$ 1.3 million for the residential sector and US$ 9.2 million for the business sector. The flood damage estimates obtained by the survey approach are lower compared to the damage scanner approach due to different ways in obtaining flood damage data and in defining the maximum flood damage per object, the different spatial levels of analysis, and uncertainties in constructing the flood damage curves that were applied in the damage scanner mode

Keywords: Flood damage, Survey, Damage scanner, Jakart

1 INTRODUCTION

Jakarta, the capital city of Indonesia, has been regarded as one of the world's most vulnerable cities with regard to climate change-related disasters. The city is prone to flooding, not only because of precipitation, but also as a result of soil subsidence, increasing sea level, and natural calamities (Firman et al. 2011). Over the past two decades, Jakarta was hit by major floods in 1996, 2002, 2007, and 2013. The 2007 flood was regarded as a national disaster, which caused a total loss of US$ 565 million.[1] The residential sector suffered most with 74% of total losses (BAPPENAS 2007). Excessive rain in Jakarta and surrounding cities caused a major flood from 17 to 19 January 2013, which inundated 124 villages in Jakarta Province with 98,000 houses, displaced 40,000 people, and killed 20 people. The flood caused in total an estimated damage of US$ 775 million[2] (BPBD 2013). These events show that flooding is a serious problem in Jakarta

The ability to assess flood damage is the key in developing effective flood policies and implementing flood protection measures to reduce expected flood damages (Green 2003; Mays 2011). Therefore, an accurate flood damage assessment is crucial for effective and efficient flood risk management (Middelmann-Fernandes 2010

The government of Jakarta conducted flood damage assessments following the 2007 and 2013 floods by using the damage and loss assessment methodology of the Economic Commission for Latin America and the Caribbean. Besides the assessment of ex-post flood events, several studies estimated the future coastal flood and river flood damages in Jakarta (e.g. PU 2011; Ward et al. 2011b). These studies, however, cover the whole Jakarta area and do not provide the detailed information that survey data of specific flooded areas provide. As detailed information is important in presenting the local flood situation and in evaluating the cost-effectiveness of flood measures for stakeholders, a detailed flood damage assessment for a specific area is needed

The aims of this paper are (1) to provide insight into the level of the flood damages; (2) to estimate the relation between flood damage and flood characteristics; and (3) to compare the results of two different approaches. The following three research questions are addressed: What is the flood damage in the residential sector and the business sector? How is this related to flood characteristics? And Are the results comparable to the results obtained by the damage scanner model for a similar flood event for the relevant area

This paper is organized as follows. Section 2 presents a literature review on flood damage assessment. Section 3 provides the methodology of this study, where the flood damage function is specified. Section 4 gives the results of the estimated flood damage from the survey and compares the results with the experts' assessment through the damage scanner model. Section 5 contains the discussion and conclusions

2 SOME CONCEPTS AND APPROACHES USED IN THE LITERATUR

Previous studies have distinguished several types of flood damage. First, there are *direct damages* and *indirect damages*. The former refers to damages caused by the physical contact of flood with humans, property, or any other objects, while the latter implies indirect damages, either caused by the flood impact or occurring after the flood event (Merz et al. 2010). Second, based on the ability to assess damages in monetary terms, these damages are categorized into tangible and intangible damages. Tangible damages mean the damages that can be valued in monetary terms, e.g. damages to buildings and contents, while intangible damages are difficult to be expressed in monetary terms, e.g. loss of life and trauma (Jonkman et al. 2008; Merz et al. 2010). Most studies only estimated direct tangible damages (see Gissing and Blong 2004; Kreibich et al. 2010), because the estimation of intangible damages is difficult.

Flood damage is often classified into actual and potential damage (Smith 1981). The former refers to the damage resulting from a particular flood event (Gissing and Blong 2004), while the latter estimates maximum possible damage, which may occur if an area without any flood protection measures becomes inundated (Messner and Meyer 2007). In addition, expected damage refers to flood risk (Gartsman et al. 2009), which is the product of the probability of flooding (which is defined as "the probability that floods of a given intensity and a given loss will occur in a certain area within a specified time period" (Merz et al. 2007, p.238)) and the damages when a flood occurs.

The spatial boundaries of a flood damage assessment influence its result (Merz et al. 2010). Based on a spatial approach, flood damage assessment can be conducted at the micro-, meso-, and macro-scales, which are estimates of the damage at the object level, land-use category, and municipal level, respectively (Messner and Meyer 2007).

Most studies estimate flood damage using damage functions, which connect the damages to particular elements at risk with the flood characteristics (Merz et al. 2010). The element at risk represents the number of social, economic, or ecological units, which are affected by flood hazard in a particular area, i.e. people, households, companies, and infrastructures, while the flood characteristics include its depth, duration, velocity, and contamination (Messner and Meyer 2007). Thieken et al. (2005) use impact and resistance factors in determining flood damage to buildings. The impact factors are similar to flood characteristics, while resistance factors are the resistance condition of the building to flood, which can be temporary (e.g. due to flood warning and preparedness) or permanent (e.g. due to building material and precautionary measures).

A flood damage function can be developed by using empirical and synthetic approaches (Merz et al. 2010). The empirical approach implies using data gathered following a particular flood event, while the synthetic approach uses data collected through "what-if-questions", i.e. the amount of expected damage in case of a certain flood condition (Merz et al. 2010; Thieken et al. 2005). However, a synthetic approach involves trade-offs between required time and precision, because it relies on fictional scenarios from which information is derived (Smith 1994).

Previous studies developed flood damage functions by relating the direct flood damage to flood depth (see Jonkman et al. 2008; Kazama et al. 2010; Scawthorn et al. 2006; Suriya et al. 2012). Such a function is called a stage-damage function or depth-damage function. Meanwhile, other studies add other parameters, such as flood duration (e.g. Dutta et al. 2003; Tang et al. 1992), velocity (Thieken et al. 2005), and contamination (Kreibich et al. 2010). The two steps to analyse flood damage include determining the impact factors or the exposure indicators, e.g. depth and duration, and assessing the damage in monetary terms (Messner and Meyer 2007).

3 METHODOLOGY

We conducted surveys on households and business units from April to June 2013 to obtain information on flood damage, flood characteristics, and socio-economics characteristics, based on the 17–19 January 2013 flood event. Based on those data, we used regression analysis to develop flood damage functions to know how flood characteristics are related to the damage. Finally, we compared the flood damage obtained from the survey with the estimated flood damage obtained by using the damage scanner model of Jakarta for the relevant area.

3.1 Study area

Our survey focused on the residential and business areas along the *Pesanggrahan* River (see Fig. 1). For the business sector, the survey also covered some parts of the *Angke* River in West Jakarta.

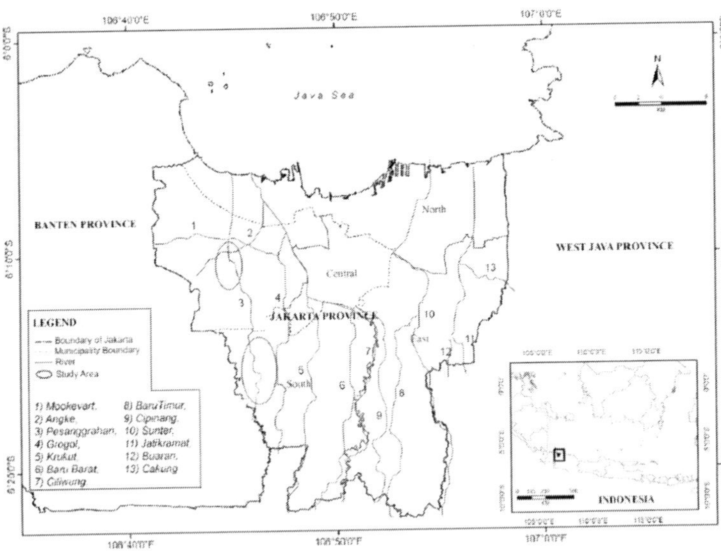

Figure.1 Map of Jakarta indicating the study area

We conducted the survey in two municipalities: South Jakarta and West Jakarta. Figure 1shows our study area, which extends from south to northwest Jakarta. The study area covers six villages with a total area of 12.3 km^2. In South Jakarta, the survey included *Pesanggrahan* and *Kebayoran Lama* Districts and covered three villages, i.e. *Ulujami, Kebayoran Lama Utara,* and *Cipulir*. While in West Jakarta, it included *Kebun Jeruk* and *Cengkareng* Districts and covered three villages, i.e. *Kedoya Selatan, Sukabumi Selatan*, and *Rawa Buaya*.

Those villages are densely populated areas, and most of the population lives in slums and works in the informal sector. During 2013, at least seven episodes of flooding on different scales occurred in these areas. The January 2013 flooding was caused by a heavy rainstorm, which resulted in the overflowing of the *Angke* and the *Pesanggrahan* Rivers (Deltares 2013) and inundated 1706 houses in those villages.

3.2 Sampling methods

Samples were selected by using a two-stage sampling method. First, villages along the river were listed and the number of flooded neighbourhoods within villages was identified. Second, the neighbourhoods for the survey were selected by purposive sampling. Household samples were then selected in various neighbourhoods by using simple random sampling. The total number of samples was about 300 households and 150 businesses.

Households and business units were eligible for the survey if they had been affected by the flood and had suffered from economic, health, and/or physical damages. The contact person at the household level was the head of household, while the contact person for businesses was the owner or the manager. If a contact person was not eligible or inhabited at the time of the survey, the next closest household or business unit was approached.

3.3 Model specification to develop flood damage function

We employed a micro-scale approach to assess damages in the residential and business sectors. The flood damage function was estimated empirically. In our model, the flood damage was explained by two types of factors, i.e. flood impact factors (or flood characteristics) and socio-economic factors. The first set of factors consist of depth, duration, and distance from buildings to a river. The depth is the over-floor depth of the flood, while the duration indicates the length of time that the water stays inside the buildings. The second set of factors include the income of the sector and the surface area of the building. For the residential sector, income was represented by monthly income of households, while the business sector was represented by monthly turnover. The surface area indicates how many square metres of the living area owns by each building.

We developed the flood damage function to indicate the economic relationship between actual flood damage (AFD) as a dependent variable, and flood characteristics and socio-economic factors as independent variables. Because data collected by the survey are not normally distributed, we tested several methods for regressions analysis and considered several alternative combinations of variables and functional forms, e.g. linear, log-linear, double log, and quadratic, by including and excluding the constant. Finally, we applied the linear functional form, which gave the best results in terms of statistical significance:

$$\mathrm{AFD}_i = \beta_0 + \beta_1 \mathrm{DEP}_i + \beta_2 \mathrm{DUR}_i + \beta_3 \mathrm{DIS}_i + \beta_4 \mathrm{INC}_i + \beta_5 \mathrm{ARE}_i + \varepsilon_i \quad (1)$$

where AFD_i, actual flood damage (IDR thousand) of respondent i; DEP_i, flood depth (cm) inside building of respondent i; DUR_i, flood duration (h) inside building of respondent i; DIS_i, distance from a river to the building (or housing) (m) of respondent i; INC_i, income (IDR thousand/month) of respondent i; ARE_i: building (or housing) area (m^2) of respondent i; ε_i, error term.

The dependent variable AFD covered the direct and indirect tangible damage. The direct damage included building structure damage and building content damage. The content meant the detailed inventories of the physical objects inside and outside the buildings. To estimate the building structure damage, we used replacement value, while to assess the building contents, we used depreciated value with a depreciation rate of 1% per year. The indirect damage in the residential sector included clean-up costs, loss of income, costs related to evacuation, and health-related costs during the flood. Meanwhile, for the business sector, the indirect damage costs were approximated by the loss of turnover during the flood, because detailed information on profit loss could not be obtained as business men are unwilling to report about profits.

The loss of income of households was calculated by multiplying the number of days missed at work due to flooding with the daily income. The costs of evacuation and temporary accommodation were the sum of the total costs for travel, food, and lodging due to evacuation. The cost due to illness is equal to the costs of visiting a doctor, staying in hospital, and buying the medicines required following a flood.

For the given obtained samples, we estimated the flood damage function by the ordinary least squares (OLS) estimator. The OLS produces unbiased and efficient estimated coefficients in the linear model (Greene 2012; Verbeek 2012). Based on Eq. (1), the null hypothesis is that each coefficient has no effect on the expected AFD. We tested the hypothesis by means of t tests and F tests, and we report R-square (R^2). In order to detect multicollinearity and homoscedasticity, we used the variance inflation factor (VIF) value and white test, respectively. Having the estimated coefficients and corresponding standard errors, we interpreted the results and used the model to estimate the annual AFD. The observed data were analysed using software STATA 10.

3.4 Damage scanner model

The damage scanner model has three major inputs, i.e. (1) flood hazard, represented by the inundation depth and extent map, (2) flood exposure, represented by the land-use value, and (3) flood vulnerability, represented by the depth-damage curve (te Linde et al. 2011). The damage scanner has been described in several studies (te Linde et al. 2011; Ward et al. 2011a). This study employed the damage scanner model for Jakarta developed by Budiyono et al. (2015), which used *Python* programming, particularly *Numpy* and GDAL libraries to produce the flood risk maps and *CorelPolyGUI add-ons* on *LibreOffice Calc* to plot the exceedance probability.

To estimate flood damage in the survey areas by the damage scanner, the initial step is to know the return period of rainfall that occurred from 15 to 18 January 2013. The return period is the expected value of the recurrence interval (time between occurrences) of rainfall events of the same magnitude (Mays 2011). The daily rainfall intensity during four days was recorded in the *Pondok Betung* weather station managed by the Indonesian agency for meteorology, climatology, and geophysics (BMKG). The sum of those rainfall data produced the accumulated rainfall intensity that was used to indicate the return period of the flood. BMKG provides the calculation of the accumulated rainfall intensity (extreme rainfall) for various return periods from 2 to 100 years, based on the generalized extreme value (GEV) method by employing rainfall data from 1981 to 2010 (Table1).

Table1 Extreme rainfall and its return period in *Pondok Betung* station, Jakarta, from 1981 to 2010 with GEV method *Source*: BMKG and KNMI (2013)

Return period	2-year	5-year	10-year	20-year	25-year	50-year	100-year
Rainfall (mm/day)	116.6	156.9	190.6	229.4	243.2	291.1	347.6

The flood hazard map was acquired from the Flood Hazard Mapping (FHM) project conducted by Deltares in 2012. It was produced by using the one-dimensional hydraulic module (1D) and two-dimensional hydraulic module (2D) of the SOBEK hydrology model to study the hazard over the greater Jakarta catchment, including the *Pesanggrahan* River. The 1D simulates one-dimensional flow (i.e. water level and discharges) through the main rivers and the main drainage system, and 2D simulates the inundation pattern over the project area from the locations where the 1D water courses are overtopped (Deltares2012). Inputs to the model characterize the 2007 flood occurrence. The hazard map acquired from FHM has not been validated on the event from the survey data, because the FHM used data obtained before January 2013.

The information on flood exposure, i.e. the value of assets that are exposed to flood, was gathered from a previous study conducted by Budiyono et al. (2015). To compare the damage scanner results with the survey results, we focused on only two land-use classes, i.e. the high-density urban village (representing the residential sector), and the commercial and business area (representing the business sector). An expert workshop, which involved experts consisting of representatives from government agencies, public organizations, and businesses, was organized to estimate the maximum extent of damages created for each land-use class.

The vulnerability curves indicate the percentage of damage that would occur at different flood depths in each land-use class (Merz et al. 2010). Here, the curves were developed in the workshop by using a synthetic stage-damage curve through "what-if-questions" on each land-use class at incremental 25 cm depths up to maximum inundation depth of 200 cm. Prior to discussion, a description on the possible structure and assets was delivered to the experts to give an overview on the assets exposed to flooding. Furthermore, experts from related fields discussed the land-use class groups to reach a consensus on the damages and losses for various scenarios of depth.

4 RESULTS

In Sects. 4.1, 4.2, and 4.3, we present the descriptive statistics of the damage data set, the actual flood damage based on the collected data from survey, and the relationship between flood damage and the flood characteristics as obtained in our regression analysis of the survey data. Finally, Sect. 4.4 compares the results of the flood damage estimates obtained from the survey with flood damage estimates for the relevant area based on the damage scanner model.

4.1 Descriptive statistics

Household included in the survey were living, on average, 84.8 m away from the river, with a maximum distance of 284 m. Their houses were mostly permanent constructions with a total area of less than 70 m^2, on average. Half the respondents lived in one-storey houses and half in two-storey houses. The second floor was deliberately built as an adaptation mechanism to flooding. Their monthly income in 2013 was on average of US$ 344, and 42% received monthly income below the standard regional minimum income, which is US$ 226. The majority of livelihoods of local people are mostly in the informal sector, and 36% of respondents claimed they earned a living from running small businesses. From April 2012 to April 2013, 53% respondents experienced 6–15 flood events. During the January 2013 flood, there was on average 87 cm of water in the house for 98 h.

Most of the business units were in charge of small- to medium-scale enterprises, e.g. small shops, groceries, cloth shops, and small restaurants. About 81% respondents employed one to four workers, and about 75% had less than US$ 205 business daily turnover. Most businesses occupied small buildings to run their activities. On average, the area including the garage was 38 m^2. Not all buildings had a basement, and 87% were permanent buildings. About 72% had an attic, and 13% had a garage or carport. Buildings were located, on average, 84.8 m away from the river, up to 123.7 m away from the river. About 51% of business activity took place in their buildings. From April 2012 to April 2013, 45% of respondents experienced 6–15 flooding events. After the January flood, three flood events occurred in the survey areas; however, the magnitude of those events was small and did not cause significant losses. During the January 2013 flood, the average inundation depth was 74 cm and there was water in the buildings for 84 h.

4.2 The actual flood damage

For the residential sector, the average flood damage per household was US$ 308. Households with more property at risk, such as vehicles and electronics, experienced the highest flood damages. For most households, the direct damage was greater than the indirect one. Table 2indicates that the content damage was the major component of losses.

The content losses varied across households from zero to an estimated US$ 863. The zero value is obtained in the very poor households, which have limited household stuffs. The higher losses were found in households with higher family income and larger houses. Households who owned a two-storey house could safely store their assets on a higher floor; hence, they could minimize the losses. After the flooding episode, households spent on average 24 h or three person-days to clear their houses of mud and to tidy up the house contents. The average number of days missed from work was four, and the income loss per household during those days varied from zero to US$ 359.

Table 2 Average actual flood damage per household for the 17–19 January 2013 flood event

Damage	Value (US$)	Percentage (%)
1 Direct		
1a Structural damage	43	14
1b Content damage (inside and outside)	193	63
2 Indirect		
2a Clean-up cost	25	8
2b Loss of income	30	10
2c Evacuation and temporary house	12	4
2d Cost of illnesses	5	2
Total	308	100

About 70% of the respondents moved to temporary places such as family houses, mosques, or evacuation camps because they were evacuated or their homes needed repair as a result of the flood. On average, they stayed in temporary shelters for about five to six days and spent about US$ 12 per day for travelling, food, water, and lodging.

Regarding the cost of illnesses, the number of family members who suffered from several diseases during and after the flood event varied from one to three people per house and most of them were children. They suffered from fever (34%), skin irritation (21%), and diarrhoea (18%). Although 89% visited a doctor, the average cost of illnesses was low for each household (not for society) because 45% visited doctors provided by the government or social organizations, which those are provided freely. Theoretically the freely provided assistance should also be counted in the costs; however, due to untraceable information this was not possible in practice. To what extent the cost of illness is due to the flood is not clear. However, the damages related to the cost of illness are relatively low, and if only a share of these damages is attributed to the flood event, the overall results would not change substantially.

For the business sector, the average flood damage per business unit was US$ 854. About 12% of the respondents had flood damage less than US$ 154, and 10% of the respondents had flood damage in excess of US$ 2054. Business units with more items in their properties experienced a higher flood loss.

For most business units, the total indirect damage was greater than the total direct damage. Total direct damages including structural and contents damages were US$ 216, whereas total indirect damages including turnover loss, temporary premises cost, labour cost, and clean-up cost were US$ 638.

In the direct costs, the structural damage was less than content damage because most owners had small buildings. Content damage was quite high, and it showed that the owners did not have enough time to evacuate their contents. Table 3 shows that the highest cost was the turnover loss, which depends on the number of days the businesses had to close due to flooding. About 93% of the respondents closed their business due to flooding with an average five-day closure. Nevertheless, only 4% of businesses set up temporary headquarters at another location.

Table 3 Average actual flood damage per business unit for the 17–19 January 2013 flood event

Damage	Value (US$)	Percentage
1 Direct		
1a Structural damage	58	7
1b Content damage	158	19
2 Indirect		
2a Turnover loss	540	63
2b Temporary quarters	17	2
2c Labour cost	40	5
2d Clean-up cost	41	5
Total	854	100

4.3 Analysis of the relationship between flood characteristics and flood damage

4.3.1 Residential sector

The flood damage function in the residential sector produces the results given in Table 7 in "Appendix". The function has R^2 0.26, and this means 26% of the variation in household AFD can be attributed to the five exogenous variables. DEP, DUR, INC, and ARE have p values lower than 0.05. We therefore reject the null hypothesis that those variables have no effect on AFD and show that those variables are good predictors of AFD. Our hypothesis was that the distance to the river (DIS) would have a negative sign. However, in the regression analysis it shows a positive sign. This might occur since households living close to the river have already taken precautionary actions, e.g. increasing floor height and constructing concrete walls in front of their houses. Furthermore, in some areas, the elevation area located close to the river is higher than that further away from the river; in that case, the overflowing water goes to the lower areas.

All coefficients have positive signs, which indicate that the change in an independent variable will increase the expected AFD, ceteris paribus. For example, increasing one additional centimetre on the flood depth will increase the expected damage by IDR 13, 040 (US$ 1.3), ceteris paribus. The *F*-statistic takes the value of 20.25, the appropriate 5% critical value being 2.10, which implies that the joint hypothesis of all coefficients being zero has to be rejected. All VIF values for five independent variables are less than 10, meaning that there is no multicollinearity or no correlation between ε_i and the independent variables. Moreover, this also indicates that the linear relationship among the independent variables leads to reliable regression estimates (Verbeek 2012). However, we rejected the null hypothesis of homoscedasticity because the NR^2 (76.83) is higher than $\chi^2(5)$ at 5% significant level (9.49). This indicates that heteroscedasticity is present in the model. Results from multicollinearity and homoscedasticity tests indicated that using OLS in the model may produce biased and inefficient estimated coefficients (see Appendix Table 7 for the statistical results). The estimated damages of flooding for the affected population are obtained by multiplying the average AFD by the number of affected households, i.e. 1706, which gives US$ 524,999.

To describe how the AFD varies under different income levels, Eq. (1) is adjusted for different income groups. Households were classified into three different income groups, i.e. low (≤US$ 226), middle (US$ 226–514), and high (>US$ 514). We used US$ 226 value as the lowest base since it is the minimum monthly wage rate for Jakarta workers based on the Decree of the Governor of Jakarta No. 182 year 2012.

We re-estimated Eq. (1) for the three different income groups and found that DEP and ARE positively and significantly affect the AFD for all three income groups (see Tables 8, 9, 10 in "Appendix"). Table 4 shows the estimated AFD for each income group, which reveals that the highest flood damage per household occurred in the high-income group.

4.3.2 Business sector

Estimation by OLS for the AFD in the business sector shows the results given in Table 11 in "Appendix". DEP, DUR, TUR, and ARE have high *t* values, which indicate that those variables are good predictors of AFD; however, variable DIS does not have a significant contribution to AFD. Using the DEP variable as an example, we can interpret that increasing one additional centimetre on the flood depth will increase the expected damage IDR 30,460 (US$ 3.1), ceteris paribus.

Table 4 Estimated actual flood damage (AFD) for the residential sector by income group

Income group	Damage per household (US$) (1)	Percentage from population (%) (2)	Households (units) (3) = (2) * 1706	Total damage (US$) (4) = (1) * (3)
Low	278	42.0	717	199,367
Middle	297	45.7	779	231,166
High	449	12.3	210	94,269
Total		100.0	1706	524,802

The *F*-statistic takes the value of 22.08, whereas the appropriate 5% critical value is 2.10, and the hypothesis of all five coefficients being zero has to be rejected. All VIF values for five independent variables are less than 10, which means there is no multicollinearity. Similar to the damage function in the residential sector, heteroscedasticity is present in the business damage function, because the NR^2 (65.75) is higher than $\chi^2(5)$ at 5% significant level (9.49). This implies that the damage function for the business sector also might be biased and inefficient.

The flood damage estimation for the population is obtained by multiplying average AFD with the number of affected business units, 816, which gives US$ 697,050. Furthermore, Eq. (1) is used to analyse AFD for two different turnover groups. The business flood damage function was analysed for two groups of businesses by turnover per year, i.e. micro (turnover ≤US$ 30,815) and small-medium (turnover between US$ 30,815 and US$ 513,584). For both models, DEP, DUR, and TUR positively and significantly influence the AFD (see Tables 12, 13 in "Appendix"). Furthermore, we calculated the total damages for each group by multiplying the coefficient of each independent variable with the means of the independent variables in each model. Table 5indicates that the flood damage per business in the "small-medium group" is 2.6 times higher than that in the "micro-group".

Table 5 Estimated actual flood damage (AFD) in the business sector by business turnover group

Business turnover	Damage per building (US$) (1)	Percentage from population (%) (2)	Number of business (units) (3) = (2)*816	Total damage (US$) (4) = (1)*(3)
Micro	465	49	400	186,000
Small-medium	1233	51	416	512,928
Total		100	816	698,928

Figure2shows the relationship between flood depth and flood damage. Note that the flood depth is only one of the five variables that were used to explain the AFD. Therefore, these scatter plots just indicate a small part of the relations in the full model, other things being equal.

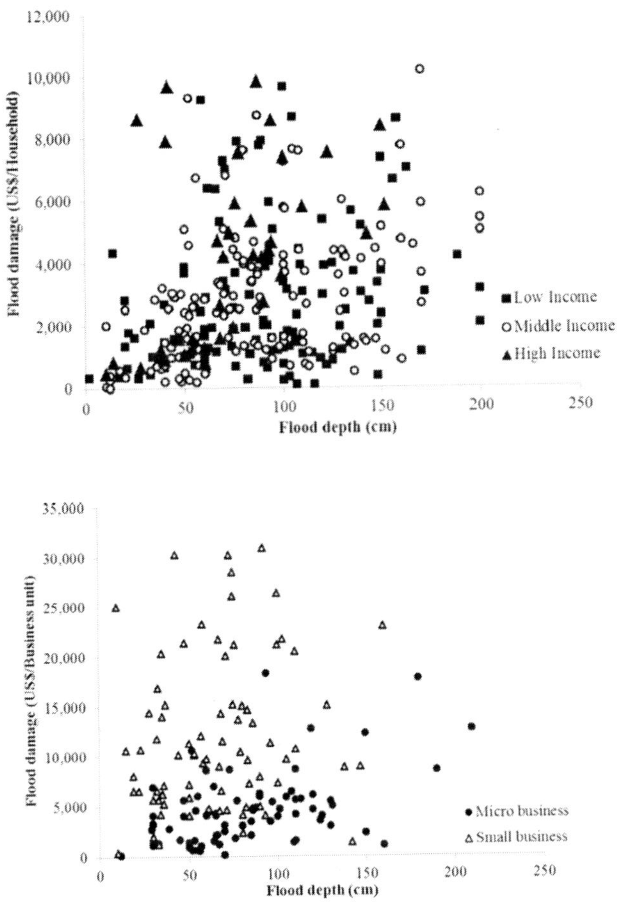

Figure. 2 Relationship between depth and flood damage in residential sector (*left*) and business sector (*right*)

4.4 Comparison of flood damage estimations by survey and by damage scanner

The daily rainfall intensity from 15 to 18 January 2013 was recorded in the *Pondok Betung* weather station as follows: 94, 27, 59, and 76.5 mm/day.

Therefore, the accumulated rainfall intensity during those four days was 256.5 mm, which is the equivalent of 30-year flood return period (FRP). To simulate the 30-year flood risk, an interpolation of damage maps of 25- and 50-year FRPs is masked using the Jakarta districts boundary map.

The 30-year FRP damage map was overlaid with the survey areas to see the locations with land-use classes for the residential sector (high-density urban village) and the small and medium business sector (business and commercial). The simulated 30-year FRP most represented the survey. The natural logarithmic function produces the flood damage model as follows:

$$Y = -353{,}230 \ln x + 91{,}745 \tag{2}$$

where$Y$ represents the flood damage (US$ thousand) and x represents the exceedance probability. The model shows $R^2 = 0.97$. The workshop estimated the maximum damage value at US$ 155,400/ha for residential sector or US$ 930 per household (assuming that the standard house area is 60 m^2 per house), and US$ 517,900/ha for business sector or US$ 5179 per business unit (assuming that the standard business unit area is 100 m^2 per building). Figure 3 shows the vulnerability curves for the two sectors, and Fig. 4 shows the flood damage curve under different exceedance probabilities.

The resulting clips taken using damage scanner show over six districts in the survey areas and the 760 inundated cells (190 ha). Total direct flood damage cost is estimated at US$ 18.2 million over 12 land-use classes. Furthermore, if the total damage is further analysed using land-use classes that were used in the survey approach, the total flood damage cost is US$ 10.5 million, in which the residential sector damage is US$ 1,318,235 and commercial and business is US$ 9,248,201.

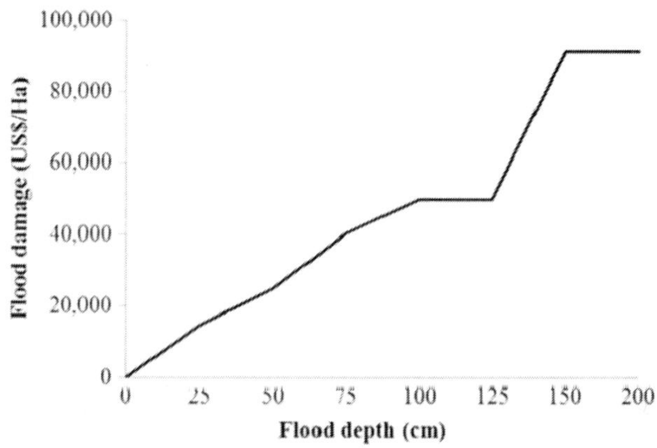

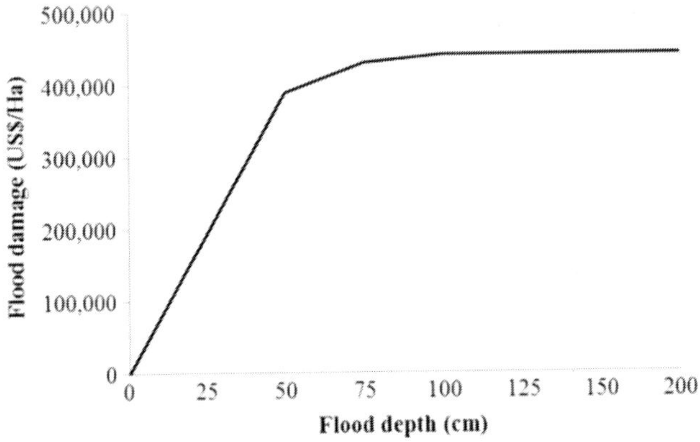

Figure.3 Vulnerability curves for residential sector (*left*) and business sector (*right*)

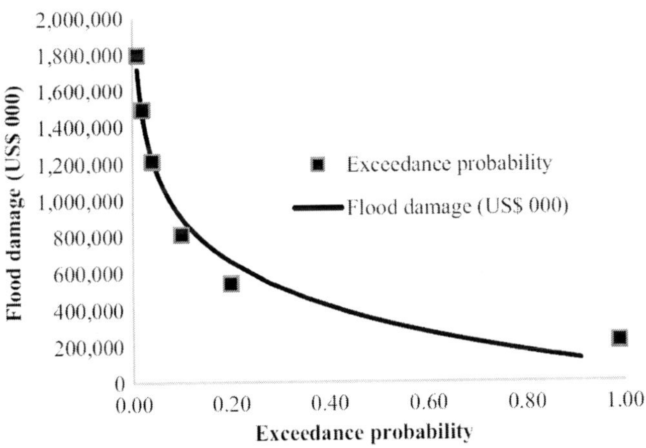

Figure.4 Flood damage in Jakarta under different exceedance probabilities in the damage scanner

Figure 5 shows an illustration of flood damage in Jakarta under the 50-year flood return period scenario. The grey and light blue areas indicate the survey areas.

The flood damage resulted by the two approaches is shown in Table 6. The estimated damage according to damage scanner is higher than the survey approach, in which the damage estimates differed by about a factor of 2.5 for the residential sector and 13 for the business sector.

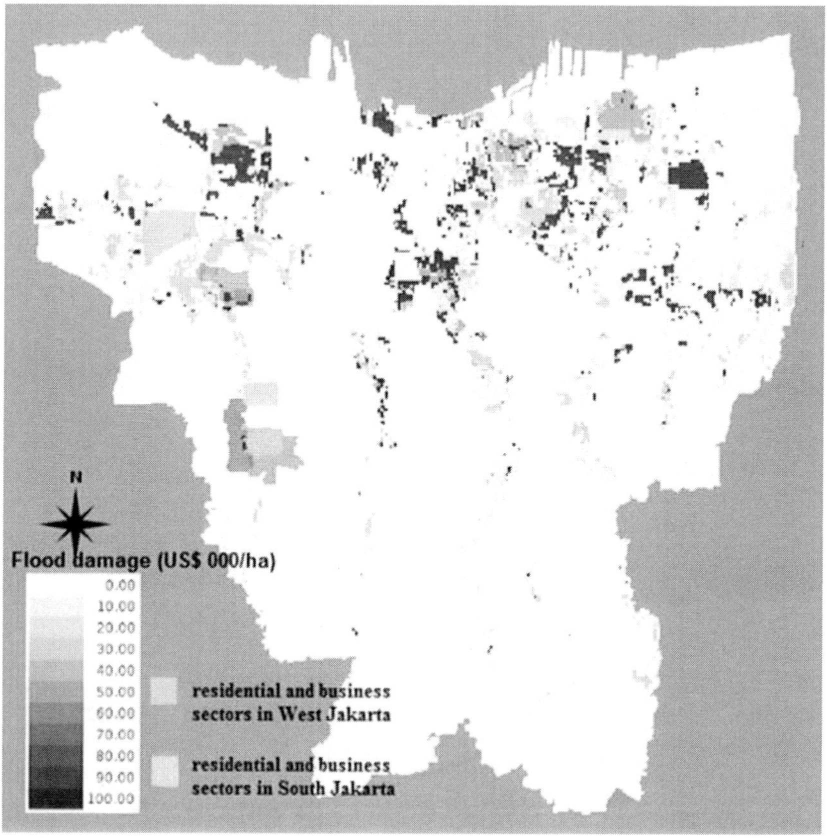

Figure.5 Jakarta flood risk map under a 50-year flood return period

Table 6 Estimated flood damage cost between 17 and 19 January 2013 under survey and damage scanner

Sector	Flood damages (US$)	
	Survey	Damage scanner
Residential	524,999	1,318,235
Business	697,050	9,248,201

5 DISCUSSION AND CONCLUSION

In this paper, we compare two different approaches of estimating river flood damage in Jakarta that are based on the survey and the damage scanner. Both approaches were conducted separately. Below, we discuss the differences between the results of the two approaches.

The regression analysis from the survey data showed significant relationships among flood characteristics, socio-economic factors, and flood damage, but only a fraction of the damages could be explained by the variables included in the model. Similar to Thieken et al. (2005) and Islam (1997), this study showed that depth, duration, income, and house area are positively correlated with the extent of flood damage in the residential sector. Nevertheless, the R^2 for the residential damage function was only 0.26. Earlier studies also found low R^2 for residential damage function, e.g. 0.25 (Islam 1997) and 0.35 (Pistrika and Jonkman 2010).

Our surveys were conducted in a small area, and the reported flood damage did not present a statistically significant relationship with flood depth. Similarly, Pistrika and Jonkman (2010) found that there is no strong one-to-one relationship between depth and flood damage for stage-damage function developed for small cluster of buildings. Flood depth by itself cannot explain the damages, unless other factors, e.g. duration, contamination, and velocity, are involved (Middelmann-Fernandes 2010; Pistrika and Jonkman 2010). However, in our study, contamination and velocity were beyond the scope of the study, because those data are not available. Furthermore, a large part of the damages may be independent of the flood depth once a certain threshold of flooding has been passed and this may explain why flood depth is not significant.

In the business sector, depth, duration, daily turnover, and business area are positively correlated with the damage. This result corresponds with Islam (1997) and Kreibich et al. (2010). The regression analysis showed that the relationship between flood characteristics and flood damage was moderate with R^2 0.43. As comparison, a previous study conducted by Tang et al. (1992) showed that depth is a significant factor for commercial flood damage with R^2 0.9. Tang et al. (1992), however, used a bigger sample which included 951 cells for the commercial sector.

The relatively low R^2 values obtained in our study may imply that the regression method is not suitable or there is no strong relationship between the variables. We applied a traditional regression model to develop flood damage functions as employed by previous studies (e.g. Appelbaum 1985; Gissing and Blong 2004; Islam 1997; Tang et al. 1992), and the scope of our survey was limited because of this traditional set-up. Merz et al. (2013) found that applying decision tree technique in selecting damage influence variables and in deriving multivariate flood damage models outperform the traditional flood damage models. Decision tree analysis is indeed promising, but it needs large data sets to identify complex relationships. Compared to Merz et al. (2013) who employed 28 predictor variables obtained by 2158 interviews, our data set only consists of 5 variables that focused on flood characteristics and socio-economic factors obtained by 450 interviews. For further research, we recommend to consider the decision tree analysis already from the design phase of the research project and to design the survey accordingly.

There are large differences in outcomes between the survey results (focussing on actual damages) and the damage scanner results (focussing on flood damage modelling), which is also found by previous studies. For example, Jongman et al. (2012) showed that reported flood damage was higher than flood damage estimation based on seven flood damage models. Chatterton et al. (2012) found the damage estimation by two different flood damage models for residential and commercial sectors in UK differed by a factor of about 5–6.

The following causes explain the differences in the results from the two approaches. First, both approaches applied different ways in obtaining flood damage data and in defining the maximum flood damage per object. The survey directly collected the damage data from the samples that really were affected by flooding, while the damage scanner represents an estimation of the damage as seen by experts. Therefore, the survey was used to create an empirical flood model, while the damage scanner was used as a synthetic model (Jongman et al. 2012; Merz et al. 2010). In detail, the survey defined the maximum damage per object as the average of total damage that is incurred by houses and micro- as well as small business units located in slump areas of Jakarta. Particularly, the survey might underestimate the flood damages in the business sector, because it entailed some difficulties in interviewing large business managers, since they often refused to give detailed information. Nonetheless, the survey conducted detailed investigation per unit sample by using depreciation in assessing the structure and contents that were exposed to the flood, and by including the reduction in the damages as a result of the early flood warning and evacuation. The damage scanner defined the maximum damage per object as the expected maximum damage corresponding with several depth scenarios that are generally incurred by houses and big business located in Jakarta. It assumed a uniform building distribution, and the assessment did not consider depreciation and damage reduction effects. Based on our survey, valuing structure and content with depreciation value significantly produced lower values (about 10–20%).

Second, the survey approach used the micro scale (local scale), while the damage scanner approach used the meso-scale (regional scale). The survey produced an actual and detailed estimation, but it is labour-intensive especially for a large-scale study. Meanwhile, the damage scanner gave a more global estimate of the potential flood damage in Jakarta, but the calculations tend to be less accurate at the detailed level.

Third, uncertainties are present in both the survey and the damage scanner approaches. Following Wagenaar et al. (2016), uncertainties in the survey arise because flood damage was assessed based on only one single event and limited to certain ranges of depth and duration. In the damage scanner model, uncertainties appear in the construction of damage curve, the asset values associated with the curve and difference in methodological framework (Jongman et al. 2012; Merz et al. 2010; Wagenaar et al. 2016). In the damage scanner, aleatory uncertainty occurs because it used average data in valuing the maximum damage per object, assumed a uniform damage per object in each land-use class and defined expected damage by expert judgement with artificial depth scenarios. These factors lead to potential errors in the damage curve. To reduce this uncertainty, the average between different functions could be use as damage function (Wagenaar et al. 2016). Additionally, Pistrika and Jonkman (2010) found that significant uncertainty arises in damage curves when the curves are applied in small case study areas.

The hazard map input of the damage scanner could be a possible explanation for the differences found between the survey and the damage scanner results. The depths obtained by the survey are represented at a different spatial scale from the depths that are represented by the hazard map. The survey samples were not designed to represent inundation at the spatial scale of the pluvial flood model that plays a role in the damage scanner, but has been based on random sampling for each village included in the study. The survey was designed to obtain a spatial damage map within an administrative scale from a socio-economics point of view. Meanwhile, the SOBEK model was designed to produce spatial inundation within the *Pesanggrahan* catchment at a different spatial scale. This can be seen as a limitation of the SOBEK hydrology model (Deltares 2012) in producing the hazard map because the spatial resolution is too aggregated.

Mapping the results of the two approaches can be used to learn more about what contribution the hazard map has to the difference found between the two approaches, but this can only be done if the spatial scales are sufficiently similar. Unfortunately, we could not develop a detailed damage map from the survey. Since the damage scanner employs aggregated land-use data instead of individual units, we suggest for future research that the survey results can be used to improve the accuracy of the damage scanner model.

Table 7 STATA output for estimating flood damage function, multicollinearity test, and heteroscedasticity test in residential sector

Source	SS	df	MS
Model	409906773	5	81981354.6
Residual	1.1904e+09	294	4048948.95
Total	1.6003e+09	299	5352166.43

Number of obs	= 300
F(5, 294)	= 20.25
Prob > F	= 0.0000
R-squared	= 0.2561
Adj R-squared	= 0.2435
Root MSE	= 2012.2

AFD	Coef.	Std. Err.	t	P>|t|	[95% Conf.	Interval]
dep	13.03926	3.237011	4.03	0.000	6.668614	19.40992
dur	4.473057	2.552958	1.75	0.081	-.5513312	9.497446
dis	2.65409	1.828448	1.45	0.148	-.9444159	6.252595
inc	.1967721	.0366812	5.36	0.000	.1245812	.2689631
are	7.961585	1.762233	4.52	0.000	4.493394	11.42978
_cons	-14.3507	358.5665	-0.04	0.968	-720.0331	691.3317

Variable	VIF	1/VIF
dep	1.35	0.741247
dur	1.30	0.767737
are	1.08	0.927431
inc	1.07	0.938577
dis	1.05	0.950729
Mean VIF	1.17	

```
White's test for Ho: homoskedasticity
         against Ha: unrestricted heteroskedasticity

         chi2(20)     =     27.17
         Prob > chi2  =    0.1306
```

Cameron & Trivedi's decomposition of IM-test

Source	chi2	df	p
Heteroskedasticity	27.17	20	0.1306
Skewness	22.72	5	0.0004
Kurtosis	7.71	1	0.0055
Total	57.59	26	0.0004

Finally, the real damage information has a greater accuracy than synthetic data (Gissing and Blong 2004), but we are also aware of the need to make assessments for larger areas in Jakarta. For further research, we recommend to also use decision tree analysis as used in Merz et al. (2013) and to design the surveys accordingly.

Table 8 STATA output for estimating flood damage function in low-income residential sector

Source	SS	df	MS
Model	79522374.6	5	15904474.9
Residual	537167094	119	4514009.19
Total	616689468	124	4973302.17

Number of obs =	125
F(5, 119) =	3.52
Prob > F =	0.0053
R-squared =	0.1290
Adj R-squared =	0.0924
Root MSE =	2124.6

AFD	Coef.	Std. Err.	t	P>|t|	[95% Conf.	Interval]
dep	10.95329	5.279813	2.07	0.040	.4987316	21.40785
dur	-.4213087	3.967084	-0.11	0.916	-8.276531	7.433914
dis	4.615639	2.94367	1.57	0.120	-1.213122	10.4444
inc	-.4368561	.4308509	-1.01	0.313	-1.289984	.4162717
are	11.81916	4.808209	2.46	0.015	2.298428	21.3399
_cons	1388.445	907.6645	1.53	0.129	-408.8212	3185.712

Table 9 STATA output for estimating flood damage function in middle-income residential sector

Source	SS	df	MS
Model	171260893	5	34252178.6
Residual	438661911	132	3323196.3
Total	609922805	137	4451991.27

Number of obs =	138
F(5, 132) =	10.31
Prob > F =	0.0000
R-squared =	0.2808
Adj R-squared =	0.2535
Root MSE =	1823

AFD	Coef.	Std. Err.	t	P>|t|	[95% Conf.	Interval]
dep	11.92706	4.262985	2.80	0.006	3.494456	20.35967
dur	9.469953	3.464846	2.73	0.007	2.616145	16.32376
dis	2.662335	2.468928	1.08	0.283	-2.221448	7.546117
inc	.4310882	.218537	1.97	0.051	-.0011997	.8633761
are	6.323553	2.376223	2.66	0.009	1.62315	11.02396
_cons	-1108.882	836.4048	-1.33	0.187	-2763.373	545.6092

Table 10 STATA output for estimating flood damage function in high-income residential sector

Source	SS	df	MS
Model	134659129	5	26931825.9
Residual	157136084	31	5068905.93
Total	291795213	36	8105422.59

Number of obs =	37
F(5, 31) =	5.31
Prob > F =	0.0012
R-squared =	0.4615
Adj R-squared =	0.3746
Root MSE =	2251.4

AFD	Coef.	Std. Err.	t	P>|t|	[95% Conf.	Interval]
dep	30.52812	12.81071	2.38	0.023	4.400497	56.65574
dur	7.122202	11.29376	0.63	0.533	-15.91158	30.15598
dis	-1.408446	6.788739	-0.21	0.837	-15.25417	12.43728
inc	.2158973	.0768904	2.81	0.009	.0590784	.3727163
are	7.665146	3.348846	2.29	0.029	.8351303	14.49516
_cons	-1408.86	1413.429	-1.00	0.327	-4291.567	1473.848

Table 11 STATA output for estimating flood damage function, multicollinearity test and heteroscedasticity test in business sector

Source	SS	df	MS
Model	3.2915e+09	5	658293192
Residual	4.2183e+09	144	29293627.8
Total	7.5097e+09	149	50400995.7

Number of obs	=	150
F(5, 144)	=	22.47
Prob > F	=	0.0000
R-squared	=	0.4383
Adj R-squared	=	0.4188
Root MSE	=	5412.4

AFD	Coef.	Std. Err.	t	P>|t|	[95% Conf.	Interval]
dep	30.46068	13.08076	2.33	0.021	4.605586	56.31578
dur	29.54769	9.019176	3.28	0.001	11.72061	47.37477
dis	1.825865	2.896997	0.63	0.530	-3.900266	7.551996
tur	2.284552	.2295142	9.95	0.000	1.8309	2.738204
are	13.17589	4.614319	2.86	0.005	4.055345	22.29644
_cons	-1245.373	1340.576	-0.93	0.354	-3895.123	1404.376

Variable	VIF	1/VIF
dur	1.28	0.780258
dep	1.25	0.801639
tur	1.18	0.850134
dis	1.05	0.954844
are	1.02	0.977558
Mean VIF	1.16	

White's test for Ho: homoskedasticity
against Ha: unrestricted heteroskedasticity

chi2(20) = 49.46
Prob > chi2 = 0.0003

Cameron & Trivedi's decomposition of IM-test

Source	chi2	df	p
Heteroskedasticity	49.46	20	0.0003
Skewness	6.99	5	0.2211
Kurtosis	4.40	1	0.0359
Total	60.85	26	0.0001

We tend to agree with Middelmann-Fernandes (2010) and Wagenaar et al. (2016), who emphasized the consideration of multiple methods in assessing flood damage. Thus, a combination of a damage assessment based on survey and flood model based on expert assessment such as damage scanner provides a better picture of flood damage in a flooding-prone area

Table 12 STATA output for estimating flood damage function in micro-business sector

Source	SS	df	MS
Model	361072171	5	72214434.2
Residual	607182946	68	8929160.97
Total	968255117	73	13263768.7

Number of obs =	74
F(5, 68) =	8.09
Prob > F =	0.0000
R-squared =	0.3729
Adj R-squared =	0.3268
Root MSE =	2988.2

AFD	Coef.	Std. Err.	t	P>|t|	[95% Conf.	Interval]
dep	30.81265	9.493503	3.25	0.002	11.86866	49.75664
dur	18.98813	6.458346	2.94	0.004	6.100697	31.87555
dis	1.557442	2.104255	0.74	0.462	-2.641534	5.756419
tur	2.711482	1.671413	1.62	0.109	-.6237697	6.046734
are	20.13589	11.23525	1.79	0.078	-2.283698	42.55547
_cons	-1760.934	1151.59	-1.53	0.131	-4058.895	537.0274

Table 13 STATA output for estimating flood damage function in small business sector

Source	SS	df	MS
Model	1.4019e+09	5	280380345
Residual	3.0428e+09	70	43468749.4
Total	4.4447e+09	75	59262855.7

Number of obs =	76
F(5, 70) =	6.45
Prob > F =	0.0001
R-squared =	0.3154
Adj R-squared =	0.2665
Root MSE =	6593.1

AFD	Coef.	Std. Err.	t	P>|t|	[95% Conf.	Interval]
dep	42.42023	24.76554	1.71	0.091	-6.973088	91.81355
dur	46.28122	18.1179	2.55	0.013	10.1462	82.41624
dis	3.310014	5.796919	0.57	0.570	-8.251577	14.8716
tur	1.80069	.3687375	4.88	0.000	1.065266	2.536114
are	24.95858	11.82158	2.11	0.038	1.381168	48.53598
_cons	-877.6077	2584.325	-0.34	0.735	-6031.882	4276.666

FOOTNOTE

1 We express values in US$ of January 2007, using an average exchange rate of US$ 1 = IDR 9125

2 We express values in US$ of January 2013, using an average exchange rate of US$ 1 = IDR 9735.5

ACKNOWLEDGEMENTS

This research was carried out with the aid of a grant from the *Economy and Environment Program for Southeast Asia of WorldFish* (PCO13-0308-012)

and the Dutch research programme *Knowledge for Climate and Delta Alliance* Jakarta Climate Adaptation Tools research project (HSINT02a)

REFERENCES

1. Appelbaum SJ (1985) Determination of urban flood damages. J Water Resour Plan Manag 111:269–283
2. BAPPENAS (2007) Laporan Perkiraan Kerusakan dan Kerugian Pasca Bencana Banjir Awal Februari 2007 di Wilayah Jabodetabek (Jakarta, Bogor, Depok, Tangerang, dan Bekasi). Badan Perencanaan Pembangunan Nasional (BAPPENAS) Indonesia, Jakarta **(in Bahasa)**
3. BMKG, KNMI (2013) Southeast Asian Climate Assessment & Dataset (SACA&D) project. Badan Meteorologi Klimatologi dan Geofisika (BMKG) and Royal Netherlands Meteorological Institute (KNMI). Accessed 1 Dec 2013
4. BPBD (2013) Kajian Kerusakan, Kerugian Dan Kebutuhan Pemulihan Dampak Bencana Banjir Jakarta Januari 2013. Badan Penanggulangan Bencana Daerah (BPBD) DKI Jakarta, Jakarta **(in Bahasa)**
5. Budiyono Y, Aerts J, Brinkman JJ, Marfai MA, Ward P (2015) Flood risk assessment for delta mega-cities: a case study of Jakarta. Nat Hazards 75:389–413. doi:10.1007/s11069-014-1327-9
6. Chatterton J, Penning-Rowsell EC, Priest S (2012) The many uncertainties in flood loss assessments. In: Beven K, Hall J (eds) Applied uncertainty analysis for flood risk management. Imperial College Press, London, pp 335–343
7. Deltares (2012) Jakarta flood management information system. FMIS and JFEWS. Towards flood early warning, planning, and design. Ministry of Public Works Indonesia, Jakarta
8. Deltares (2013) Jakarta flood January 2013. Royal Embassy of the Netherlands and Ministry of Public Works Indonesia, Jakarta
9. Dutta D, Herath S, Musiake K (2003) A mathematical model for flood loss estimation. J Hydrol 277:24–49. doi:10.1016/S0022-1694(03)00084-2
10. Firman T, Surbakti IM, Idroes IC, Simarmata HA (2011) Potential climate-change related vulnerabilities in Jakarta: challenges and current status. Habitat Int 35:372–378
11. Gartsman B, Nooyen RV, Kolechkina A (2009) Implementation issues for total risk calculation for groups of sites. Phys Chem Earth 34:619–625. doi:10.1016/j.pce.2008.12.001
12. Gissing A, Blong R (2004) Accounting for variability in commercial flood damage estimation. Aust Geogr 35:209–222. doi:10.1080/0004918042000249511
13. Green C (2003) Handbook of water economics: principles and practice. Wiley, Chichester
14. Greene WH (2012) Econometric analysis, 7th edn. Prentice Hall, Upper Saddle River
15. Islam K (1997) The impacts of flooding and methods of assessment in urban areas of Bangladesh. Middlesex University, London
16. Jongman B et al (2012) Comparative flood damage model assessment: towards a European approach. Nat Hazards Earth Syst Sci 12:3733–3752. doi:10.5194/nhess-12-3733-2012
17. Jonkman SN, Bočkarjova M, Kok M, Bernardini P (2008) Integrated hydrodynamic and economic modelling of flood damage in the Netherlands. Ecol Econ 66:77–90
18. Kazama S, Sato A, Kawagoe S (2009) Evaluating the cost of flood damage based on changes in extreme rainfall in Japan. In: Sumi A, Fukushi K, Hiramatsu A (eds) Adaptation and mitigation strategies for climate change. Springer, Tokyo, pp 3–17
19. Kreibich H, Seifert I, Merz B, Thieken AH (2010) Development of FLEMOcs—a new model for the estimation of flood losses in the commercial sector. Hydrol Sci J 55:1302–1314. doi:10.1080/02626667.2010.529815CrossRefGoogle Scholar

20. Mays LW (2011) Water resources engineering, 2nd edn. Wiley, Hoboken
21. Merz B, Thieken AH, Gocht M (2007) Flood risk mapping at the local scale: concepts and challenges. Springer, Dordrecht
22. Merz B, Kreibich H, Schwarze R, Thieken A (2010) Review article "assessment of economic flood damage". Nat Hazards Earth Syst Sci 10:1697–1724. doi:10.5194/nhess-10-1697-2010
23. Merz B, Kreibich H, Lall U (2013) Multi-variate flood damage assessment: a tree-based data-mining approach. Nat Hazards Earth Syst Sci 13:53–64. doi:10.5194/nhess-13-53-2013
24. Messner F, Meyer V (2007) Flood damage, vulnerability and risk perception—challenges for flood damage research. In: Schanze J, Zeman E, Marsalek J (eds) Flood risk management—hazards, vulnerability and mitigation measures, earth and environmental science, vol IV. Springer, Dordrecht, pp 149–168
25. Middelmann-Fernandes MH (2010) Flood damage estimation beyond stage-damage functions: an Australian example. J Flood Risk Manag 3:88–96. doi:10.1111/j.1753-318X.2009.01058.x
26. Pistrika AK, Jonkman SN (2010) Damage to residential buildings due to flooding of New Orleans after hurricane Katrina. Nat Hazards 54:413–434. doi:10.1007/s11069-009-9476-y
27. PU (2011) ATLAS: Pengamanan Pantai Jakarta. *Kementerian Pekerjaan Umum* (PU) Indonesia, Jakarta **(in Bahasa)**
28. Scawthorn C et al (2006) HAZUS-MH flood loss estimation methodology. II. Damage and loss assessment. Nat Hazards Rev 7:72–81. doi:10.1061/(asce)1527-6988(2006)7:2(72)
29. Smith DI (1981) Actual and potential flood damage: a case study for urban Lismore, NSW, Australia. Appl Geogr 1:31–39. doi:10.1016/0143-6228(81)90004-7
30. Smith D (1994) Flood damage estimation—a review of urban stage-damage curves and loss functions. Water S A 20:231–238
31. Suriya S, Mudgal BV, Nelliyat P (2012) Flood damage assessment of an urban area in Chennai, India, part I: methodology. Nat Hazards 62:149–167. doi:10.1007/s11069-011-9985-3
32. Tang JCS, Vongvisessomjai S, Sahasakmontri K (1992) Estimation of flood damage cost for Bangkok. Water Resour Manag 6:47–56. doi:10.1007/bf00872187
33. te Linde AH, Bubeck P, Dekkers JEC, de Moel H, Aerts JCJH (2011) Future flood risk estimates along the river Rhine Nat Hazards Earth. Syst Sci 11:459–473. doi:10.5194/nhess-11-459-2011
34. Thieken AH, Müller M, Kreibich H, Merz B (2005) Flood damage and influencing factors: New insights from the August 2002 flood in Germany. Water Resour Res 41:1–16. doi:10.1029/2005wr004177
35. Verbeek M (2012) A guide to modern econometrics -, 4th edn. Wiley, West Sussex
36. Wagenaar DJ, De Bruijn KM, Bouwer LM, De Moel H (2016) Uncertainty in flood damage estimates and its potential effect on investment decisions. Nat Hazards Earth Syst Sci 16:1–14. doi:10.5194/nhess-16-1-2016
37. Ward PJ, de Moel H, Aerts JCJH (2011a) How are flood risk estimates affected by the choice of return-periods? Nat Hazards Earth Syst Sci 11:3181–3195. doi:10.5194/nhess-11-3181-2011
38. Ward PJ, Marfai MA, Yulianto F, Hizbaron DR, Aerts JCJH (2011b) Coastal inundation and damage exposure estimation: a case study for Jakarta. Nat Hazards 56:899–916. doi:10.1007/s11069-010-9599-1

CHAPTER 4

Investigating the presumed causal links between drought and dzud in Mongolia

Troy Sternberg[1]

1 University of Oxford, Oxford, UK

ABSTRACT

Extreme weather events present environmental and social challenges across the Eurasian steppe. In Mongolia much attention is given to drought and dzud (severe winter conditions) impact on rural livelihoods, landscapes and governance. A link between the two events, fostered by international and state agencies, speculates that drought leads to dzud; this has become the widely accepted doctrine. However, the relationship between the two events is assumed rather than analysed. Whilst there may be natural links between climate events, causality is more difficult to establish yet often claimed post-event. This paper stresses Mongolia's destructive dzuds of 1999–2001 and 2009–2010 in examining drought frequency before dzud events. Findings question the hazard connection as just 3 of 32 examined dzud events were preceded by drought. Investigation did not document a relationship between the disasters; linkages between extreme events were implied rather than established. The human role in disaster also needs to be assessed as preparation, and response are key factors for mitigation. Study results identified a lack of causality between the disasters, suggesting more assiduous investigation of hazards is needed in Mongolia. This can clarify causal factors, identify risk and improve disaster mitigation strategies in Mongolia.

Keywords: Drought, Dzud, Causality, Mongolia, Natural hazards

1. INTRODUCTION

Natural hazards disrupt physical processes, disrupt social systems and dislocate livelihoods and communities. Identifying and understanding endemic natural hazards is essential to reducing risk, mitigating disasters and strengthening adaptation to extreme climate events. This is critical in societies where residents are directly dependent on the environment for their livelihoods. Across the Eurasian steppe pastoralism exemplifies the exposure and vulnerability of livestock, households and governance to extreme climate conditions. For instance, in Mongolia, drought and dzud (severe winter conditions) create the greatest threats to animal mortality and herder well-being (IPCC 2012; Fernandez-Gimenez et al. 2012; Shinoda 2012). The destructive national impact of dzud events reached crisis proportions in 1999–2001; the 2009–2010 disaster killed ~25% of the national herd, led to significant displacement as 3% of the population lost livelihoods and reduced GDP by −4.4% (Batima et al. 2005; Sternberg 2010) whilst 1.1 million animals died in the 2015–2016 event (UN 2016). Numerous academic, government, international NGO publications and news reports highlight how concurrent or sequential natural hazards threaten society by stressing causality between drought and dzud that creates significant disasters (IPCC 2012; UNOCHA 2010).

A presumed connection between drought and dzud has become the standard narrative in research, policy and mitigation strategies. Whilst there has been notable research on dzud impacts (Tachiiri et al. 2008; Middleton et al. 2015; Rao et al. 2015), the drought–dzud link is often inferred rather than investigated. As a cold, dry environment limited rainfall and severe winter conditions (temperatures to −40 °C) are threats to pastoral livelihoods. The Government of Mongolia (2001) stressed the impact of a prolonged summer 2000 drought on the 2001 dzud, Fernandez-Gimenez et al. (2012) identified several drought–dzud events and Shinoda (2012) found the two events associated. The World Bank (2010), FAO (2010) and general media, such as the New York Times (2016), identified summer drought as antecedent to severe dzuds.

Dzuds, particularly the 2010 event, are regarded as the country's worst natural disaster (Sternberg 2010); their immediacy consumes attention, aid and resources. Direct and antecedent events are acknowledged yet receive limited examination or integration into dzud strategies. Indeed, what constitutes a dzud is debated and reflects the lack of clarity. Table 1presents dzud definitions, whilst Table 2 reflects diverse perceptions of dzud. Herders' traditional ecological knowledge contributes local perspectives on dzud, from 'drought causes dzud' (Fernandez-Gimenez et al. 2012) to celestial signals (position of stars, shape of the moon), animal behaviour (marmots sealing burrows, sheep leave mountains early in autumn, livestock fight in autumn) and plant signs (yellowing of leaves from top of tree) (Middleton et al. 2015). The dominant focus is on natural disaster and meteorological events though social and livestock parameters are also considered. This reflects how dzuds and their contributory factors can be perceived differently. Thus perspectives on causality, correlation or circumstance vary and are often not examined. When the FAO/Suttie (2006) or UN/Mongolian National Emergency Management Agency (2010) report on dzud and link events to drought, the drought is not quantified, a process that repeats throughout dzud studies (Table 2). This is a major research gap and failure in comprehension. The 2010 dzud decimated livestock and livelihoods to catastrophic levels yet work evaluating and distilling cause and effect remains negligible. This paper frames the challenge through concise investigation of drought–dzud interaction to examine potential causality, relation or independence of the two natural disasters.

Table 1 Dzud definition

White dzud deep snow with density >0.20 g/cm^3
Black dzud lack of snow, freezing temperatures 5.0 °C below average
Iron dzud snow cover melts and refreezes to form an impenetrable ice cover that prevents livestock from foraging
Cold dzud severe cold with temperatures lower than monthly mean for several days
Hoof dzud extreme dry weather leading to depletion of grass due to heavy grazing and trampling
Storm dzud heavy snow and high winds
Combined dzud multiple dzud conditions occurring simultaneously

FAO/Suttie (2006), Middleton et al. (2015), Rao et al. (2015)

Table 2 Perceptions of dzud, with sources

Complex, long-lasting natural disaster	IFRC (2016)
Extremely harsh winters characterised by heavy snow and low temperature	Miao et al. (2016)
Meteorological phenomenon	Mayer (2016)
Complex social and ecological phenomenon	Murphy (2011)
Incidences of mass livestock mortality	Rao et al. (2015)
Unique weather phenomenon	World Bank (2011)

Themes reflect key policy, planning and donor country perspectives on disaster in Mongolia where agendas, government action and international assistance are based on conventional paradigms of climate–environment–human interactions (see above, World Bank 2011). Thus, for effective disaster mitigation and risk reduction in Mongolia, clear comprehension of natural hazards, causal factors and their interaction is essential for disaster preparation and response.

2. DISASTER IN MONGOLIA

There is a long, established dzud record in Mongolia with growing literature on several aspects of dzud (IPCC 2012; Shinoda 2012; Fernandez-Gimenez et al. 2012). Dzud is particularly damaging to rural livelihoods and mobile pastoralism that provide income for 30% of the population. Work by the UN/Mongolian National Emergency Management Agency (2010) identifies extreme events in 1945, 1999–2002 and 2009–2010 (Table 3). The important condition is restricted forage and animal exposure to extreme cold that leads to mortality, particularly in spring after the long winter. Coping strategies include adequate grain and fodder storage, livestock offtake in autumn, otor (long distance movements) and reduced pregnancy rates to improve animal survival in harsh winters. International relief efforts offer limited dzud relief; for example, after the 2010 disaster >$18 million in aid was promised yet only $3 + million was received (World Bank/Benson C 2011). Response to the 2010 dzud highlights both the limited domestic mitigation capacity and the country's dependence on international assistance for dzud relief (Sternberg and Batbuyan 2013).

Table 3 Dzud relation to drought, with sources

Natural disaster—drought followed by heavy snow, low temperatures	UNOCHA (2010)
Stems from summer droughts	Murphy (2011)
Prolonged summer drought impacts dzud	Government of Mongolia
Consequence of a summer drought and extreme overgrazing leading to insufficient hay for winter grazing	IFRC (2016)
Heavy drought in late summer followed by intense snow storms	World Bank (2011)
Triggered by prolonged drought	FAO (2010)
Dzud follows drought	Herder saying, in Fernandez-Gimenez et al. (2011), p. 14

Climate research highlights drought prevalence in Mongolia. Whilst Pederson et al. (2014) suggest drought in the late twelfth century may have influenced Chinggis Khan's formative years; Bayarjargal et al. (2006),

Sternberg et al. (2011) and Nandintsetseg and Shinoda (2013) examined multifaceted drought impacts on environment on society. Little work has explicitly addressed drought/dzud interaction, although Sternberg et al. (2009) disaggregated drought and dzud in Omnogovi Province from 1999 to 2002. Though identified, the theme has not become part of the policy or mitigation discourse. The outcome has been a presumed link between drought and dzud (see Table 4) ascribed to physical factors, whilst concurrently the socio-economic nature of dzud has gained wide recognition (Murphy 2012; Fernandez-Gimenez et al. 2012). As causal factors for dzud become a major concern, clarity is needed on the complex interplay of contributory factors to address detrimental outcomes. This highlights how human action, such as decisions that affect the number of livestock, mobility, use of water sources and intensified herding practices, affects dzud impact. Through examination of historical drought and recorded dzuds, we evaluate the potential link between drought and dzud disasters. The outcome can identify natural relationships, clarify causality, suggest event independence or reflect a more nuanced relation between the two hazards.

Table 4 Historical dzud record established by the Government of Mongolia (UN/National Emergency Management Agency 2010); international agencies (World Bank/Benson C 2011) and academics (Fernandez-Gimenez et al. 2012; Middleton et al. 2015)

Year	Event	Livestock mortality (millions)
1944–1945	Dzud + Drought	8.6
1954–1955		1
1956–1957		1.8
1967–1968	Dzud + Drought	3.8
1976–1977	Dzud	4.1
1986–1987	Dzud	0.8
1993–1994	Dzud	0.8
1996–1997	Dzud	0.7
1999–2002	Dzud + Drought	10.8
2009–2010	Dzud + Drought	10.3
2015–2016	Dzud + Drought	1.1

3. CAUSALITY

An understanding of causality is essential to identify possible connections between natural events (e.g. drought and dzud) and attempt to separate biophysical risk from social capacity and response (Cutter et al. 2008). In evaluating forecasting, planning and response the antecedent structures and follow-on consequences are key; causality can be seen as a flow of processes. However, correlation does not *prima facie* identify a cause–effect relationship but rather indicates an important path of further investigation (Rajsekhar et al. 2015). This is a significant point in evaluating major hazards in Mongolia and reflects the ability of original research to highlight critical associations that may be missed or ignored at first glance. This is particularly salient to in a discussion of drought, an event which remains elusive when identifying onset, extent and impact due to its variegated spatio-temporal scale and ensuing effects. As the most widespread hazard affecting the greatest number of people globally, drought is routinely perceived as a common causal and detrimental episode in multiple socio-environmental contexts; research can verify whether the conventional perception is valid at local and regional levels (Rajsekhar et al. 2015; Sternberg et al. 2011).

Relevant to research on two potentially related hazards, Below et al. (2007) stressed how a disaster often represents a cumulative attribution of losses across sectors, including physical and social domains. In Mongolia, this sees drought attributed to climate change and global warming with impacts as diverse as drying of water sources, reduced pasture biomass, lowered animal weight gain and degradation of the steppe landscape. Dzud is ascribed to changing weather patterns, severe cold, shifts in pastoralism, poor policy or lack of preparedness, socio-economic forces, the 'market-economy' and most frequently drought (UN/National Emergency Agency 2010). The challenge of disaster attribution is key to the Mongolian context where policymakers and practitioners seek to assign disaster causality to reduce uncertainty, attribute liability to physical factors that are beyond state control and determine reference conditions which can then be addressed through intervention, new policy and often by international aid or development projects (i.e. World Bank's Sustainable Livelihoods Program, the UN's Strengthening Disaster Mitigation Strategy).

Appropriate identification and attribution of Mongolia's drought and dzud encompasses several factors. As a developing nation limited research capacity, few alternate sources of funding (aid, relief), low to moderate intervention capacity and physical limitations (distance, infrastructure, timeliness, effective delivery mechanisms) combine to frame disaster risk and strategy. If initial presumptions about causal relationships are adequate and established, then policy can build on real-world circumstances as a foundation. A Mongolian example was widespread pasture degradation in 2005 attributed to the tsar-tsai (grasshopper/insect) invasion that focused response on containing the insect population rather than assigning culpability to overgrazing or climate (Hauck et al. 2008). Similar clarity is needed for the dominant hazards that consume attention, focus and resources of the government and herders. For example, pastoralists identify hazard mitigation through mobility, fattening of animals, reserve pastures and reciprocal land use patterns (Fernandez-Gimenez 2000). The questions surrounding drought and dzud make clarity key to system behaviour and livelihood improvement in the country. As several citations show (FAO 2010; Murphy 2011; IFRC (International Federation of the Red Cross) 2016), definitive causative understanding of the country's major disasters has been marginalised in favour of long-standing assumptions. This paper addresses the knowledge gap through investigating cause and attribution between drought and dzud.

4. METHODS

To investigate the relationship between drought and dzud in Mongolia, diverse geographical sites were chosen. This provides a prima-facie assessment of drought–dzud at a national scale through a focus on four different regions of the country. Sites were selected to represent landscape variability, present distribution in areas identified as vulnerable to dzud and drought and had working meteorology stations (World Bank 2011). The vast country, 1.56 million km^2, covers several ecotones and much environmental variety. The study uses documented dzud events identified as natural disasters at the national level (Fernandez-Gimenez et al. 2012).

Study sites with meteorological stations include Choibalsan (N 48.067, E 114.5; elevation 756 m) in the eastern Mongolian steppe grasslands with a cold, semi-arid climate; Dalanzadgad (N 43.583, E 104.417; elevation 1470 m) in the arid desert-steppe of south-central Mongolia; Hovd (N 48.017, E 91.650; elevation 1406 m), a dry steppe zone near the Altai Mountains in the west; and Ulaan Baatar (N 47.933, E 106.983; elevation 1316) in central-northern Mongolia at the edge of the steppe-mixed forest eco-zone (Fig. 1). Ulaan Baatar is the national capital; the other sites are provincial capitals. Using climatological data from Mongolia's National Agency for Meteorology, Hydrology and Environment Monitoring (NAMHEM) and NOAA (National Oceanic and Atmospheric Administration) and UK's Climate Research Unit (Harris et al. 2014), drought was calculated with the Standard Precipitation Index. Findings were then correlated with documented dzud events (see Table 4). The objective was to stress connections between:

1) dzud events preceded by drought

2) drought events followed by dzud

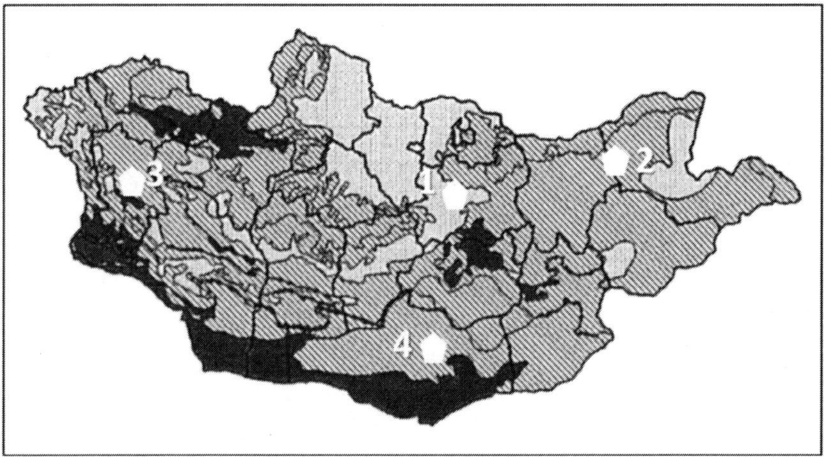

Figure. 1 2016 pasture vulnerability map indicating dzud hotspots—darker colour indicates greater risk. Meteorological stations: *1* Ulaan Baatar; *2* Choibalsan; *3* Hovd; *4* Dalanzadgad. *Source* Mongolian Livestock Early Warning System (2016)

This was done by identifying summer drought events at each site that preceded winter from 1970 to 2015. To do this, drought during the maximum 6-month spring-summer plant growth season was examined at the sites. This captures reduced and negative precipitation anomalies from annualised precipitation records for the April–September potential vegetation growth period in the country. We note that in Mongolian the word 'gan' is translated as drought yet may also signify high temperatures that affect livestock. Results quantify moisture deficiencies which may then contribute to dzud events in the following winter. Data were then analysed to determine association between hazards; the outcome is a record of drought–dzud interaction. This addresses the link between the two hazards with significance for disaster policy, planning and mitigation in Mongolia. Detailed evaluation at province or district levels, or with other specific dzud criteria (socio-economic factors, livestock mortality; see Table 1) may lead to varied results.

4.1 Standardised precipitation index

The standardised precipitation index (SPI) documents anomalous and extreme precipitation and drought events (McKee et al. 1993; Sonmez et al. 2005; Sternberg et al. 2009). Using >30 years meteorological data, the SPI gives a numeric value to precipitation surplus or deficiency; this enables drought monitoring at different time scales and facilitates comparison across regions and varied climatic zones. The index calculates drought initiation, magnitude, duration, and frequency; results can be interpolated to provide spatial and temporal drought parameters. The SPI is determined by fitting data to a probability distribution defining a series of aggregated monthly precipitation observations. The cumulative probability of a precipitation amount is calculated with the inverse normal function applied to the probability (Guttman 1999). This enables analytical comparison of exceedance probabilities and comparison of locations at selected timescales (1, 3, 6 ... months). The resulting function gives the cumulative probability for precipitation at a specific station for a given month and timescale; drought intensity reflects the negative precipitation anomaly (McKee et al. 1993; Sternberg et al. 2011). The SPI provides statistical consistency, identification of short and long-term drought episodes and robustness in comparison with other drought indices (Keyantash and Dracup 2002). Additional methodological details are presented in cited work.

Using meteorological data from NAMHEM, NOAA and the UK's Climate Research Unit (Harris et al. 2014) 6-month drought was assessed from 1970 to 2015. The peak biomass and plant growth period, from the 1st of April through the 30th of September, was assessed. The SPI identifies positive/negative anomalies from zero that give a probability indication of event severity; a result of −1 reflects moderate drought, −1.5 severe drought and <−2 extreme conditions. With event probability based on a 100 year scale, an extreme drought would be expected to occur 2.3 times per century (Sonmez et al. 2005). As the data record in this study is 45 years, drought frequency is extrapolated rather than a reflection of long-term (unavailable) data.

5. RESULTS

Historical drought at Choibalsan, Dalanzadgad, Hovd and Ulaan Baatar Mongolia documents variable moisture levels, fluctuating precipitation patterns over time and disparate drought events across the sites during the 6-month summer plant growth season (Fig. 2a–d). In general, the 1980s were part of a wetter phase, whilst the 2000s were a drier period.

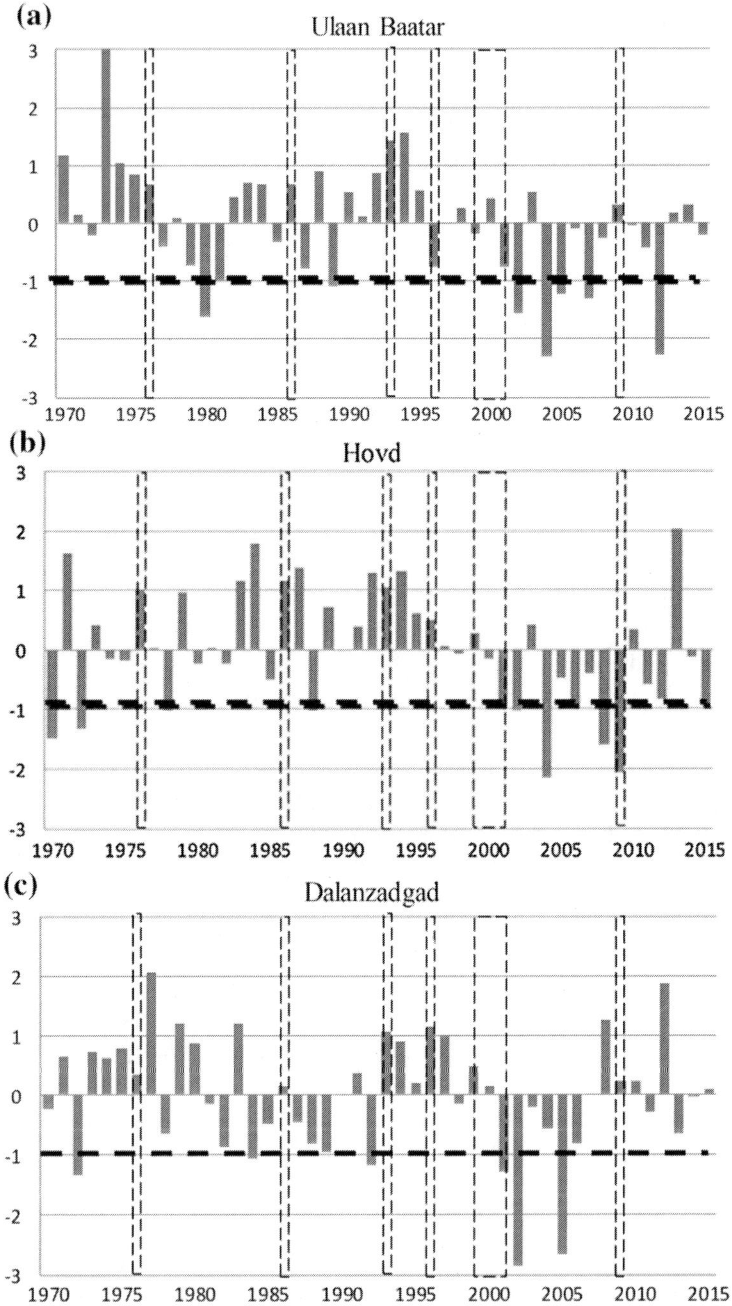
(a)
Ulaan Baatar
3
2
1
0
-1
-2
-3
1970
1975
1980
1985
1990
1995
2000
2005
2010
2015
(b)
Hovd
3
2
1
0
-1
-2
-3
1970
1975
1980
1985
1990
1995
2000
2005
2010
2015
(c)
Dalanzadgad
3
2
1
0
-1
-2
-3
1970
1975
1980
1985
1990
1995
2000
2005
2010
2015

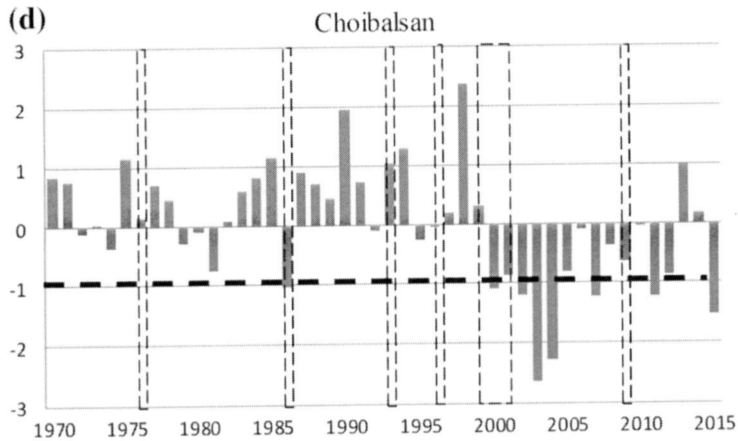

Figure. 2 A–D Drought records for study sites from 1970 to 2015; **A** Ulaan Baatar, **B** Hovd, **C** Dalanzadgad, **D** Choibalsan. Levels of −1 reflects moderate drought, −1.5 severe drought and <−2 extreme events. Boxes (*dashed lines*) denote dzud years; below the *dotted line* at −1 indicates the drought threshold

Severe and extreme droughts varied across time and sites with drought episodes concentrated in the 2000s. The percentage of assessed months in drought conditions, at 10–14%, was lower than SPI probability (15.9%) When related to dzud, data finds limited association between the two hazards. One extreme drought event was correlated with dzud over 45 years; the link between drought and dzud—at a rate 9%, suggests a random connection. The clear finding is that the two events occur independently at multiple sites in Mongolia.

Given established perceptions in the country, matching the drought history and dzud record presents remarkable findings. Since 1970, 3 of 32 dzud events were preceded by drought (Table 5). Thus the two hazards were seldom connected in this sample of high-risk winter hazard regions. Examining quantified drought identified 18 events from moderate to extreme severity at the selected timescale. Performing the reverse analysis—droughts followed by dzud—shows that 1 of 11 severe or extreme droughts were linked to dzud whilst 3 of 7 moderate droughts were followed by dzud (Table 6). The extreme droughts in Dalanzadgad (2002, 2005) were at 1-in-100 year levels yet were not associated with dzud. Investigation found little relation or causality between drought and dzud in Mongolia over 45 years. Data highlight a random connection between dzud and drought rather than establishing a causal link.

Drought events themselves were seldom related between sites. Only Ulaan Baatar and Choibalsan showed a correlation regarding when precipitation deficits occurred (P = 0.005). Other site combinations varied in drought episodes over 45 years. This stresses how Mongolia's environment cannot be characterised by country-wide events. Thus, linkages between drought and other events or inferences need to be identified at more local scales. Analysis points out the inability to draw conclusions on physical parameters at a national scale.

Absence of a notable physical connection between these two natural hazards calls into question how climate variability and hazards interact in the steppe environment. Regional climate dynamics suggest that summer meteorological conditions (such as drought) are decoupled from winter temperature and precipitation events. Findings also highlight how disaster events are presumed and perceived by society. The abundant literature on drought and dzud suggest a strong appeal and intent to connect vegetation productivity, as affected by drought, and exposure to severe winter episodes, factors not evident in this study. Table 7 documents that the majority of dzuds (63%) were preceded by wetter-than-average climatological conditions.

In winter, reflected in average January temperatures, there was interannual fluctuation but slight variation over the study time (Fig. 3). Temperature was independent of disaster cycles, not having an apparent effect on drought or dzud events. Sites, excepting Ulaan Baatar, showed notable oscillation in average temperature with Hovd having most divergent temperatures.

Table 5 Dzud events across sites that were preceded by an identified drought

Year	Disaster	Dzud preceded by drought			
		Dalanzadgad	Hovd	Ulaan Bataar	Choibalsan
1976–1977	Dzud	–	–	–	–
1986–1987	Dzud	–	–	–	Yes
1993–1994	Dzud	–	–	–	–
1996–1997	Dzud	–	–	–	–
1999–2000	Dzud	–	–	–	–
2000–2001	Dzud	–	–	–	Yes
2001–2002	Dzud	Yes	–	–	–
2009–2010	Dzud	–	Yes	–	–

Table 6 Droughts across sites that were followed by dzuds. Table is ranked from most extreme (−2.86) to moderate (−1.02) events

Drought level	Year	Location	Followed by dzud
−2.86	2002	Dalanzadgad	No
−2.66	2005	Dalanzadgad	No
−2.65	2003	Choibalsan	No
−2.31	2004	Ulaan Bataar	No
−2.27	2004	Choibalsan	No
−2.27	2012	Ulaan Bataar	No
−2.16	2004	Hovd	No
−2.07	2009	Hovd	Yes
−1.63	1980	Ulaan Bataar	No
−1.60	2008	Hovd	No
−1.55	2002	Ulaan Bataar	No
−1.48	1970	Hovd	No
−1.28	2001	Dalanzadgad	Yes
−1.22	2005	Ulaan Bataar	No
−1.20	2002	Choibalsan	No
−1.08	2000	Choibalsan	Yes
−1.02	2002	Hovd	No
−1.02	1986	Choibalsan	Yes

Table 7 Wetter-than-average (+) versus drier-than-average (−) 6-month SPI episodes preceding dzud events

Year	Disaster	Dalanzadgad	Hovd	Ulaan Bataar	Choibalsan
1976–1977	Dzud	+	+	+	+
1986–1987	Dzud	+	+	+	+
1993–1994	Dzud	+	+	+	+
1996–1997	Dzud	+	+	−	+
1999–2000	Dzud	+	−	−	+
2000–2001	Dzud	+	−	+	−
2001–2002	Dzud	−	−	−	−
2009–2010	Dzud	−	−	+	−

Summer temperatures, measured in July, vary yet show gradual increasing trends over 45 years. Dzud years did not experience the coldest temperatures just as drought episodes were not marked by intense heat. Rather, there were not clear associations between temperature and hazards. More intra-site difference was found in winter, whilst there was less range in summer.

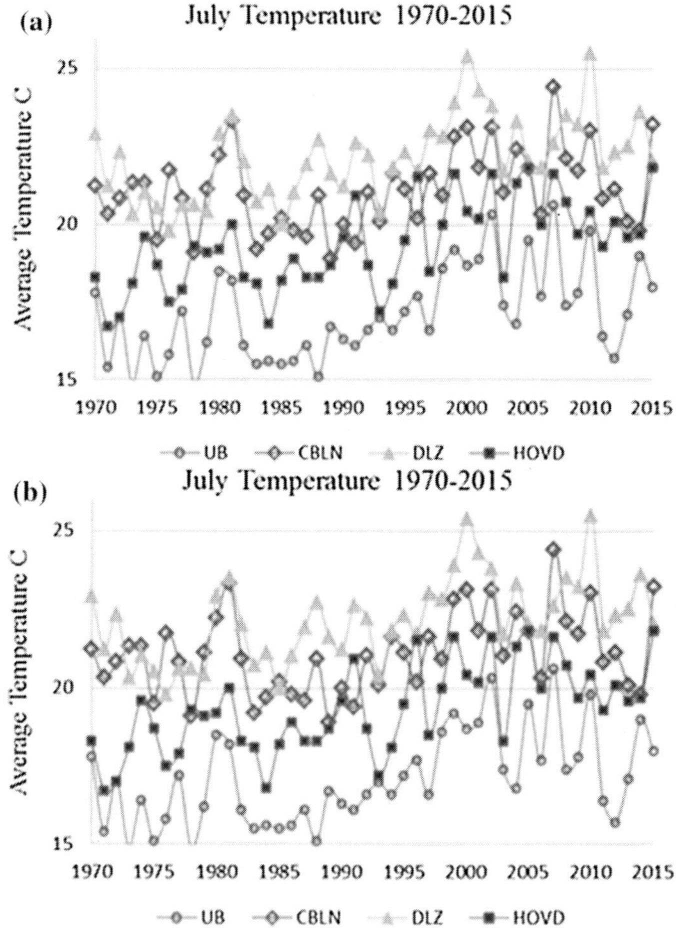

Figure. 3 July and January average monthly temperatures at the sites. January data showed low r^2values. An increase in July temperatures reflected a gradual warming trend, Ulaan Baatar—$r^2 = 0.28$; Choibalsan—$r^2 = 0.11$, Dalanzadgad—$r^2 = 0.28$, Hovd—$r^2 = 0.41$

6. DISCUSSION

The findings of this study—that in just 3 of 32 events analysed was dzud preceded by drought—contravene entrenched opinion, bedrock policy foundations, international agency presumptions and development practices. Questioning the drought–dzud assumption and contesting the environment-driven disaster scenario can invigorate research, policy and disaster mitigation. The ability to better understand hazards has increased significantly over time yet the Mongolian process reflects how the 'drought leads to dzud' inference became an unchallenged link. A new paradigm better suited to on-the-ground realities may do more to reduce disaster impact that increased funding based on past attitudes. Long-standing folk wisdom has an important place in thought and discussion, but it should not limit scientific insight that can ameliorate livelihoods. That other research has not chosen to isolate drought is perhaps not surprising given the predominance that the drought–dzud hypothesis has enjoyed. The study objective was to investigate standard assumptions; results encourage reassessment of physical relationships in Mongolia and the greater region.

The clear outcome of the study is that past and present presumption of drought causality of dzud requires re-examination. Findings suggest more varied and nuanced research that addresses non-drought factors that contribute to dzud is essential (Sternberg 2010; Fernandez-Gimenez et al. 2012). This can commence with further investigation of winter snow depth, timing of freeze–thaw cycles, separating summer and winter climate patterns, anomalous events and early and late-onset disaster. Factors contributing to animal weight gain, mobility, pasture degradation livestock mortality, could be reviewed. Examining disparities between dzud intensity, livestock health and pasture conditions at regional and local levels will further understanding. Greater recognition of the role of herder decision-making and actions, acknowledgement of their unique ecological insight, the contributory socio-economic forces and government policy is vital. Choices, selection and behaviour have antecedent motivation and commonly, rational reasons if regarded from the herder perspective. Dzud relief efforts, emergency fodder at local and national levels and international aid are part of the context. The extant literature identifies the conflicted, dual nature of dzud and implicitly asks the question 'is it a physical or social disaster?' (Murphy 2012). The body of literature can advance, delve into the bifurcated nature of cold disaster in livestock systems and move beyond the concept that 'drought causes dzud' which has been a poor explanation of complex physio-social processes. Mongolian and international research in the country has progressed greatly in the last 20 years; it remains a puzzle why as technical capacity has improved more effort has not been spent establishing baseline knowledge about endemic hazards. Thus 17 years after the first severe post-soviet era dzud, we continue to rely on Batjargal (1997) and Natsagdorj and Dulamsuren (2001) early assessments of dzud and environmental conditions.

In part reliance on the imprecise drought–dzud relation was born of exigency in the 1990s as Mongolia abruptly moved from communism to a market system through much turmoil. At the same time, advising international agencies had little knowledge of Mongolian natural and social systems. Limited capacity and funding, severe dzuds from 1999 to 2002 and great human misery meant basic response and rapid reaction was needed. With time the concepts of the era became entrenched; now science can reassess the country's hazards and to further engage and test the rather messy intricacies of dzud. We see this in new approaches by Shinoda (2017), Middleton et al. (2015) and Ahearn (2016); now more than previously the research agenda should be wide open from macro- to microscales.

The government role is highlighted through reports written in Ulaan Baatar (e.g. MARCC 2009; National Action Programme on Climate Change 2011; annual National Statistical Yearbook) yet the state maintains a modest presence in the drought and dzud impacted countryside dynamics (Sternberg and Batbuyan 2013). The key perspective is that as government interaction with herding changed from communism's top-down state control to today's laissez faire approach, the assumption of responsibility also changed. Thus, the effective district hay barns of the Soviet era are now empty and derelict; today the disaster response scenario is remote, cumbersome and of limited value to herders. For example, a herder experiencing dzud in Hovd Province, 1800 kilometres from the capital, has little prospect of fodder, relief or aid from the government as exemplified in 2010 (Middleton et al. 2015). Previously response mitigated disaster to a degree; now Mongolia depends on a fickle international community that promises aid but seldom delivers in full (World Bank/Benson C 2011). What does arrive faces severe infrastructure and distribution problems so as to be of little timely, immediate relief. The gradual change in the government's disaster perception, role and performance promotes a drought–dzud focus where the message is of unforeseen, uncontrollable natural hazards beyond the government's remit to address or ameliorate. An updated assessment can directly address the government's limited ability to reduce disaster threat. To herders, it may be preferable to have matters presented clearly rather than foster unreasonable hope of support. This also encourages herders to acknowledge that they are effectively responsible for mitigation.

Data from the Mongolian Statistical Office (2016) provide a way to stimulate new questions at relevant scales. Figures 4 and 5 present changes in livestock numbers at the national level and in the four selected provinces. Patterns show much annual variability with decreases after major dzud events. Examining livestock numbers and hand-collected fodder harvests (reflecting herder effort) show the strong drivers that can affect extreme events (Fig. 5). One is positive—the collection of fodder by hand for emergency winter supply (r^2 = 0.8) whilst a doubling of livestock in less than 15 years increased pressure on pastoral systems. The livestock curve is impressive as animal numbers increased from 22 million in post-dzud 2002 to 51 million in 2014 (r^2 = 0.6) with similar patterns of livestock change in the four study sites (Fig. 5). Over the same time the amount of fodder has distinctly fluctuated. The 1990s were low-harvest years. After the 1999–2002 dzud era fodder collection increased, only to decline through 2009 as drought but not dzud occurred. Drought affects the amount of vegetation cover and plant biomass, thus potentially limiting forage for animal weight gain and reducing fodder harvests. Following the 2009–2010, dzud fodder harvest increased 26% in the immediate year and 42% the next, to then drop 15%. The national and provincial data provides a lens to conceive of the wide range of potential human factors affecting pastoral vulnerability. The motivations reflect both economic—increase herd size for revenue and practical reasons—improve veterinary nutrition and provide fodder to keep animals from starving. Throughout processes decision-making, key players (herder households, government, NGOs) and their actions are divergent yet impacted by natural hazards.

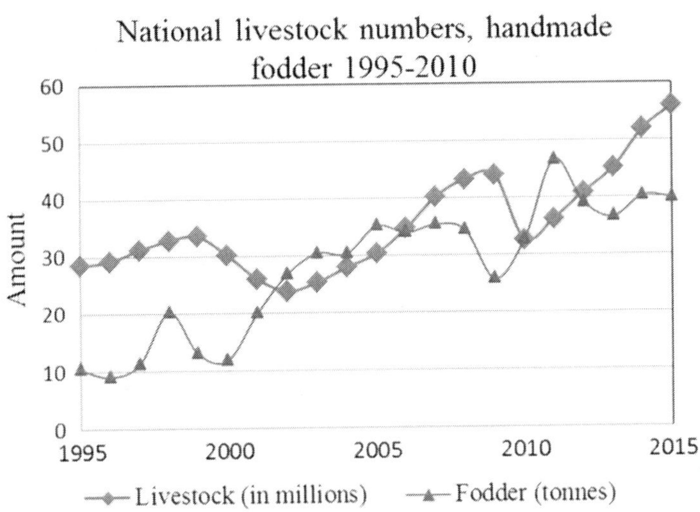

Figure. 4 Livestock numbers in the four study provinces, 1990–2015. This examines change in animals, not livestock mortality. National Statistical Office 2016

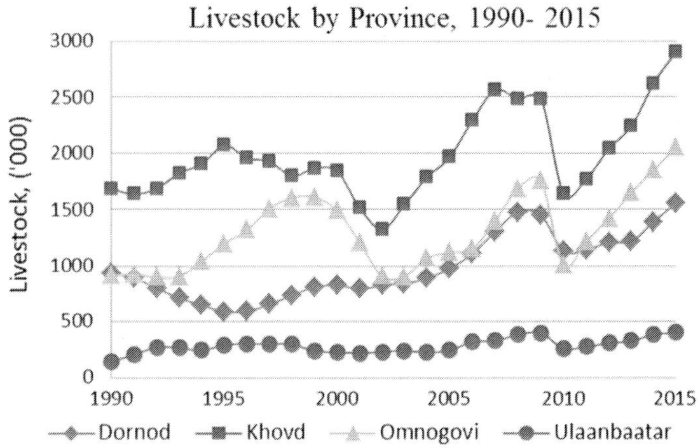

Figure. 5 Two trends affecting disaster vulnerability that reflect human influence in natural disaster. The first presents herder collection of fodder by hand for emergency winter supply. In the second, livestock numbers show a doubling of animals in less than 15 years. National Statistical Office 2016

Increasing livestock numbers, variable fodder resources, intermittent drought events and less-remarked yet significant allied practices (e.g. mobility, animal intensification, markets, competition for water) raise additional concerns that are reflected in research literature. This centres on physical resources, degradation, overgrazing, land tenure etc. and their implications for herding practices, assigning vulnerability and criticising herders when rational livelihood decisions (increased livestock) may be regarded as negative processes environmentally. This apparent conundrum, a version of Hobson's choice—selecting environmental vs economic benefit—can potentially be a sustainable option depending on context and resources. Additional livestock may be feasible (and not detrimental) if they are accompanied by adequate pasture, mobility, fodder, lack of resource competition—factors that enable increased levels of productivity. The disaggregation of multiple factors is difficult in the strictures of an academic article, yet this paper raises the point that pastoralism has several approaches as different as the herders practicing it. Their inherent local wisdom and ecological knowledge may be more reliable than desk-based drought or dzud analysis (this paper included). Yet to influence policy, practice, funding and mitigation at national or international scales data, justification and explanation is required. This article challenges one of the established pastoral axioms in questioning the relationship between drought and dzud.

The study data present four meteorological stations over a 45-year period. The country's large size and disparate landscapes encourages more in-depth study both at the national level and within regions and localities. One challenge is the limited number of stations in the country. NOAA identifies 40 stations nationwide, or one station per 39,000 km^2 (compare this to nearby Kazakhstan, with one station for every 8200 km^2 or Iceland, one per 1470 km^2). This means much detail is lost due to scale issues and recognises that results may differ in a detailed province or district level study. A potential new method would be to use extracted global data sets, such as from the UK's Climate Research Unit at the University of East Anglia that generates gridded data sets from 1900 (Harris et al. 2014). The caveat is that for much of the period Mongolia had limited weather monitoring, thus extrapolated data would lack accuracy and value for much of the twentieth century. Other technical approaches, such as satellite vegetation, temperature or moisture monitoring, can provide a short-term perspective on climate and pasture fluctuation in the country. The UNESCO Remote Sensing Lab at Mongolian National University and the National Remote Sensing Office are two capable centres that could engage in remotely sensed environmental monitoring, whilst NAMHEM can undertake climate monitoring.

An interesting factor is that dzud is a particularly Mongolian phenomenon, at least in its perception as a major disaster. The word dzud is Mongolian; across the border in northern China there is no local equivalent. The closest term is 'snow disaster' that lacks the impact or urgency of dzud. Equally, there is little research on dzud in Kazakhstan, perhaps in part due to the decrease in pastoralism and sedentarisation of recent decades. Both countries share the steppe landscape and herding traditions yet do not perceive or experience major dzuds nor focus on a drought–dzud link. This suggests the dynamics are country-specific and that the disasters reflect unique domestic factors. These may include the importance of pastoralism as a livelihood in Mongolia, the high proportion of rural residents, limited non-herding opportunities in the countryside and strong cultural links to pastoralism. Climate threats may also highlight the decline in effectiveness of the government to mitigate hazards over recent decades, or the concurrent dramatic growth in livestock numbers that increases pasture stress and herder vulnerability. These key factors reflect social influences rather than physical parameters. The principal message becomes that disaster is significant in the country, that they are multifaceted and result from physical forces and human action. 'Drought causes dzud' becomes an ingenuous assumption that was not supported by investigation—in 29 of 32 events analysed, dzud was not preceded by drought. Clarity on causal complexity will strengthen disaster strategies with significant benefit for pastoralists at the centre of the hazard tempest.

SUMMARY

Whilst natural disasters are endemic in Mongolia, a clear understanding of hazard links and contributory factors has been elusive. The country's chaotic transition away from communism led to a generation of social and development processes and thought that excused society from looking too closely at hazard causality. Through times of uncertainty Mongolians' approach to hazards has been settled in logic, presumption and the status quo. Yet it serves poorly if on-the-ground reality diverges from long-held belief. This succinct study of drought and dzud in diverse areas of the country makes this important point as findings identify a lack of causality between the disasters. An event-by-event examination at Choibalsan, Dalanzadgad, Hovd and the capital Ulaan Baatar stresses the independence of dzud and drought. This is case whether evaluating dzud preceded by drought or the inverse, drought followed by dzud.

Findings that drought preceded dzud in just 3 of 32 analysed events suggest more assiduous investigation of disasters is needed in the country. Efforts could focus at larger scales to highlight variegation, then distil details at regional and local levels and by environmental zones. This special journal issue on Eurasia provides a format to intensively examine disasters from different angles. Hazards instigate physical and social processes; in Mongolia their impact reverberates across several research fields as diverse as livestock science, ecology, economics, development, migration and climatology to sociology and governance. Assessment should be encouraged to use contemporary techniques in combination with knowledge generated through intensive study to better serve individuals and communities. Only then will disaster strategies improve their ability to mitigate Mongolia's unusual hazard scenarios.

ACKNOWLEDGEMENTS

Research was supported by the John Fell Fund, University of Oxford, Mongolia's National University and Academy of Sciences and China's International Science and Technology Cooperation Program of China (No. 2013DFA91700). Discussion at the 3rd Oxford Desert Conference (UK) and the University of Nagoya's 4D–'Multi-Hazard Approach in Mongolia' workshops encouraged conceptual development. Special thanks goes to Masato Shinoda, Ariell Ahearn and Batbuyan for their insights and conversations.

REFERENCES

1. Ahearn A (2016) The role of kinship in negotiating territorial rights. Inner Asia 18:245–264
2. Batima P, Natsagdorj L, Gombluudev P, Erdenetsetseg B (2005) Observed climate change in Mongolia. Assess Imp Adapt Clim Change Work Pap 12:1–26
3. Batjargal Z (1997) Desertification in Mongolia. RALA Report 200:107–113
4. Bayarjargal Y, Karnieli A, Bayasgalan M, Khudulmur S, Gandush C, Tucker C (2006) A comparative study of NOAA–AVHRR derived drought indices using change vector analysis. Remote Sens Environ 105:9–22
5. Below R, Grover-Kopec E, Dilley M (2007) Documenting drought-related disasters a global reassessment. J Environ Dev 16:328–344
6. Cutter S, Barnes L, Berry M, Burton C, Evans E, Tate E, Webb J (2008) A place-based model for understanding community resilience to natural disasters. Glob Environ Change 18:598–606
7. FAO (2010) FAO's role in the Mongolia dzud appeal. fao.org/fileadmin/user_upload/emerg encies/docs/app_mongoliadzud2010.pdf. Accessed 26 Nov 2016
8. FAO/Suttie J (2006) Country Pasture and Forage Resource Profiles–Mongolia. FAO report
9. Fernandez-Gimenez M (2000) The role of Mongolian nomadic pastoralists' ecological knowledge in rangeland management. Ecol Appl 10:1318–1326
10. Fernandez-Gimenez M, Batkhishig B, Batbuyan B (2011) Mongolia - Understanding resilience in Mongolian pastoral social-ecological systems: adapting to disaster before, during and after 2010 dzud. World Bank. http://documents.worldbank.org/ curated/en/946711468061500091/Mongolia-Understandingresilience-in-Mongo lian-pastoral-social-ecological-systems-adapting-to-disaster-before-during-and-after-2010-dzud
11. Fernandez-Gimenez M, Batkhishig B, Batbuyan B (2012) Cross-boundary and cross-level dynamicsincrease vulnerability to severe winter disasters (dzud) in Mongolia. Glob Environ Change 22:836–851
12. Government of Mongolia Appeal for International Assistance for Mongolia—Winter Disaster—Dzud (2001) reliefweb.int/report/mongolia/un-and-government-mongolia-appeal-inter national-assistancemongolia-winter-disaster. Accessed 26 Jan 2016
13. Guttman N (1999) Accepting the standardized precipitation index: a calculation algorithm. J Am Water Resour As 35:311–322
14. Harris I, Jones Osborn T, Lister D (2014) Updated high-resolution grids of monthly climatic observations–the CRU TS3. 10 Dataset. Int J Climatol 34:623–642
15. Hauck M, Dulamsuren C, Heimes C (2008) Effects of insect herbivory on the performance of Larix sibirica in a forest-steppe ecotone. Environ Exp Bot 62:351–356
16. IFRC (International Federation of the Red Cross) (2016) http://www.ifrc.org/en/news-and-media/pressreleases/asia-pacific/mongolia/red-cross-launches-emergency-appeal-as-extreme-winter-dzud-threatensmongolian-herders-/. Accessed 26 Nov 2016
17. IPCC (2012) Managing the risks of extreme: vvents and disasters to advance climate change adaptation. In: Field et al. Cambridge Press, Cambridge
18. Keyantash J, Dracup J (2002) The quantification of drought: an evaluation of drought indices. Bull Am Meteorl Soc 83:1167–1180. doi:10.1175/1520-0477(2002)083
19. Mayer B (2016) Climate migration and the politics of causal attribution: a case study in Mongolia. Migr Dev 5:234–253
20. McKee T, Doeskan N, Kleist J (1993) The relationship of drought frequency and duration to time scales. In: Eighth conference on applied climatology, American meteorological association, Anaheim, CA, pp 179–184
21. Miao L, Fraser R, Sun Z, Sneath D, He B, Cui X (2016) Climate impact on vegetation and animal husbandry on the Mongolian plateau: a comparative analysis. Nat Hazards 80(2):727–739
22. Middleton N, Rueff H, Sternberg T, Batbuyan B, Thomas D (2015) Explaining spatial variations in climate hazard impacts in western Mongolia. Landsc Ecol 30:91–107

23. Mongolian Livestock Early Warning System (2016) Forage Deviation www.mongolialews.net/index.php/map-products/english/. Accessed 22 Feb 2016
24. Murphy DJ (2011) Going on otor: disaster, mobility, and the political ecology of vulnerability in Uguumur Mongolia. PhD thesis, University of Kentucky
25. Murphy DJ (2012) Encountering the franchise state: dzud, otor, and transformations in pastoral risk. In: Knauft BM, Taupier R (eds) Mongolians after socialism. Admon Press, Ulaanbaatar, pp 68–80
26. Nandintsetseg B, Shinoda M (2013) Assessment of drought frequency, duration, and severity and its impact on pasture production in Mongolia. Nat Hazards 66:995–1008
27. National Statistics Office/Mongolian Statistical Yearbook www.en.nso.mn. Accessed 06 Dec 2016
28. Natsagdorj L, Dulamsuren J (2001) Some aspects of assessment of the dzud phenomena. Pap Meteorol Hydrol 23:3–18
29. New York Times (2016) Mongolia's Livestock in danger after drought, harsh winter 2016/02/26 http://www.nytimes.com/aponline/2016/02/26/world/asia/ap-as-mongolia-harsh-winter.html?_r=0. Accessed 28 Jan 2016
30. Pederson N, Hessl A, Baatarbileg N, Anchukaitis K, Di Cosmo N (2014) Pluvials, droughts, the Mongol Empire, and modern Mongolia. Proc Natl Acad Sci 111:4375–4379
31. Rajsekhar D, Singh V, Mishra A (2015) Multivariate drought index: an information theory based approach for integrated drought assessment. J Hydrol 526:164–182
32. Rao M, Davi N, D'Arrigo R, Skees J, Nachin B, Leland C, Lyon B, Wang S, Byambasuren O (2015) Dzuds, droughts, and livestock mortality in Mongolia. Environ Res Lett 10:074012
33. Shinoda M (2012) Proactive Management of Drought and Its Derived Disasters in Mongolia. In: Tran P, Shaw R (eds) Environment disaster linkages. Emerald Group Publishing, Bingley, pp 61–78
34. Shinoda M (2017) Evolving a multi-hazard focused approach for arid Eurasia. In: Sternberg T (ed) Climate hazard crises in Asia. Routledge Publishers, Abingdon
35. Sonmez K, Komuscu A, Erkhan A, Turgu E (2005) An analysis of spatial and temporal dimension of drought vulnerability in Turkey using the standard precipitation index. Nat Hazards 35:243–264. doi:10.1007/s11069-004-5704-7
36. Sternberg T (2010) Unravelling Mongolia's extreme winter disaster of 2010. Nomadic Peoples 14:72–86
37. Sternberg T, Batbuyan B (2013) Integrating the Hyogo Framework into Mongolia's disaster (DRR) policy. Int J Disaster Risk Reduct. doi:10.1016/j.ijdrr.2013.05.003
38. Sternberg T, Thomas D, Middleton N (2009) Pressurized pastoralism in South Gobi Province, Mongolia: what is the role of drought? Trans Br Geogr IBG 34:364–377
39. Sternberg T, Thomas D, Middleton N (2011) Drought dynamics on the Mongolian steppe, 1970–2006. Int J Climatol 31:1823–1830
40. Tachiiri K, Shinoda M, Klinkenberg B, Morinaga Y (2008) Assessing Mongolian snow disaster risk using livestock and satellite data. J Arid Environ 72:2251–2263
41. UNDP/NEMA (2010) Dzud national report 2009–2010. UNDP-NEMA, Ulaanbaatar
42. United Nations (2016) Mongolia: dzud. United Nations Office of the Resident Coordinator. www.unmongolia.mn
43. UNOCHA (2010) Mongolia dzud appeal unocha.org/cap/appeals/mongolia-dzud-appeal-2010. Accessed 26 Nov 2016
44. World Bank (2010) http://blogs.worldbank.org/eastasiapacific/dzud-a-slow-natural-disaster-kills-livestockand-livelihoods-in-mongolia. Accessed 10 Dec 2010
45. World Bank (2011) Exploring options to institutionalize the dzud disaster response product in Mongolia. No. 12760. World Bank, Washington
46. World Bank/Benson C (2011) Dzud disaster financing and response in Mongolia. World Bank, Washington

CHAPTER 5

Communication and Preparedness Issues on Various Scales from Extreme Atmospheric Hazards

Robert M. Schwartz

ABSTRACT

Most federally declared disasters are from atmospheric hazards. These could be from floods, tropical cyclones, tornadoes, or winter storms. Some of these hazards are events with relatively short warning times such as tornadoes or have sufficient warnings from tropical cyclones. This research examined communication of weather information and personal preparedness following the Florida landfall from Tropical Storm Debby in 2012. Another case study examined emergency management issues such as preparedness and response after the 2011 tornado in Tuscaloosa, Alabama. The concentration was on emergency management agencies at the County, University, State, and Federal levels. A sample of elderly residents of Pinellas and Pasco Counties in Florida completed a self-administered survey to examine various means of receiving weather information along with hurricane preparedness actions. In-depth interviews were conducted with representatives of various agencies on different scales in regard to preparedness and response following the Tuscaloosa tornado. The elderly used television as the primary means of receiving weather information, thus stressing the importance of utilizing both traditional and newer forms of communications to reach all citizens. One of the major issues on all levels following the Tuscaloosa tornado is related to communications such as resource allocations and response actions.

Keywords: disaster communications, tornado and hurricane preparedness, personal and community emergency management preparedness, methods of receiving weather information, emergency management response

1. INTRODUCTION

Most of the federally major declared disasters are due to atmospheric hazards. Of the 3477 total declarations, from 1953 to 2015, 2449 or 71% are weather related [1]. There were 293 major disaster declarations from 2011 to 2015. Only 18 of these major declarations were non-weather related such as earthquakes [2]. Meteorological disasters could be from events such as but not limited to floods, tropical cyclones, tornadoes, or winter storms. Some of these hazards are quick fuse events with relatively short warning times such as tornadoes or have sufficient warning such as a tropical cyclone. This research examined two different atmospheric hazards that were declared major disasters. The first studied means of communicating weather information and personal preparedness of elderly citizens following Tropical Storm Debby in 2012 that made a landfall in Florida. An additional case study examined emergency management issues such as preparedness, response, recovery, and mitigation after the tornado in Tuscaloosa, Alabama, in 2011. The research covered emergency management organizations at the county, university, state, and federal levels.

2. BACKGROUND

2.1. Tropical cyclones

Some of the most destructive storms are tropical cyclones. Based on the geographic location, tropical cyclones are known as typhoons over the western Pacific Ocean, cyclones over the Indian Ocean, and hurricanes over the Atlantic and eastern Pacific Oceans. Hurricane season in the North Atlantic basin (which impacts the United States) starts 1 June and ends 30 November, with the peak being around mid-September. The North Atlantic basin includes the Gulf of Mexico, Caribbean Sea, and Atlantic Ocean [3,4].

Tropical cyclones are classified by organization of thunderstorm clusters, circulation patterns, and wind speeds. First there is a tropical wave which is an unorganized cluster of thunderstorms with a weak surface circulation.

An organized cluster of thunderstorms is generally 150–350 miles (250–600 km) in diameter with no closed circulation and maintains an identity for 24 hours is a tropical disturbance. A tropical depression has an identifiable pressure drop, closed circulation, and wind speed less than 39 mph (34 kts); the system is then assigned a number by the National Hurricane Center. If the storm continues to organize with wind speeds greater than or equal to 39 mph (34 kts) to less than 74 mph (64 kts), it is classified as a tropical storm and given a name. Continued development of the system becomes a hurricane with wind speeds exceeding or equal to 74 mph (64 kts) [3,4].

There are five major environmental factors which determine successful tropical cyclone development. These are sea surface temperature, surface layer of warm water, weak vertical wind shear, sufficient moisture in the middle troposphere, and a location at least 5° north or south of the equator. Sea surface temperature must be greater than 80°F (26.5°C) as this supplies the heat and moisture released into the atmosphere. The layer of warm water is usually around 200 ft (60 m) to ensure there are enough warm water and energy as the ocean gets mixed in a process known as upwelling. This depth keeps the warmer water rising and cooler water in deeper regions that will not impede the energy source. Wind shear (winds from opposite directions or too high a speed) needs to be relatively weak to form the vortex. If speeds are too high, the vortex will be torn apart and moves downstream. Dry air in the middle troposphere can weaken the storm by reducing latent heat and increasing downdrafts through evaporative cooling; hence, a tropical cyclone requires enough moisture to ensure formation. Rotation is another important aspect of tropical cyclones and thunderstorm clusters. These clusters need to be 5° north or south of the equator so the Coriolis force is strong enough to help develop rotation; the Coriolis force is zero at the equator [4].

The Saffir-Simpson Scale classifies hurricanes based on wind speed on a scale of 1–5 with 5 the strongest. Before 2010, the scale predicted storm surge height and barometric pressure along with wind speeds. The wind scale was revised again in 2012 (**Table 1**) due to the rounding and conversions from mph to km/h; only category 4 and 5 storms were impacted with this update. Previous storms will not have their categories changed with the new scale. Expected damage generally increases with higher category levels. Category 3 and higher hurricanes are known as major hurricanes [4,5].

Tropical cyclones are some of the most destructive storms. Some of the impacts are due to storm surge, heavy rain, inland flooding, high winds, and potential tornadoes. Storm surge refers to the rise sea level as the hurricane makes landfall. Onshore winds and the barometric effect (rise in sea level due to low pressure) cause storm surge. Other factors are the wave height, tides, and shoreline shape. Storm surge can cause extensive damage to the landscape and structures. If the tropical cyclone is slow moving or stalled, heavy rains can hit an area. Heavy rain can produce inland flooding and causes the most fatalities along with property destruction. High winds can cause damage, especially to structures not built to withstand higher tropical force winds. There can also be hurricane spawned tornadoes which are often highly concentrated in the right, front quadrant (when looking at the perspective of the hurricane approaching the shore). These tornadoes are usually in the EF0–EF2 range and can also cause property damage [4].

Table 1. Saffir-Simpson scale [5].

Category	Wind speed	Damage type
1	74-95 mph	Some
2	96-110 mph	Extensive
3	111-129 mph	Devastating
4	130-156 mph	Catastrophic
5	157 mph or higher	Catastrophic

2.2. Tropical storm debby

During the 2012 hurricane season, Tropical Storm Debby originated in the south-central Gulf of Mexico after a surface low developed near the Yucatan peninsula and propagated eastward toward an area where the prevailing subtropical ridge had weakened. There was a northern edge of a tropical wave in the Caribbean Sea that merged with the disturbance near the Yucatan peninsula that eventually became Tropical Storm Debby on 22 June [6]. Hurricane Hunter aircraft determined circulation was well defined and the winds were tropical storm strength on 23 June (**Figure 1**). Over the next 24 hours, Tropical Storm Debby moved slowly north to northeastward without a well-defined trajectory forming a rain shield over the northeastern Gulf of Mexico on 24 June. On 25 June, Debby approached the Big Bend area of Florida (area where the Panhandle and Florida peninsula curve on the Gulf of Mexico) and made landfall near Steinhatchee, Florida, on 26 June. Peak winds were estimated at 63 mph (55 kts) and minimum surface pressure was 990 mb [6].

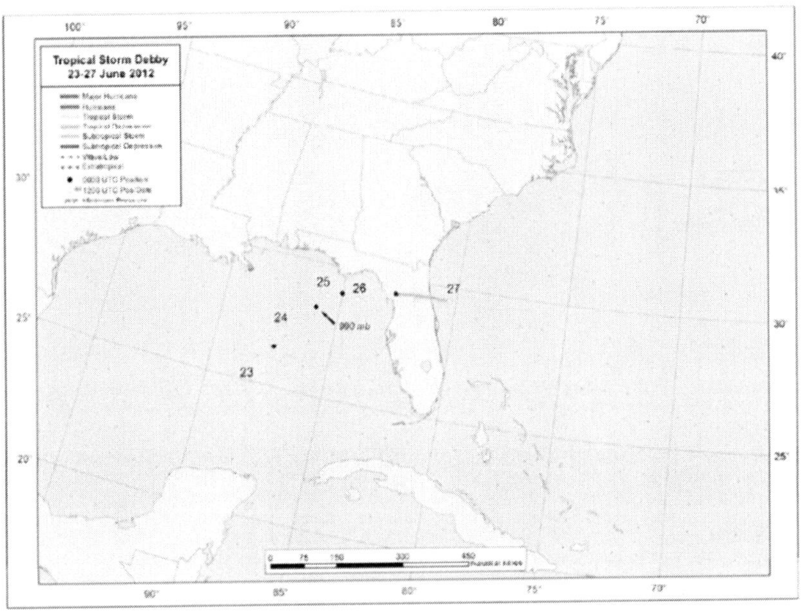

Figure 1. Track positions of Tropical Storm Debby [6].

Winds were not the major impact from Debby but the torrential rains caused major impacts along with inland flooding in various parts of the Florida peninsula. A local observer in Wakulla County measured 29 in. (731 mm) during the event and there were other reports greater than 20 in. (508 mm) in the same region (**Figure 2**). Wakulla County is northwest of the track line near the Great Bend with the light blue shading. There were several totals greater than 10 in. (254 mm) over western and northeastern Florida (see **Figure 2**, south of the purple-shaded track line) [6].

Besides inland flooding, storm surges from 2 to 4.5 ft (0.6–1.4 m) occurred from the Florida Panhandle to the southwestern coast of Florida. This resulted in inundation 1–3 ft (0.3–0.9 m) aboveground level with the highest surge reported between Apalachicola and Cedar Key [6] (see **Figure 2**) on and northwest of the purple-shaded track line) [6]. In addition to the heavy rain and flooding, the rain bands east of the center produced a number of tornadoes. NOAA's Storm Prediction Center (SPC) recorded 24 tornadoes primarily rated at EF0 in central Florida on 23 June. On 24 June, tornadoes hit over the southern and central Florida with several rated at EF1 and EF2 (**Figure 3**) [6].

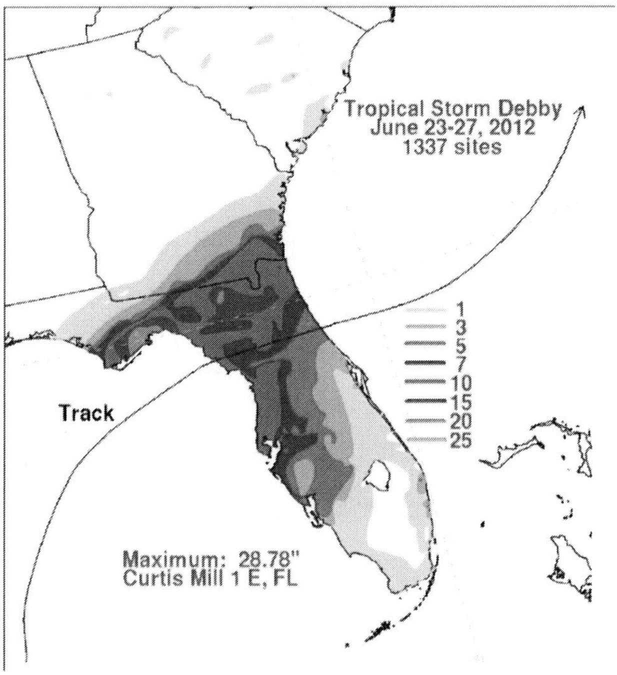

Figure 2. Rainfall totals associated with Tropical Storm Debby [6].

Five direct fatalities and three indirect deaths were recorded with Debby. A mother was killed in Venus, Florida (west of Lake Okeechobee, the largest lake in Florida), after a tornado hit a mobile home. One drowned in heavy surf in Pinellas County, Florida (west-central Florida on the Gulf of Mexico), and another in Orange Beach, Alabama. A canoe capsized near Lake Dorr, Florida (central Florida north of Orlando), and the person drowned and another man presumed to drown from the storm was found near Anclote Key, Florida (west-central Florida). The indirect fatalities were a man wading in floodwaters in Pinellas County, Florida, and two others were in automobile crashes on wet roads from Debby [6].

Most of the damage was from inland flooding from the heavy rains in parts of northern and central Florida. The Sopchoppy River in Wakulla County (northwest of track line in **Figure 2**) crested at 36.8 ft (11.21 m) and affected 400 structures. There was flooding in Pasco County (south of track line near the Gulf of Mexico in **Figure 2**) along the Anclote and Pithlachascotee Rivers damaging 106 homes. The Suwanee River had observations highest since Hurricane Dora in 1964. Roads such as US Highway 90 and Interstate 10 (north of track line in **Figure 2**) were closed due to floodwaters. US 90 was closed for almost two weeks and Interstate 10 for two days [6]. In addition, there were several roads closed in many counties and sinkhole problems in Marion County (south of track line in **Figure 2**) due to the heavy rains [7].

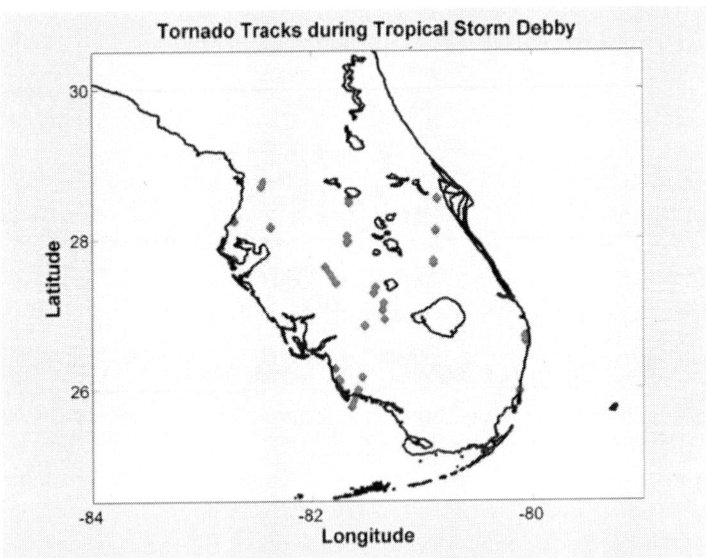

Figure 3. Tornado tracks during Tropical Storm Debby [6].

Coastal areas were affected from storm surge in the Panhandle, Big Bend, and along US Highway 19 in Hudson, Florida, on the west-central coast of Florida. Some roads were submerged for days and others were washed out. There was extensive beach erosion form Pinellas County southward to Charlotte County with the worse erosion in Treasure and Anna Maria Islands [6].

Preliminary insured losses in Florida were $105 million according to the Property Claims Services with $40 million in flood damage covered by the National Flood Insurance Program. Total damage could be $250 million since insured values are doubled along with flood losses [6]. The Federal Emergency Management Agency (FEMA) approved 6758 Individual Assistance applications in the amount of $27,800,267.48 and Public

Assistance for communities for $52,197,352.72. These figures do not include flood insurance claims [8].

2.3. Us tornado background, activity, and measurement

2.3.1. Tornado background

Tornadoes are rapidly rotating columns of air extending from the cloud to the ground [4]. According to the American Meteorological Society [9], a tornado is "a violently rotating column of air, pendant from a cumuliform cloud or underneath a cumuliform cloud, and often (but not always) visible as a funnel cloud." By definition, tornadoes are invisible but visible once debris is in the funnel cloud [4].

All tornadoes come from thunderstorms but all thunderstorms do not form tornadoes [10]. Most tornadoes are formed in supercell thunderstorms but can develop from hurricane thunderstorms, squall lines, and regular thunderstorms. Some of these are non-supercell tornadoes, landspouts, waterspouts, mesovortices, and gustnadoes [4]. Supercells are rotating thunderstorms and consist of a mesocyclone (circulation that is detected on radar) and can spawn a tornado along with hail, high winds, lightning, and heavy rain [10].

The typical life cycle or tornadogenesis begins in the dust whirl stage, followed by the organizing stage that reaches full damage at the mature stage, the weakening stage, and finally the rope stage. Environmental factors necessary for formation include vertical wind shear and horizontal rotation [4]. Uplift is also necessary with temperature and pressure differentials.

Tornado widths are commonly 150 ft to 0.5 miles (50–800 m) with wind speeds ranging from 65 mph to greater than 200 mph (57 kts to over 174 kts) [4]; however, greater tornado diameters can occur. Most tornadoes are only on the ground for a short time such as less than 10 minutes [10] but can remain over an hour with damage paths over 30 miles (50 km) [4].

A tornado can occur at any time of the year in the United States. The United States averages around 1000 tornadoes a year. Peak season depends on the location. On the Gulf of Mexico coast, it is the early spring, while May and early June generally have more tornadoes in the southern plains, while the northern plains and upper Midwest have more in June and July [10].

Other countries also get tornadoes but most are reported in the United States. Some of the regions associated with tornado activity are also main agricultural areas that do not include the tropics [4]. Countries with reported tornadoes are Canada, United Kingdom, Bangladesh, Mexico, Argentina, Brazil, and Russia [10].

2.3.2. Tornado measurement and activity

Tornadoes are classified by the Enhanced Fujita Scale (EF). Prior to 2007, the Fujita Scale was used based on damage. The original Fujita Scale was developed by Dr. Theodore Fujita in 1971 as a forensic examination of structural damage. Since 2007, the EF takes into account the type of building materials and construction, damage level, and estimated sustained wind speeds. EF categories range from EF0 to EF5 [4,10,11].

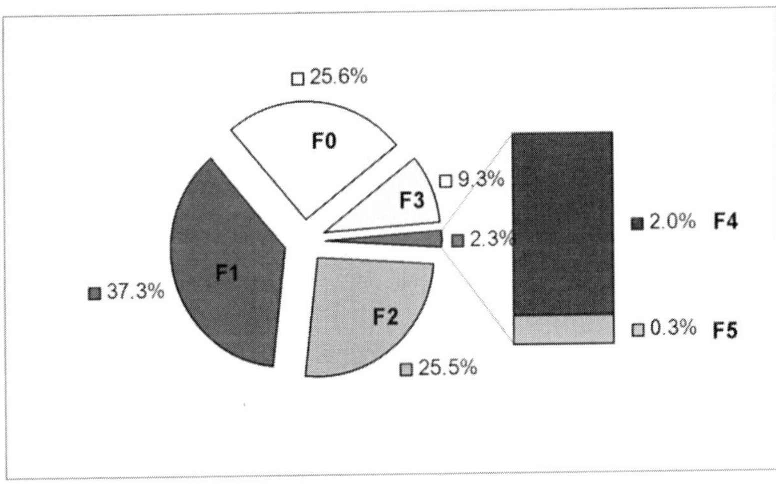

Figure 4. Percentage of all US tornadoes that occurred in each EF-scale category [4].

The majority of tornadoes in the United States are EF0 (25.6%) and EF1 (37.3%). EF4 and EF5 tornadoes are the rarest at 2.0% and 0.3% of all tornadoes, respectively (**Figure 4**) [4]. Wind speeds with EF0 range from 65 to 85 mph (105–137 km/h), EF1 86 to 110 mph (138–177 km/h), EF2 111 to 135 mph (178–217 km/h), EF3 136 to 165 mph (218–266 km/h), EF4 166 to 200 mph (267–321 km/h), and EF5 >200 mph (322 km/h) (**Table 2**). Damage associated with tornadoes is from property damage to structures from high wind speeds and heavy rain from the thunderstorms which can lead to flash flooding and hail [4].

Table 2. Enhanced Fujita Scale [11].

EF number	3-second gust (mph)	3-second gust (kph)
0	65–85	104.58–136.77
1	86–110	138.38–176.99
2	111–135	178.60–217.22
3	136–165	218.83–265.49
4	166–200	267.101–321.81
5	Over 200	Over 321.81

2.4. Tuscaloosa alabama tornado

A tornado outbreak consisting of 353 tornadoes in 21 states occurred between 25 April and 28 April 2011 [4]. This outbreak had more tornadoes than the 1974 Super Tornado Outbreak (148 tornadoes) and more fatalities than the outbreak that occurred on Palm Sunday in 1965. There were also 2400 injuries and over $4.2 billion in damages associated with this outbreak. In the southeast United States, there were 122 tornadoes that resulted in 313 fatalities in the afternoon and evening of 27 April. Tornadoes that had hit before dawn on the 27th added three more deaths for a total of 316. States affected on the 27th include central and northern Mississippi, central and northern Alabama, eastern Tennessee, northern Georgia, and southwestern Virginia. There were 15 violent tornadoes (EF4 or EF5) and eight had paths longer than 50 miles (or about 80 km) (**Figure 5**). Two of these tornadoes in Alabama, one in the northern part of the state and the other that struck Birmingham and Tuscaloosa, each had more than 60 deaths [12].

The outbreak was forecast by the SPC five days before the event. Weather forecast offices (WFOs) in the area were also preparing for the threat of severe convective weather and tornadoes five days in advance. Some of the activities from the WFO to emergency managers included discussions and tools such as "Hazardous Weather Outlooks, Web images, prerecorded multimedia briefings, and webinars that discussed the potential impacts [12]."

All of the tornadoes occurred in tornado watch box and warning areas. Lead time from watches to warning averaged 2.4 hours while watch time to the first significant tornado ranged from 3 to 6 hours in each area. Warning lead time for tornadoes was 22.1 minutes. The fatalities were all in watch and warning boxes [12].

Even with the forecasts from the SPC and WFOs, there were a high number of fatalities and injuries for several reasons. The tornadoes hit urban, suburban, and rural areas and were long-track violent tornadoes. The storms damaged warning sources, such as NOAA Weather Radio Transmitters. Human behavior was also a major factor as many individuals did not respond to warnings without additional confirmation or waited for visual confirmation before taking action. Furthermore, the storms moved 45–70 mph (or about 72–112 km/h) which gave less time for those who waited to seek shelter, and for some, adequate shelter was not readily available [12].

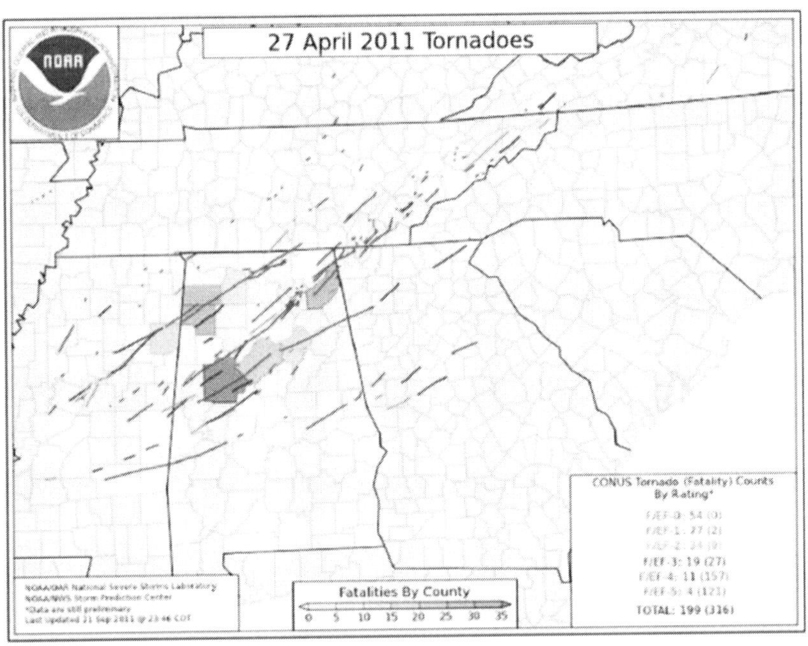

Figure 5. Tornado tracks from 27 April 2011 outbreak [12].

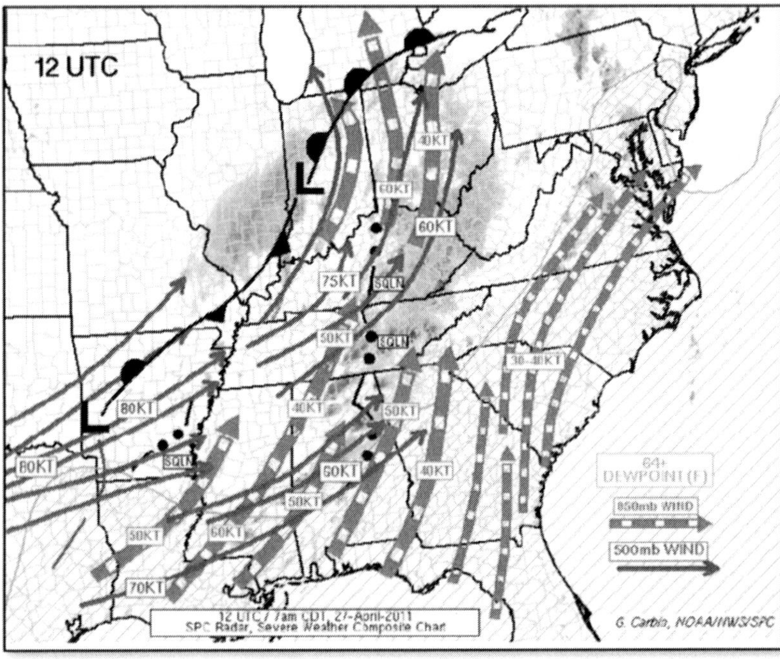

Figure 6. Composite weather analysis 1200 UTC 27 April [12].

The active April severe weather event in the Southern Plains and southeastern United States peaked on 27 April. Conditions were ripe for atmospheric instability due to colder air (compared to previous systems) with an upper-level storm moving east out of the southern Rocky Mountains. The upper-level storm continued east while a strong low-pressure system formed in western Arkansas (**Figure 6**). "As this low formed in the morning, southerly winds increased dramatically in the lower portion of the atmosphere, from around 15 mph at the surface to 45 mph approximately 3000 ft aboveground level. The change of wind direction and speed with height, known as vertical wind shear, helped create highly organized storms that could develop strong rotation in the lower and mid-levels. The approaching upper-level storm brought strong westerly winds at high altitudes, helping ensure that long-lived thunderstorms would occur [12]."

A line of severe thunderstorms hit central and northern Mississippi, central and northern Alabama (including Tuscaloosa County), and southern middle Tennessee before dawn and produced over 24 tornadoes and caused three fatalities and over 40 injuries (**Figure 7**). Besides causing widespread power outages, several NOAA Weather Radio All-Hazards Transmitters were out of service; this would be a factor later in the day regarding warnings in some of these areas [12].

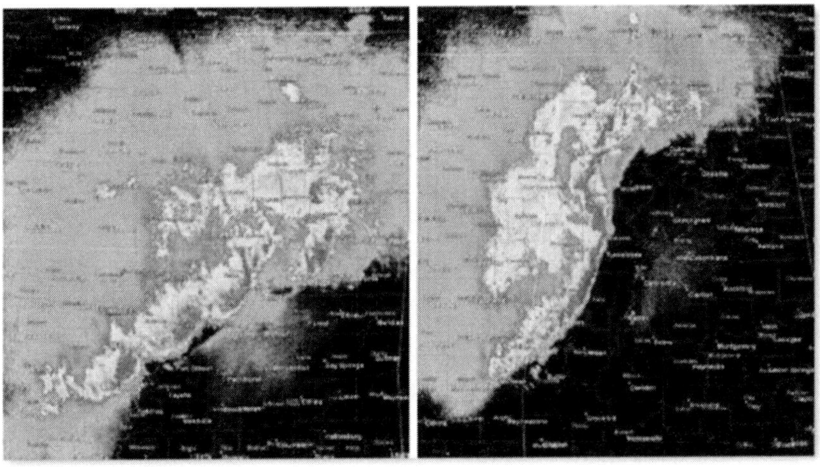

Figure 7. Radar images of early morning storms 27 April 2011 from Jackson Mississippi (left) and Birmingham Alabama (right) WFOs [12].

The early morning storms left a strong low-level jet along with a lot of moisture to assist in more atmospheric instability. An outflow boundary in northern Mississippi and northern Alabama brought more severe storms later in the morning. Further south heating occurred from the sun resulting in more heated low-level air resulting in more destabilization. Vertical wind shear increased, from 20 mph (32 km/h) at the surface to 70 mph (112 km/h) at 3000 ft (0.9 km) to over 100 mph (160 km/h) near the tropopause (34,000 ft or 10 km). These conditions gave "an extraordinary high potential for strong low-level rotation in the storms." In addition, the upper-level wind speeds helped to produce long-lived storms, an environment suggesting severe, long-lived supercells capable of producing violent tornadoes (**Figure 8**) [12]. There were 62 confirmed tornadoes in Alabama and 29 in central Alabama 27 April [13].

Storm Data [14] named the Tuscaloosa-Birmingham EF4 Tornado the "outstanding storm of the month." This tornado was not the strongest (there were EF5s) or the longest track of the outbreak, but it had significant impacts due to the population affected and social impacts. There are 194,656 people in Tuscaloosa County and 90,468 in the city of Tuscaloosa according to the US 2010 Census [15]. The 27 April 2011 outbreak was well forecast by the SPC and local WFOs with most severe weather parameters such as instability and shear on the higher ends of the scale.

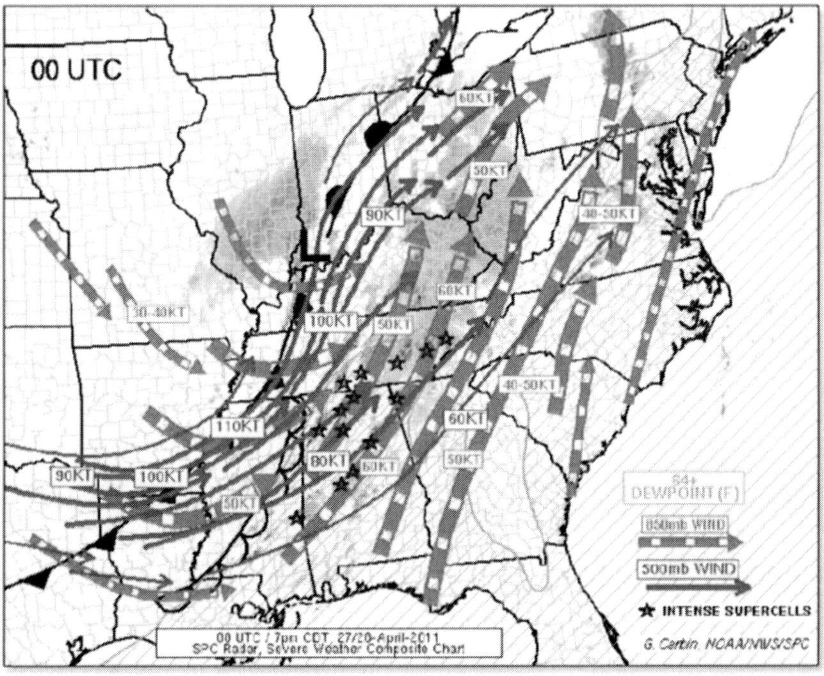

Figure 8. Composite weather analysis 0000 UTC 28 April [12].

The first afternoon tornado touched down at 3:05 pm CDT (2005Z), in Marion County, Alabama, was an EF5. Another supercell would produce the Tuscaloosa-Birmingham tornado started around 3:00 pm in Newton County, Mississippi (110 miles or 176.99 km southwest of Tuscaloosa), and took about an hour and 45 minutes to develop a tornado. This tornado dropped in Greene County, Alabama, and then to Tuscaloosa County into the populated city of Tuscaloosa. As the tornado continued to move east-northeast, it moved into Jefferson County, Alabama, and finally weakened around north of downtown Birmingham, Alabama, and lifted around 4 miles northeast when the parent supercell and another supercell merged. This supercell would produce other tornadoes and was tracked into western North Carolina. The National Weather Service (NWS) damage survey said the tornado caused 64 direct fatalities and over 1000 injuries along with impacting more than 36,000 people [14].

3. METHODS AND CASE STUDIES

These two atmospheric hazards had two separate case studies and different research questions and methods. Both of these studies examined preparedness along with communications on different levels and scales.

3.1. Tropical storm debby case study

3.1.1. Research objectives

The research objectives of this study examined various means of communicating weather information among the elderly. One of the major themes was comparing traditional media (i.e., television and newspapers) to newer forms of social media such as Facebook or Twitter. In addition, questions were asked in regard to preparedness actions for a tropical storm or hurricane.

3.1.2. Methods

Self-administered questionnaires were given on voluntary basis to elderly citizens. Since American Association Retired Persons (AARP) considers membership from age 50 and older, this was the threshold for elderly. Surveys consisted of both close-ended and open-ended questions. Questions dealt with methods of weather communications prior and following Tropical Storm Debby. Additionally, questions dealt with other potential disaster situations, evacuations, and any changes in actions following this event. A few respondents were interested in elaborating in interviews but the majority only completed the surveys. Subjects resided in Pinellas and Pasco Counties, part of the Tampa Bay area in Central West Florida. There were 30 participants in the 40 question survey. All of the responses were kept confidential with no identifying characteristics of the respondents.

3.1.3. Results

The first question asked on the primary method of receiving weather information. Some of the choices were local television, newspaper, radio, cable television news (i.e., CNN, MSNBC, and Fox), specialized cable news (Weather Channel), local cable news, the Internet, National Weather Service, cell phone, or personal communication with friend or relative. Local television was the overwhelming choice with 60.2% selecting this option. The second choice was local cable news at 13.3% (there is a 24-hour local news channel with weather updates every 10 minutes). Third was the National Weather Service at 10.0%. All of the other choices totaled 16.5%.

Question two asked if respondents utilized a second method of receiving weather information. Of those responding, 83.3% were affirmative and 16.7% did not go to a secondary source. For those who went to another source, 23.3% used the Internet, 13.3% the National Weather Service, 13.3% Weather Channel, local television and cable news each 10%, newspaper and local cable news each 6.7%, and cell phones and personal communication 3.3% each.

In regard to technology owned, 90% owned a computer, 6.7% did not own a computer, and 3.3% did not respond. Almost as many respondents owned a cell phone at 86.7% while 10% did not own a cell phone, and 3.3% did not respond to the question. For those who owned a cell phone, 23.3% indicated it was a smartphone while 66.7% did not have a smartphone and 10% did not respond.

Besides owning technology, the survey inquired how it was used by the respondents. A large majority of 90% use e-mail, while 6.7% do not and 3.3% did not respond to the question. The majority of respondents who use e-mail only on a computer were 66.7%, none used a phone only, 23.3% used both the computer and phone, and 10% did not respond. A minority utilized texting at 36.7% but 60% did not text, and 3.3% did not respond. Similar to texting, only 26.7% of the respondents engaged in social media with an overwhelming majority of 70% not using services such as Facebook or Twitter. The nonresponse rate was 3.3%.

Those who used social media primarily were for personal reasons, such as communicating with family. Only 4.6% used social media to get information. Some of the reasons stated for not using social media from the respondents include "don't like it, security issues, not interested in public exposure, privacy concerns, do not need it, don't feel comfortable with it, no time, and prefer to use the telephone."

Only five respondents indicated they had any damage from Tropical Storm Debby and all five noted it was minor damage. No insurance claims were filed or requests for disaster assistance.

The majority of respondents (36.7%) did not think it was applicable to prepare for Tropical Storm Debby. However, 33.3% did and 26.7% did nothing while 3.3% did not respond. Preparedness actions included close hurricane shutters if threatened, have an evacuation plan, follow instructions from local authorities (such as emergency management officials), have equipment and supplies on hand, secure property if threatened, have no special actions, minimize mulch around back of house to avoid a dam and cause flooding, alert the Citizens Emergency Response Team (CERT), and have more water (for drinking and other purposes).

Another question asked if respondents owned a NOAA All-Hazards or Weather Radio. A large majority of 63.3% owned a radio with 36.7% not owning a weather radio. A question was asked regarding the perception of FEMA with 6.7% answering excellent, 3.3% good, and 90% stating not applicable.

Besides questions on weather communications and preparedness, demographic information was requested from the respondents. The gender breakdown was almost equal with 46.7% female, 50% male, and 3.3% not responding. A question inquiring to primary employment status indicated the majority (76.7%) were retired, 13.3% employed full-time, 3.3% employed part time, and 3.3% not responding. There was a range of incomes with categories of $20,000–34,999 (16.7%), $35,000–49,999 (13.3%), $50,000–74,999 (16.7%), $75,000 or above at 20%, and 33.3% preferred not to answer the question.

Ages were in categories with 50–54 (6.7%), 55–60 (13.3%), 60–64 (3.3%), 65–70 (16.7%), 71–74 (20%), 75–80 (16.7%), 81–85 (10%), and 86 or above (10%). Only 3.3% did not respond to the question.

3.2. Tuscaloosa, alabama, tornado case study

3.2.1. Research objectives

Research objectives of this study examined fundamental emergency management issues on four different levels. These levels are Tuscaloosa County, the University of Alabama at Tuscaloosa, the Alabama Emergency Management Agency, and FEMA. Besides the emergency management issues, lessons learned were noted for each level.

3.2.2. Methods

In-depth interviews were conducted with various emergency management officials at each level. An open-ended survey/interview was utilized with the respondents. Interviews were with personnel from the Tuscaloosa County Department of Emergency Management, University of Alabama Department of Public Safety, Alabama Emergency Management Agency, and FEMA. Field observations were also completed by the author. Topics included existing and revised preparedness plans, major components of planning, lessons learned from plans, mutual aid, recovery progress, and lessons learned. Other issues included citizen and survivor experiences, mitigation, and interactions among the various agencies at the four levels.

3.2.3. Results: county level

Tuscaloosa County utilized various preparedness actions. Preplanning for events was more common in the post-Katrina era (since 2005). Emergency management personnel had plans for events such as tornadoes, hurricanes, flooding, and winter storms. Besides having planning, exercises were practiced with various community stakeholders. One of the plans and exercises was for mass casualties. Procedures were established for damage assessment such as the type of tags to be issued based on damages and having architects and engineers ready to inspect structures to determine safety or be habitable. Price gouging laws were enacted to keep those from profiteering, especially for necessary items such as gasoline and hotel rooms. The city of Tuscaloosa and Tuscaloosa County had budgeted for financial reserves in case of disaster. Local media was also a partner with the city and county to help disseminate severe storm and tornado awareness to the citizens.

Response was challenging as the tornado made a direct hit and damaged the Emergency Operations Center (EOC) for Tuscaloosa County. The EOC was relocated to the University of Alabama campus. Prior to the tornado, emergency management personnel were aware of the severe weather and tornado potential and responded to the early morning tornadoes. There was a weather briefing with the Birmingham WFO at 2:00 pm and emergency management officials used tools such as Weather Messenger and Emergency Management Weather Information Network (EMWIN) Injects to stay apprised of current and future weather. Various responders (i.e., firefighters, law enforcement, and medical personnel) were also prepared and ready to respond if necessary. The tornado cut through the city and there was a lot of structure damage in addition to the high number of fatalities and injuries. A lock down and curfew was imposed in the city of Tuscaloosa. Some looting occurred but the offenders were not from Tuscaloosa County.

Recovery included tasks such as cleaning debris, power restoration, and working to bring the community back to normal conditions. Citizen needs such as food, water, clothing, housing, and recharging cell phones were attended to by various groups. Some were private or faith-based and others were governmental organizations such as FEMA. Four Disaster Resource Centers (DRCs) were opened by FEMA in Tuscaloosa to assist survivors with their needs. In addition, one Small Business Administration (SBA) Center was available to assist with low-interest loans; SBA works with both individuals and businesses after disasters.

Tuscaloosa County worked with a few mitigation methods. One mitigation method was establishing an active group of Skywarn volunteers, a group of citizens that spot storms and report observations to the National Weather Service. Skywarn training is an annual event with many citizens either taking new or refresher training. Community shelters were established as many do not have storm shelters in their homes or apartments. Some of the structures to serve as shelters are schools and recreation centers. There have been discussions on stricter building codes and recommendations to tie down objects such as air conditioning compressors and water heaters so they are not projectiles in a tornado.

The loss of the EOC in Tuscaloosa County brought up some issues that serve as lessons learned from the incident. First, when the building was damaged, the fire suppression systems went off since the water was on. The backup generator was damaged and needed protection from the elements and water. When a generator is hooked up to the building, circuits on the generator need to be well marked to save time trying to find the live outlets. Spare COAX cable (used in communications) and antennae are necessary for setting up networks and communications. The internet was backed up with a virtual private network (VPN) that went to the University of Alabama that allowed workers to be on the county computer network. A lot of handheld radios were used and issues were found getting a signal in substantial buildings due to the interference from construction materials; hence, multiple methods for communications are needed. Besides the issues at the EOC, there was a lot of interaction within the community that involved both the private and nonprofit sectors. Some events could have gone smoother such as the Governor's Summit (meeting with emergency management and political officials) that was held too soon after the event to cover response and recovery strategies. The FEMA housing event (for citizens impacted by the tornado on rebuilding options) was not well marketed to the community and had low attendance.

3.2.4. Results: university level

Preparedness at the University of Alabama was evident with their established Emergency Operations Plan (EOP) (Tuscaloosa County had one too), marked tornado shelters, an Emergency Call Center Plan (to get out information), preestablished mutual aid agreements, and following the Incident Command System (ICS). The university also had an Emergency Notification/Crisis Communication Plan. These systems utilized texts, e-mails, phones, signage, and public address system to get out warnings and information. Additionally, the university was designated by the National Weather Service as StormReady, meaning the university meets standard communication protocols for informing their community on present and future hazardous weather conditions. Other plans were established for using resources such as the campus buses along with shelter plans with dormitories and continuity plans dealing with dining, facilities, payroll, human resources, and public safety.

The tornado path came within 1200 ft (366 m) of the campus and impacted many students and faculty. There was a lack of power in Tuscaloosa and cell phone communications were severely hampered. This was due to tower damage, lack of power, and heavy usage. The university was able to offer mutual aid to the community in a number of ways for response. First, the

EOC was used by Tuscaloosa County for two days and university law enforcement officers assisted the city for a month. Many university vehicles and equipment such as trucks, vans, gators (small all-terrain vehicles for transporting people and materials), and forklifts were used in the community. The campus was used as a staging area for several groups including Urban Search and Rescue (USAR) teams. Some university personnel were used as translators for the non-English-speaking population in shelters. Classes were canceled and dorms used for responders such as utility workers, National Guard, Red Cross, and law enforcement personnel. Still, some students and employees also needed shelter along with meals for responders and key personnel. A "Seek and Find" website was established to track missing people. University resources were used for power and communications, especially with towers. With classes canceled, many students volunteered to help in the community and the university offered food assistance to citizens in need. Other university resources included medical personnel working at the hospital along with the Incident Command Center that was open for 17 days, 24/7. A Joint Information Center (JIC) was established to communicate with citizens and the media. Participants in the JIC included the City of Tuscaloosa, Tuscaloosa County, the University of Alabama, and the State of Alabama.

Since there was no damage to the campus, recovery was more financial than physical. The university was reimbursed by FEMA for expenses assisting the city and county. They also worked with the city recycling services with the large amount of debris. An "Acts of Kindness" program was established to assist students and employees who needed financial aid for recovery and helped manage donations. Another mitigation effort was a Hazard Mitigation Grant Request to FEMA for additional community shelters and generators on campus. The decision was not known at the time of research visit and it was not listed as a funded grant by FEMA. A Damage Assessment Response Plan was established to coordinate personnel to assess and restore buildings.

The University of Alabama learned several emergency management lessons through this event. For example, the university needed to evaluate the generator fuel supplier as the university was in direct competition with the City and, if conditions warranted, both could run low on fuel. Generators were found to be in short supply and increased power capacity was needed. The incident reinforced the need for redundant internet and network pathways; this could be accomplished with multiple ingress/egress core routers. Volunteers are usually forthcoming in disasters and the university needed to establish and coordinate their use in an effective manner. Once Incident Command was established, all requests for resources should go through the Incident Commander to avoid miscommunication issues along with utilizing volunteers. Another lesson learned was to expect traditional communications to be overwhelmed or unavailable. Good working relationships (networking, training, and exercises) and mutual aid agreements with other partners such as the city, county, and state are essential during a crisis. Finally, having large events (i.e., football games) with partners and utilizing Incident Command is important in a real disaster. Although a non-crisis situation, large numbers of citizens in a concentrated setting provide community experience in dealing with multiple agencies and stakeholders, a situation similar to a disaster.

3.2.5. Results: state level

Preparedness for the Alabama Emergency Management Agency also involved planning, exercises, and training. The state agency also has a strong relationship with Voluntary Organizations Active in Disasters (VOAD) and the various organizations that are members. Faith-based groups along with civic organizations are among the members that are prepared to respond when needed at a disaster. The State Emergency Management Agency partners with the National Weather Service in educating the public on Severe Weather Awareness Day. Alabama Emergency Management Agency was very aware and communicating with the National Weather Service before and during the tornado outbreak.

Response from the state was to ensure citizens and communities received necessary aid and resources. Mutual aid compacts worked in the majority of counties with no problems in Tuscaloosa County (there were some issues in other counties). State subject matter experts coordinated and worked well with FEMA. Personnel from the state were working in the EOC at the University of Alabama within three hours of the tornado.

One of the major recovery objectives was to give support to the local level. The process for the citizens who needed assistance can be described as "bottom-up" (citizens starting the process and working up through the various agencies), whether to rebuild or obtaining information. The state made a big effort to prevent those "from falling through the cracks" or not following through with individual and community cases. This is due to the bureaucracy and regulations for receiving aid and resources – both to individuals and communities. Another major task was assisting with debris removal and reimbursements.

The State Emergency Management Agency was involved with a few items regarding mitigation. There were grants available to individual homeowners for tornado shelters. They received hundreds of calls along with almost 4200 applications regarding these grants for a shelter or safe room. Hurricane straps and safe rooms were encouraged to be included in the rebuilding process; however, these items were not required items per the building code. Other mitigation issues were related to historic structures. The state advocated slowing down the process and be practical in regard to regulations dealing with historic preservation. Several individuals in Tuscaloosa are not fond of government regulations and wanted to proceed quickly with recovery and not spend too much time regarding older damaged structures that needed to be repaired or totally rebuilt.

A Tornado Recovery Action Council of Alabama (TRAC) was conferred by the Governor and aided in lessons learned from the outbreak. The major sections of the report were a summary of the tornadoes and prepare, warn, respond, recover, and forum reports [14]. Other variables included the importance of ICS training and that all resources and requests should go through the Incident Commander which helps with proper tracking. This is important for getting reimbursements. Like the other levels, the state found issues with communications, coordination, and managing resources.

3.2.6. Results: federal level

FEMA was following the whole community approach of collaboration in the Post Craig Fugate (Administrator since May 2009 through present time of 2016) era for the entire disaster cycle including preparedness. Personnel in the agency were aware of the severe weather potential and had strike teams ready to activate to work with partners at all levels of government along with the private and nonprofit sectors.

There was a quick response by FEMA with Community Affairs workers on the scene within 24 hours. A Joint Field Office was established in Tuscaloosa along with DRCs. In addition, a Federal Coordinating Officer was dispatched to Tuscaloosa to aid with response and recovery functions. At the peak, 3000 FEMA employees were assisting in the disaster. It was imperative for qualified personnel and a strong command and control structure to assist in getting a quick awareness of the situation and evaluate what resources would be needed to help the community.

Some of the recovery tasks involved debris removal resources and wholesale rebuilding of structures. The Army Corps of Engineers also worked with homeowners on debris removal. A program called Tuscaloosa Forward was established consisting of town hall meetings to inform residents of assistance and procedures in this phase. One of the first priorities for survivors was temporary housing assistance while public assistance was infrastructure restoration. There were around 30,000 applications submitted to FEMA for assistance. Those with private insurance used their policies for recovery first and then FEMA assistance would come into aid if they were eligible.

Mitigation measures included education of current building codes and suggestions of strengthening structures (even if not enforced by building codes). This included tie downs and safe rooms. There was also a priority of establishing community shelters, especially with the number of apartments and manufactured housing in the community.

Lessons learned by FEMA went along with the other levels of government, such as the success of using the ICS and the cooperation/knowledge of community leaders with the process. All levels expressed a great working relationship with the local, state, and federal government personnel. The importance of a Public Information Officer (PIO) was useful with the different agencies along with communicating information to the media and public; however, there were communication and coordination glitches. Additionally there was a housing shortage due to the tornado hitting a populated area with housing needs for residents, volunteers, and responders. Due to the number of damaged vehicles, transportation shortages existed. Finally, there were challenges with commercial sites and insurance policies in regard to debris management.

4. SUMMARY

4.1. Tropical storm debby summary

Research indicated that the elderly use local television to get their primary weather information. As with most individuals, they also want to verify their information with a secondary source. The majority used the National Weather Service and the Weather Channel for their sources of secondary information. However, results indicated the elderly use technology in their lives. A large majority own computers and cell phones which are used for personal use such as e-mail and phone calls instead of receiving weather information. Another technology used by the respondents was NOAA Weather Radios. This could be due some of them being members of CERT teams along with previous tropical storm experiences. The majority of respondents do not text or use social media. These results indicate the importance of using both traditional means of communicating weather information and newer methods such as social media in order to reach all citizens. Only a third indicated preparedness actions with Tropical Storm Debby compared to around a quarter who took no action. More than a third said it was not applicable and did not feel the need with this tropical system.

4.2. Tuscaloosa alabama tornado summary

The research examined emergency management issues on different scales ranging from county to federal levels. All of the respondents indicated the importance of using Incident Command in the response of a disaster. Other commonalities emphasized the usefulness of planning and exercises along with networking. Mutual aid agreements for entities are also important for assistance and resources. Tornado-specific findings were having more tornado shelters, especially community shelters and tougher building codes. Common issues on all levels were with communications and obtaining and utilizing resources.

4.3. General themes

These two case studies examined communication and emergency management issues along with disaster preparedness on different levels. Elderly citizens were surveyed to their preferences for receiving weather information along with preparedness actions before Tropical Storm Debby. Seniors are often vulnerable populations in disasters. They use more traditional methods of receiving weather information and will listen to authorities such as local media and the National Weather Service. Elderly preparedness actions follow practices established and communicated by the media and local emergency management agencies. Tropical systems such as tropical storms or hurricanes are usually well forecast and receive a lot of media attention. However, most of the respondents did not have any impacts from Tropical Storm Debby but there were citizens who were flooded or had tornado damage.

Tornadoes do not usually have the lead time compared to a tropical storm or hurricane. However, the Tuscaloosa tornado had lead time but previous severe weather damaged some communication methods such as NOAA Weather Radio. This outbreak was forecast days in advance and the media communicated its progress live to listeners and viewers. Emergency management agencies were prepared and utilized existing networks for response and recovery actions. There was cooperation and mutual aid demonstrated on the various levels which ranged from the local community to the federal. While there were some issues and glitches, the objective of reducing social vulnerability was primarily successful. Most citizens were able to survive even with the large amount of property damage and high number of fatalities and injuries.

These case studies demonstrate the necessity of efficient communications, from a personal level of receiving weather information to all the levels of emergency management. In addition, preparedness actions and procedures need to be explicit and able to be understood by all, again from a personal level to working with agencies. Being prepared will not stop a tropical storm or hurricane but can help mitigate an event and if disaster strikes makes response, recovery, and resiliency better for those impacted and hopefully decreases fatalities, injuries, and property damage. Effective communications can help achieve this goal. No matter what the scale is, it is important to remember that all disasters are local.

ACKNOWLEDGEMENTS

The author would like to thank Dr. Jason Senkbeil from the Department of Geography at the University of Alabama for his assistance with arrangements and fieldwork.

REFERENCES

1. Federal Emergency Management Agency. Data Visualization: Summary of Disaster Declarations and Grants. [Internet]. 2015. 44325–4304 Available from: http://www.fema.gov/data-visualization-summary-disaster-declarations-and-grants [Accessed: 2016-01-13].
2. Federal Emergency Management Agency. Disaster Declarations by Year. [Internet]. 2016. Available from: http://www.fema.gov/disasters/grid/year [Accessed: 2016-01-13].
3. National Hurricane Center. Tropical Cyclone Climatology. [Internet]. 2016, Available from: http://www.nhc.noaa.gov/climo/ [Accessed: 2016-01-16].
4. Rauber R, Walsh J, Charlevoix D. Severe & Hazardous Weather: An Introduction to High Impact Meteorology. 4th ed. Dubuque: Kendall Hunt; 2012. 612 p. ISBN: 9781465250704.
5. National Hurricane Center. Saffir–Simpson Hurricane Wind Scale. [Internet]. 2016, Available from: http://www.nhc.noaa.gov/aboutsshws.php [Accessed: 2016-01-16].
6. Kimberlain T. Tropical Cyclone Report Tropical Storm Debby. [Internet]. 2013. Available from: http://www.nhc.noaa.gov/data/tcr/AL042012_Debby.pdf [Accessed: 2016-01-15].
7. National Oceanic Atmospheric Administration. Storm Data and Unusual Weather Phenomena. Asheville, NC: National Climatic Data Center; 2012. 680 p. Vol. 54 Number 6 [Internet]. Available from: http://www.ncdc.noaa.gov/IPS/sd/sd.html [Accessed: 2016-01-15].
8. Federal Emergency Management Agency. Florida Tropical Storm Debby (DR-4068). [Internet]. 2016. Available from: http://www.fema.gov/disaster/4068 [Accessed: 2016-01-18].
9. American Meteorological Society. Glossary of Meteorology. Tornado. [Internet]. 2013. Available from: http://glossary.ametsoc.org/wiki/Tornado [Accessed: 2016-01-16].
10. Edwards R. The Online Tornado FAQ Frequently Asked Questions about Tornadoes. [Internet]. 2015. Available from: http://www.spc.noaa.gov/faq/tornado/#The%20Basics [Accessed: 2016-01-19].
11. Storm Prediction Center. The Enhanced Fujita Scale (EF Scale). [Internet]. 2014. Available from: http://www.spc.noaa.gov/efscale/ [Accessed: 2016-01-19].
12. Hayes J. Service Assessment the Historical Tornadoes of April 2011. Silver Spring, MD: National Oceanic and Atmospheric Administration, National Weather Service; 2011. 76 p. [Internet]. Available from: http://www.nws.noaa.gov/om/assessments/ pdfs/historic_tornadoes.pdf [Accessed: 2016-01-16].
13. BMX Webmaster. Historic Tornado Outbreak April 27, 2011. [Internet]. 2013. Available from: http://www.srh.noaa.gov/bmx/?n=event_04272011 [Accessed: 2016-01-16].
14. Tornado Recovery Action Council of Alabama. Cultivating a State of Readiness, Our Response to April 27, 2001. Alabama; 2012. 117 p. [Internet]. Available from: http://ema.alabama.gov/filelibrary/TRAC_Report.pdf [Accessed: 2014-03-29].

15. United States Census Bureau. Tuscaloosa County, Alabama. [Internet]. 2010. Available from: http://factfinder.census.gov/faces/nav/jsf/pages/community_facts.xhtml [Accessed: 2016-02-18].

CHAPTER 6

The perceptions of social responsibility for community resilience to flooding: the impact of past experience, age, gender and ethnicity

Robby Soetanto[1], Aaron Mullins[2] and Nebil Achour[3]

[1] *School of Civil and Building Engineering, Loughborough University, Loughborough, Leicestershire LE11 3TU, UK*

[2] *School of Energy, Construction and Environment, Coventry University, Priory Street, Coventry CV1 5FB, UK*

[3] *Department of Medicine and Healthcare Science, Anglia Ruskin University, Cambridge CB1 1PT, UK*

ABSTRACT

Community resilience to flooding depends, to a large extent, on the participation of community members to take more responsibility for enhancing their own resilience. The perception of social responsibility (SR) which is argued to be one of the antecedents influencing individual's willingness to undertake resilient behaviours can significantly contribute to community resilience through individual and collective actions. Understanding of factors influencing the perceptions of SR of individuals within community might help with developing strategies to increase the perceptions of SR. This research explores perceptions of SR in relation to flooding for householders and local businesses and establishes their relationships with experience of flooding and demographic factors of age, gender and ethnicity. The data were obtained via a questionnaire survey of three communities in Birmingham and one community in South East London, UK, three with experience of flooding and one without. A total of 414 responses were received and used in the multiple regression analysis. The analysis identified 'experience of flooding', 'age' and 'South Asian' ethnic group as significant variables, suggesting that older individuals from South Asian ethnic groups with previous experience of flooding are likely to be more socially responsible than others without these attributes.

Keywords: Community Resilience, Flooding, Perceptions, Social responsibility

1 INTRODUCTION

Flooding is considered one of the biggest natural hazards, frequently experienced by communities in the UK (Fowler and Kilsby 2003; Harries 2013). Structural measures, such as flood defences, can fail if extreme floods exceed their design capacity and therefore cannot completely protect communities. The structural measures should be complemented by other measures which would require participation of members of community to take more responsibility for enhancing their resilience (DEFRA 2005; Pitt 2008). Community resilience, defined broadly, as the ability of a community to cope with the impacts, return to normal functioning, learn and adapt after a flood event (adapted from Adger 2000; Cutter et al. 2008; Norris et al. 2008), could be regarded as collective resilient behaviours of individuals and involves complex interrelationships between community members within a local social context. Exploring resilient behaviour in its context allows a better understanding of the perceptions that lead to these behaviours. The understanding is even more important as several studies have found that majority of individuals in European countries (including the Netherlands, Germany, Norway, Sweden and the UK) had not taken protective actions to prepare for floods (Krasovskaia 2005; Terpstra and Gutteling 2008; Kievik and Gutteling 2011). Specifically, Joseph et al. (2015) found that flooded homeowners in the UK were unclear about their responsibility and not confident with the ability of resistance/resilience measures to prevent flooding. Further, the nature of resilient behaviours and the perceptions that lead to these behaviours in the community are not well understood (Smit and Wandel 2006). From a thorough review of literature from a number of academic fields and policy research, the research reported in the paper proposes the concept of social responsibility

(SR) as an alternative way of understanding the behaviours of individuals to contribute to collective actions for enhancing community resilience to flooding.

SR refers to a generic ethical framework relating to the relationships between the economic, environmental and social aspects of an individual or organisation activities that endeavour to benefit society (ISO 2010). The US Federal Emergency Management Agency (FEMA) recognised personal responsibility as being the key to building a resilient community (Colten et al. 2008). Personal responsibility for taking action is also an important factor in several psychological models for preparedness behaviours (Karanci 2013). Further, individual perception of their roles and responsibilities is important to climate change adaptation and climate change (Nicholson-Cole 2005). The term SR encompasses a broader individual's responsibility to contribute to collective actions for enhancing the resilience of community which they are part of. This definition is appropriate for this research given that achieving community resilience is considered as a concerted effort by individuals within that community. Ajzen (1991) stated that behaviour is determined by intention (i.e. the decision to engage in a particular behaviour). Intention itself is understood to be determined by an individual's motivational factors (Sinatra et al. 2012; Armitage and Conner 2001). It is recognised that perceptions are one of these key motivational factors, with their relevance varying between individuals and contexts of the behavioural intention (Sinatra et al. 2012; Collins and Chambers 2005; Corraliza and Berenguer 2000; Ajzen 1985). Thus, in terms of individual cognitive decision-making, behaviour intention is an intermediate link between perceptions and behaviour. It is important to note that perceptions of SR are only one of a number of perceptions which may affect behaviour in relation to community resilience to flooding. For example, the effect of perceptions of risk has been noted across a wide range of extreme weather events, with higher perceived risk found to increase pro-environmental behaviour (Whitmarsh and O'Neill 2010; Floyd et al. 2000; Neuwirth et al. 2000) and lower perceived risk leading to a lack of pro-environmental behaviour (Whitmarsh 2011; Spittal et al. 2005; Johnston 1999; Harris 1996). This research is orientated towards the perceptions of SR and argues that individuals with higher perception of SR would display increased resilient behaviour, but individuals with lower perception of SR would display a lack of resilient behaviour.

Investigating the perceptions of SR would benefit from an understanding of factors influencing the perceptions of SR of individuals within community. This understanding might help with developing strategies to increase the perceptions of SR, which ultimately create a more desirable mindset within the key community groups that leads to the adoption of resilience measures. This understanding would also help to identify groups of individuals who tend to have lower perceptions of SR, so that appropriate targeted strategies could be developed and deployed. This is supported by Terpstra and Gutteling (2008), who highlighted the importance of increasing responsibility of, and motivating individuals who do not think that they are responsible to take protective actions. For example, demographic factors may help explain the reasons why two people from the same geographical community hold different perceptions of SR and display different behaviours when presented with the same flooding event. From reviews of research, Lindell and Perry (2000) and more recently, Lindell (2013) concluded that demographic factors can influence resilient behaviours to reduce the risk of environmental hazards and are included as influencing factors in a number of community resilience models (Tierney and Bruneau 2007; Cutter et al. 2008).

This paper reports an investigation on the effect of experience of flooding and the demographics of age, gender and ethnicity on the perceptions of SR in three communities in Birmingham and one in South East (SE) London, UK. In these four communities, householders and small businesses were invited to participate in a questionnaire survey as part of the research. 'Householder' refers to an individual who resides within the community, and 'small business' refers to either the owner, manager or a person of senior standing within the local area. Involvement in community actions by these community groups is vital for successful disaster management (Buckle et al. 2001; Tapsell et al. 2010). The questionnaire collated respondents' individual self-rated perceptions of SR and their demographic data. The responses were then analysed statistically. The following sections describe the concept of SR within the context of this investigation, the influence of disaster experience and demographic factors on the perception of SR, before research methods and findings are explained. They are followed by a discussion on the findings and implications, a conclusion and research limitations and areas for future research.

2 THE PERCEPTIONS OF SOCIAL RESPONSIBILITY AND COMMUNITY RESILIENCE

2.1 The concept of social responsibility and community resilience

As the focus of the research is to explore the relationship between social aspects (SR) and environmental aspects (resilience to flooding), it is important to note different conceptualisations of SR for different application contexts. For example, conceptualisations of SR within the majority of literature have largely been business based, exploring corporate social responsibility (CSR) (Hahn 2012; Waddock 2008; Banerjee 2007). The rise in awareness of CSR emanated from the public demanding access to more information about how companies were working, and the public are now being recognised by companies as a key stakeholder group (Waddock 2008; Horgan 2005; Clark 2000). SR has long been an important field of research for both academics and business practitioners and continues to provide a valuable research area for those wishing to investigate modern societal issues (Hahn 2012; Peterson and Jun 2007; Gorte 2005). SR has been the focus of research that investigated business SR by exploring and comparing the perspectives of business employees and social workers (Boehm 2009), investigated the relationship between perceptions of personal and SR and intrinsic motivation in the field of education (Li et al. 2008) and explored SR as a factor when investigating genetic and environmental components of pro-social attitudes (Rushton 2004). Some studies have explored the impact of socio-demographic factors on the perception of CSR from investors (Cheah et al. 2011), state residents (Panwar et al. 2010) and consumers (Wells et al. 2011); however, none has been found to examine the impact of the same on perception of SR in relation to flooding. Overall, these studies indicate that responsibility for behaviour is related to the perceptions that people hold. This adds further support to the idea that perceptions of SR and their effect upon decision-making and behaviour is an important area to explore, in relation to resilience to flooding. Exploring perceptions of SR for flooding will provide an excellent platform from which to investigate community resilience.

Protection against the impact of flooding is often perceived to be mainly the responsibility of government, with very little participation from members of community (Krasovskaia 2005; Terpstra and Gutteling 2008). Increasing impact of climate change and tightening public spending on physical flood protection measures have called individuals to be responsible for improving their own resilience by taking protective actions. Pitt's (2008) review highlighted the importance of community groups (householders and small businesses) to work together with the government in resilience-enhancing actions, for example, via local resilience forums. These initiatives can help to clarify what individuals are expected to do by themselves and what the government can support. Enhancing community resilience should be a collaborative effort between community groups and government. Thus, it is important to recognise that members of community can enhance their own resilience and, by doing so, can enhance the resilience of community. These individuals have a responsibility to increase their own resilience, and they can do so through the decisions they make about being aware of the risks, accepting these risks, seeking relevant information and taking protective actions. It is argued that individuals may differ in their perceptions of SR, based on their unique set of intrinsic factors (i.e. past experience of flooding and demographic variables of age, gender and ethnicity). The discussion of literature has so far highlighted age, gender and ethnicity as the most recurring and prominent of these factors. Here, the research is not stating that the other factors are not relevant, but simply that they were not the most relevant, given the review of literature conducted. In relation to socio-economic status, Whittaker et al. (2005) and Dunlap and Mertig (1995) suggested that affluence is not always a determining factor for environmental concern and a conservation ethos. Further, Bethel et al. (2013) found that the difference in disaster preparedness (measured by medication supplies) in eight states in the USA was not due to factors related to socio-economic status. Particularly for this investigation, socio-economic status was not considered to be one of the most relevant factors. This is because the participants within each case study area lived within the same radius of a flooding event. This proximity of living accommodation, the majority of which is similar in house type and therefore cost, was considered as a natural social leveller for socio-economic status amongst participants. It is recognised that socio-economic status may not only play a part in what people are able (or feel they are able) to achieve in relation to resilience measures, but also impact on long-term mental health of flooded households (Lamond et al. 2015). However, given that there are not expected to be great differences between the socio-economic statuses of participants, then it is reasonable to suggest that this would not be the most important factor to explore within the research. The following section explains factors that may influence individual perceptions of SR included in this research.

2.2 Past experience of flooding

Research has found that experiencing a flooding event has long-term impacts upon people's lives (Tapsell et al. 2003, 2010; Tapsell 2000; Fordham 1998). A body of research suggests that personal experience of flooding can influence their awareness of risk, response and preparedness (Nicholson-Cole 2005; Tapsell et al. 2010; Delanty 2003; Valentine 2001; Pain et al. 2001), which in turn suggest that they may show higher perceptions of SR than people who have not experienced flooding. However, there has also been research which found that there was little difference between the climate change perceptions of individuals who had experienced flooding and those who had not (Whitmarsh 2008). One potential reason for this is that research has indicated that past behaviour can give an indication of future behaviour, as people are unwilling to deviate from regular routines (Quimby and Angelique 2011; Ouellette and Wood 1998). This is supported by Whitmarsh (2008) who noted that not a single participant in their study explicitly mentioned strategies to adapt to climate change and flooding. This suggests that experience of a flooding event may not be enough to instigate long-term behavioural change. This poses the question as to whether, in the same way that experience of a flooding event may alter an individual's perception of flood risk, would experience of flooding alter perceptions of SR? Furthermore, would these perceptions be altered in a positive manner, with a clear distinction between the perceptions held by those individuals that have experienced flooding and those who have not?

2.3 Age

Hakes and Viscusi (2004) found that older people estimated risks more precisely than younger people, i.e., their perceived estimation of risk is closer to the actual level of risk they are exposed to. This view is also supported by research which found that increasing age was related to greater protective behaviour in both seismic hazard adjustments (Lindell and Whitney 2000) and preparations for El Nino (Siegel et al. 2003). Wells et al. (2011) explored consumer responsibility and found that increasing age was related to an increase in general environmental responsiveness. Given these findings, it is reasonable to suggest that older participants would be more likely to engage in resilient behaviour than younger participants.

In contrast, Tanida (1996) indicated that younger people were more likely to survive an earthquake. This suggests that younger people are more resilient to extreme events. It may also be interpreted in another way. If younger people are more likely to survive an extreme event, then they may be less likely to take measures to become more resilient to extreme events. This is because people who perceive themselves to be most at risk from extreme events are more likely to take measures to counter that risk (Armas et al. 2003). This may explain the increased interest and uptake of resilient behaviour displayed by older people. There is also the third view that age does not affect climate change risk perception at all, as found by Safi et al. (2012). These contrasting findings require further exploration of the effect of age on perceptions of SR related to flooding.

2.4 Gender

Early research indicated that females (particularly pregnant females) are more vulnerable to the effects of extreme weather events than males (Balbus and Malina 2009; Granger and Hayne 2001). Several researchers found that females are more concerned about the impact of climate change than men (Semenza et al. 2008; Sundblad et al. 2007; Leiserowitz 2006). This line of reasoning is also supported by the facts regarding the vulnerability of females during extreme events. The Women's Environment and Development Organization reported that women and children are 14 times more likely to die than men during disasters (Araujo and Quesada-Aguilar 2007). These gender influences have also been found to be present after two flooding events in Scotland, with females being affected more than males by changes within the community (Fordham 1998). Bartlett (2008) supported this early finding, indicating that females are more vulnerable to anger, frustration and violence associated with the upheaval during and after an extreme event. However, Adelekan and Asiyanbi (2016) found no significant effect of gender on feelings of worry and perceived vulnerability to flooding in Lagos. Similarly, Harvatt et al. (2011) found little evidence of gender differences in householder responses to flooding and sea level rise in three high-risk locations in England. Despite these contrasting findings, this suggests that, given the difference in responses and social reactions to flooding, there may also be gender influences in perceptions of SR, which may then affect decision-making and behaviour.

2.5 Ethnicity

Culture, generally defined as a set of beliefs, moral and values that are shared within an ethnic group, gives meaning of ethnicity to individuals and groups and is therefore considered as a basic building block of ethnicity (Nagel 1994). Potential ethnic differences in perceptions of SR related to flooding are supported by many researchers who found that perceptions of environmental and technological hazards can vary between cultures due to different perceptions (or world views) held by individuals within each culture (Kahan et al. 2010; Poortinga et al. 2002; Steg and Sievers 2000; Gyawali 1999; Slovic 1999; Ellis and Thompson 1997). Perceptions of risk can vary between different cultures because cultural norms shape which dangers are feared and what risks are taken (Douglas and Wildavsky 1982). These cultural norms influence this by entering into an individual's cognitive and social risk identification processes (Kahan et al. 2010). Cultural orientation has an impact on how people respond to climate change issues, and determines whether they are willing to take protective actions (Larson et al. 2011; Kahan et al. 2010; Myers 1994). This suggests that the willingness of individuals to improve resilience to flooding which is associated with high perception of SR may be affected by cultural differences. Thus, it is argued that cultural differences may create differences in the perception of SR between individuals of different ethnicities.

There are contrasting findings of the ethnicity influence on the perceptions of risk. In a large-scale opinion survey of risk perceptions conducted between June and September 2004 in the USA, Kahan et al. (2007) indicated that member of Black ethnic groups (i.e. African Americans) perceives risk higher than members of White ethnic groups. Hakes and Viscusi (2004) indicated that members of the White ethnic group would estimate the risks more precisely than other ethnic groups, suggesting that they would have had to take a greater interest in those risks, which Myers (1994) suggested is linked to higher take-up of protective actions. However, Kahan et al. (2007) suggested that individuals from White ethnic group consider environmental risk less seriously than those from African Americans. This finding is linked to reduce take-up of protective actions. More specific ethnic differences related to flooding can be seen in research which suggests that individuals who regard themselves as belonging to the Asian ethnic group may hold different perceptions of a community's response to and recovery from an extreme flooding event (Tapsell 2000; Tapsell et al. 1999). However, despite the importance of ethnicity as a predictor of risk perception, Bourque et al. (2012) found that perceived risk was a weak predictor of preparedness behaviour. Mullins and Soetanto (2013) found ethnic differences in the perceptions of SR and suggested three different levels of resilience and their association with different ethnic groups. However, the statistical importance of ethnicity to SR score compared to the other intrinsic factors of experience of flooding, age and gender was not investigated and therefore is still not well understood. There is scant evidence that previous research in environmental risk communication for hazard adjustments has sufficiently assessed the importance of demographic variables compared to other variables in this manner (Lindell and Perry 2004). This paper significantly extends the previous work in Mullins and Soetanto (2013) and is intended to make a novel contribution by considering these intrinsic factors simultaneously in a multivariate analysis. Thus, the research question addressed in this paper is:

What are the extent and direction of the relationships between these intrinsic factors and the perception of SR?

The research intends to establish their relationships using multiple regression analysis so that the relative importance of the factors on the perception of SR could be ascertained. Case study areas and methods employed to investigate this relationship are explained in the following section.

3 CASE STUDY AREAS

These two largest UK cities, Birmingham and (South East/SE) London, were chosen because they contain communities which had experience of flooding. It is important to note that this research is not suggesting that either of these areas are more susceptible to flooding than other areas of the UK. However, these urban areas do contain a number of characteristics that make them appropriate as case study areas for the investigation. Four communities were chosen because they allowed comparisons between communities in different locations who faced different levels of risk, as well as between communities who had experienced flooding and those who had not.

3.1 Birmingham communities (Witton, Selly Park and Digbeth)

Birmingham is the second largest city in the UK with over 1 million people. Historically, Birmingham has been particularly susceptible to flooding (Be Birmingham 2011). Although Birmingham is around 500 feet above sea level, flooding still occurs, particularly in the area around the course of three main rivers (River Tame, River Rea and River Cole). Witton and Selly Park were chosen because they were urbanised areas that lie close to water sources within tributaries of River Tame and River Rea, respectively. Furthermore, both Witton and Selly Park have frequently experienced flooding including major flooding in summer 2007. Witton was highlighted as being one of the worst affected communities, with multiple issues related to the flooding they experienced (Kotecha et al. 2008). It was reported that around 300 homes within the Brookvale Road of Witton (data collection area) were affected by extreme flooding in 2007 (Dayani 2007). To reduce flood risk, Environment Agency (2014) recommended to build and improve flood defences in the Brookvale Road area. There were estimated to be around 70 flood-affected properties in Selly Park during the 2007 flood, the largest flooding event in the area since 1927 (Clayton 2008). There are over 400 properties (in Southern and Northern areas of Selly Park) at risk of flooding from the River Rea, with several recorded flood events in this area, most recently in 2012 and June 2016. Flood defence schemes are being planned for both areas with implementation from Autumn 2016 (Rea Catchment Partnership 2016). Digbeth was chosen as a control group area as it had not experienced flooding in recent years, but theoretically still contains the potential to do so as it is also an urbanised area that is close to the River Rea. It should be noted that the community of Digbeth was not flooded during the summer 2007 floods and has not been known to have experienced a flooding event within the last 30 years.

3.2 London community (Thornton Heath)

The study chose Thornton Heath community in the London Borough of Croydon as a case study area. Croydon is the 4th highest ranked borough out of 4215 settlement areas with around 21,100 properties predicted to be at risk from surface water flooding (DEFRA 2009). It is also noted that there are a number of areas within Croydon which are at risk of sewer flooding from the Thames water and there have been a number of incidences of surface water flooding, in particular within the communities of Thornton Heath, Upper Norwood and Broad Green (Croydon SFRA 2009). The majority of flooding within London during the 2007 summer floods was from surface water flooding (City of London Corporation 2014). Thornton Heath contains the water source of Norbury Brook, whose river level is monitored by the Environment Agency. The community of Thornton Heath was flooded during summer 2007, with the highest ever recorded river level at that time. This flooding also caused disruptions to two of the four rail lines within South Croydon, due to landslips (Bannerman 2007). This experience of flooding in 2007 within the Thornton Heath community will allow direct comparison with the Birmingham communities of Witton and Selly Park. However, Thornton Heath have been identified as the most susceptible to groundwater flooding, as demonstrated by the summer 2007 floods located there (Croydon SFRA 2009). In addition, this was combined with a large amount of sewer flooding in Thornton Heath, due to the long lengths of culverted sewer in the borough (Croydon SFRA 2009). Furthermore, Thornton Heath receives greater amounts of water run-off due to the local topography of steep slopes in Coulsdon, Kenley and Upper Norwood, which channel water into the area (Croydon Council 2011). All these elements combined with the potential heavy precipitation have posed high level of flood risk in Thornton Heath.

4 METHODS

4.1 Questionnaire design

Berkowitz and Lutterman's (1968) SR scale was modified and subsequently adopted in the research. The reason of adopting this SR scale was because it allowed the research to explore an individual's attitude in relation to SR and compare attitudinal SR scores between individuals and communities. The original SR questionnaire has provided a valid and reliable basis for researching SR since its creation. The original SR scale measured an individual's acceptance of the traditional values of their society. The aim of this questionnaire survey is to collate the perceptions of community groups in relation to flooding and as such the original questionnaire was extended and the attitudinal statements were modified. Table 1 presents a comparison between original and modified statements for the survey. The modified statements made specific reference to 'flooding', 'local community', 'neighbours', 'protection against floods' and 'climate change', thus ensuring their relevance to the investigation. A number of SR studies have adopted modified versions of the original scale (Reed et al. 2005; Megicks et al. 2005; Chacon et al. 1998; Frieden and Downs 1986). The final version of the questionnaire included 12 attitudinal statements.

The respondents were asked to indicate their agreement against these statements on a 4-point scale ranging from 'Strongly Agree' (4) to 'Strongly Disagree' (1). In the analysis, several statements were reverse scored to counter acquiescence bias. Each respondent had a potential total score ranging between 12, indicating a very low SR, and 48, indicating a very high SR. These attitudinal questions provide information about how each respondent perceives their own SR. Please note that because the questions are attitudinal, then simply examining the scores of individual questions could display too much bias, or not tell us very much when analysed individually. Thus, the scores for all 12 statements were aggregated. The reliability of the scores was confirmed by Cronbach's alpha coefficient of 0.801. Before this SR questionnaire, the respondents were asked to provide demographic data of age, gender and ethnicity. The respondents were to state their age in space provided, which represents ratio data. Gender and ethnicity were considered as nominal data with multiple choice answers provided in the questionnaire. Ethnicity groups were based on the 2001 UK Census, including 'White', 'Black', 'South Asian', 'Chinese', 'Mixed—White/Black', 'Mixed—White/South Asian' and 'others' answers (ONS 2001).

4.2 Data collection and respondents

The questionnaires were distributed and collected by hand in the communities of Witton, Selly Park and Digbeth in Birmingham and Thornton Heath in SE London. The sampling approach was used to designate target population (i.e. flooded or non-flooded communities) based on the geographical area of the case studies. For targeted flooded communities, the questionnaires were distributed to respondents in and around the worst affected areas of the community, ensuring that as many of the target population had direct experience of flooding. However, practical considerations of the sampling approach would mean that the responses from flooded communities may not necessarily be from individuals whose homes or businesses have been flooded. This could be considered a potential limitation of the sampling approach. Nevertheless, floods affect communities in many ways, and an individual is considered by this investigation to have experienced a flood within their community, regardless of whether they were directly or indirectly affected by the flood water within their own homes or businesses (their street, transport links, shops they use or friends and relatives may be flooded). Further, local groups in flooded communities (such as Selly Park South Neighbourhood

Forum) which raise awareness of flood risk and promote resilience-enhancing actions could positively impact individual's perception of SR.

Table 1 Comparison of original and modified statements for social responsibility questionnaires

No.	Original statements	Modified statements
1	It is no use worrying about current events or public affairs; I can't do anything about them anyway	It is no use worrying about flooding within the community as I can't do anything about it anyway
2	Every person should give some of his time for the good of his town or country	Every person should give some of their time for the good of their local community
3	Our country would be a lot better off if we didn't have so many elections and people didn't have to vote so often	Our country would be a lot better off if we didn't have so many rules
4	Letting your friends down is not so bad because you can't do good all the time for everybody	Letting your neighbours down is not so bad because you can't do good all the time for everybody
5	It is the duty of each person to do his job the very best he can	It is the duty of each member of a community to do the very best they can to increase their protection against floods
6	People would be a lot better off if they could live far away from other people and never have to do anything for them	People would be a lot better off if they could live far away from other people and have less interaction with each other
7	At school I usually volunteered for special projects	I would like to take part in a community volunteering project
8	I feel very bad when I have failed to finish a job I promised I would do	I feel very bad when I have failed to finish a job I promised I would do
9	–	I feel it is important to always tell the truth to others
10	–	I feel it is important to get on well with your neighbours
11	–	I do not feel that climate change is an important issue that will affect me
12	–	I feel that it is important that people should always obey the law

A total of 1200 questionnaires were distributed (300 for each community), and 414 completed questionnaires were received, representing a response rate of 34.5%. The responses are detailed in Table 2. The last two columns of Table 2 display gender distributions with 58% male and 42% female respondents. Table 3 shows the ethnic distribution for communities in Birmingham and SE London. Although this result is comparable to the ethnic distribution of communities within the UK based on UK Census in 2001 (ONS 2001), the result should be interpreted with caution due to non-probabilistic sampling of the respondents. There are sufficient numbers of 'White', 'Black' and 'South Asian' ethnic groups in order to investigate the influence of each on the perceptions of SR. The numbers of 'Chinese', 'White/Black', 'White/Asian', and 'other' ethnic groups are low. The mean, median and standard deviation of respondents' age were 34.62, 33.00 and 10.125, respectively. This near equal mean and median and observation of histogram suggest that distribution of respondent age is normal. The minimum and maximum ages were 19 and 64 years old.

Table 2 Number of participants and gender distribution for each community

Communities	No. of participants	Gender	
		Male	Female
Witton			
Householders	81	32	49
Small businesses	23	17	6
Selly Park			
Householders	94	38	56
Small businesses	28	18	10
Digbeth			
Householders	49	33	16
Small businesses	27	22	5
Thornton Heath			
Householders	89	61	28
Small businesses	23	19	4
All communities (%)	414 (100%)	240 (58%)	174 (42%)

Table 3 Ethnic distribution in Birmingham and London communities

Community	White	Black	Asian	Chinese	White/ Black	White/ Asian	Other
Witton	83	3	15	1	0	1	1
Selly Park	98	10	6	3	0	1	4
Digbeth	53	1	22	0	0	0	0
Thornton Heath	67	19	19	1	4	0	2
All communities (%)	301 (72.7%)	33 (8.0%)	62 (15.0%)	5 (1.2%)	4 (1.0%)	2 (0.5%)	7 (1.7%)

4.3 Data analysis

An initial analysis of SR scores was conducted to obtain descriptive statistics (mean, median, standard deviation) and to explore whether there were differences between SR scores of different locations (Witton, Selly Park, Digbeth, Thornton Heath) and type of respondents (householders, small businesses). Two-way analysis of variance (ANOVA) tests were used because it allowed a comparison of the means of more than two samples at a time. When only two samples were being compared, *t* tests were applied. The analysis of SR scores confirmed that the parametric assumptions were met.

The primary aim of the analysis was to determine the effect of independent variables (experience of flooding, age, gender and ethnicity) on dependent variable (SR score). This was achieved by establishing relationships between the independent variables and dependent variable, using multiple regression analysis which allowed variations in the dependent variable to be explained by variations in the independent variables. It was not the intention to predict the SR score, but to understand the relative importance of the independent variables against the dependent variable.

Multiple regression can include any combination of quantitative (i.e. interval) and qualitative/nominal independent variables (Hardy 1993). Experience of flooding and demographic variables of gender and ethnicity were considered as nominal variables. These variables were transformed into dummy variables in which one nominal variable is represented by several binary dummy variables in the form of 0 or 1 ('no' or 'yes'). Each dummy variable captures one piece of the categorical information from the original measure. Here, the information in the original variables is not fundamentally altered, instead an alternative form of representing that information is chosen (Hardy 1993). Therefore, these variables do not cause the regression estimates to lose any of their desirable properties (Lewis-Beck 1993) and are legitimately employed in the analysis (Bryman and Cramer 2011). Table 4 presents a list of independent variables with their category and measurement.

The analysis of residual confirms that regression assumptions of normality, constant variance and linearity have been met (Lewis-Beck 1993; Norusis 1995). The analysis was conducted using IBM SPSS Statistics version 22.

5 RESULTS

Table 5 presents means, medians and standard deviations of SR scores in different locations for householders and small businesses. An observation of means shows that respondents from Witton, Selly Park and Thornton Heath (flooded communities) appear to have higher perceptions of SR than those from Digbeth (non-flooded community). In flooded communities, the SR scores of respondents from Thornton Heath were lower than those of Witton and Selly Park. Near equal means and medians is an indication that the data are normally distributed around a central point. The proportion of respondents from flooded communities (excluding Digbeth) is 81.6%. The two-way ANOVA results indicate that there is a significant difference ($p = 0.004$) between the means of SR scores reported by householders (Witton = 35.25, Selly Park = 35.17, Digbeth = 29.92, Thornton Heath = 32.03) and small businesses (Witton = 36.87, Selly Park = 36.86, Digbeth = 30.33, Thornton Heath = 33.39), although it was considered a 'small' effect (partial eta squared of 0.021). Furthermore, there is a significant difference ($p < 0.0005$) between the means of SR scores reported in different locations with a 'large' effect (partial eta squared of 0.241). However, there is no significant interaction ($p = 0.721$) between community groups and locations. This indicates that small businesses have significantly higher perception of SR than householders, although this difference is not as large as the difference between the perceptions of SR of communities in different locations. That flooded communities (SR scores' mean of 34.50) had higher perceptions of SR than non-flooded community (SR scores' mean of 30.07) was confirmed with a strong evidence by the result of t test ($p < 0.0005$).

Table 4 Independent variables

Variable	Category/measurement
Age of respondent	Ratio variable: years
Experience of flooding	One binary variable: 1 = 'flooded'; 0 = 'never flooded'
Gender	One binary variable: 1 = 'male'; 0 = 'female'
Ethnicity	Six binary variables: assigned 1 if respondent chose 'Black', 'South Asian', 'Chinese', 'Mixed—White/Black', 'Mixed—White/South Asian', and 'Others'. 'White' was used as a reference (0)

Table 6 presents four multiple regression models which include more independent variables as the models progressively advance. Hierarchical multiple regression analysis allows an examination of contribution of each independent variable to the explanatory power of the model (Akerlof et al. 2013; van der Linden 2015). Model 1 evaluates the contribution of 'age of respondent' to the dependent variable (i.e. SR score). 'Age of respondent' is a significant variable and explains 18% (Adjusted R^2) of the variation of the perception of SR. In addition to 'age of respondent', model 2 includes 'experience of flooding' which markedly increases the explanatory power of the model by 24% (Δ adjusted R^2), bringing a total Adjusted R^2 of 42%. Model 3 suggests that 'gender' is not a significant variable and does not increase the explanatory power of the model (Δ adjusted R^2 of −0.001). The final model 4 includes several binary variables, resulting from transformation of ethnic group variable. 'White' ethnic group was not included as it was used as a reference variable to examine the impact of non-White ethnic groups. 'South Asian' ($\beta = 0.165$, $p < 0.0005$) and 'Chinese' ($\beta = 0.075$, $p = 0.042$) ethnic groups are significant variables. This suggests that people from 'South Asian' and 'Chinese' ethnic groups are likely to have higher perception of SR than those from 'White' ethnic group. People from 'Black' ethic group are likely to have a lower level of perception of SR than those of 'White' ethnic group, although it was not highly significant ($\beta = -0.057$, $p = 0.128$). The overall contribution of ethnic group variables to the explanatory power of the model is relatively low (with Δ adjusted R^2 of 0.029) in comparison with the contributions of 'age of respondent' and 'experience of flooding'.

Table 5 Descriptive statistics of SR scores in different locations for householders and small businesses

Location	Community group	SR scores			Respondents	
		Mean	Median	SD	Number	%
Witton	Householders	35.25	36.00	4.335	81	
	Small businesses	36.87	36.00	2.897	23	
	Total	35.61	36.00	4.104	104	25.1
Selly Park	Householders	35.17	35.00	3.600	94	
	Small businesses	36.86	36.50	3.240	28	
	Total	35.56	36.00	3.579	122	29.5
Digbeth	Householders	29.92	30.00	4.541	49	
	Small businesses	30.33	31.00	3.648	27	
	Total	30.07	30.00	4.225	76	18.4
Thornton Heath	Householders	32.03	32.00	3.383	89	
	Small businesses	33.39	33.00	3.041	23	
	Total	32.31	32.00	3.348	112	27.0
Overall	Householders	33.48	33.00	4.398	313	
	Small businesses	34.33	35.00	4.238	101	
	Total	33.68	33.50	4.370	414	100.0

The final model explains 45% (Adjusted R^2 of 0.447) of the variation in the perception of SR. The model indicates that 'experience of flooding' is the most important variable ($\beta = 0.502$), followed by 'age of respondent' ($\beta = 0.446$), 'South Asian' ($\beta = 0.165$) and 'Chinese' ($\beta = 0.075$) ethnic groups.

Standardised beta coefficients (β) of 'age of respondent' and 'experience of flooding' are relatively stable across models 1–4. Higher β of these two variables in comparison with those of 'South Asian' and 'Chinese' confirms their strength as the main predictors of SR score. Due to small number of 'Chinese' respondents (n = 5), the significance of 'Chinese' ethnic group would need further investigation before it is included in the final model. The final model suggests that older individuals from South Asian group with previous experience of flooding are likely to be more socially responsible than those without these characteristics. This scenario potentially provides the highest perception of SR. Other scenarios with different combination of independent variables will bring lesser SR scores. For instance, an individual who is from White ethnic group, in their early twenties, and has experienced flooding, is likely to have a lower SR score than those with the characteristics in the previous scenario. However, it further acknowledges that the aim is not to predict the SR score, but to enhance understanding of the contribution of individual intrinsic factors as themselves, and in combination with the other significant factors. Nevertheless, the model has helped to settle some contrasting findings of the factors influencing risk/climate change-related perceptions and their potential relationship with a new concept of SR for community resilience to flooding.

Table 6 Perception of social responsibility regression models

Independent variable	Model 1		Model 2		Model 3		Model 4	
	β	p	β	p	β	p	β	p
Age of respondent	0.424	<0.0005	0.442	<0.0005	0.442	<0.0005	0.446	<0.0005
Exp. of flooding			0.492	<0.0005	0.489	<0.0005	0.502	<0.0005
Gender					−0.012	0.767	−0.035	0.360
Black							−0.057	0.128
South Asian							0.165	<0.0005
Chinese							0.075	0.042
White/Black							0.006	0.879
White/Asian							−0.003	0.927
Other							0.033	0.373
Adjusted R^2	0.178		0.419		0.418		0.447	
Δ adjusted R^2			0.241		−0.001		0.029	

β is standardised beta coefficient; p is significant level of independent variable, with t test

6 DISCUSSION

This research has provided a better understanding of factors influencing the perceptions of SR to flooding. There have been some research on social aspects of flooding, but there is very little knowledge on the perceptions of SR of key local community groups. Enhancing SR perception is important because higher perception of SR can lead to improved technical, organisational and economic resilience. The research explored not only the perceptions of SR of householders and local businesses in several communities, but also established their relationships with key intrinsic factors. The finding suggests that older community members with previous experience of flooding and from South Asian ethnic group are more likely to have higher perceptions of SR. Although it might be cases that they could be the first generation of older people with South Asian background in the UK, who were more likely to have experienced flooding in their home countries, the research did not investigate the origin of the respondents as the questionnaires were distributed to communities regardless of their ethnic background and origin, and therefore, this possible explanation cannot be ascertained.

The finding of experience of flooding supports previous research which has shown that (1) experience of a disaster can often have an influence upon an individual's motivation to cope with future risks (Siegrist and Gutscher 2008; Siegel et al. 2003), (2) personal experience can have a positive effect upon people's ability to visualise climate change risk and can alter perceptions of its importance, as well as perceptions of their ability to enhance their own resilience to it (Nicholson-Cole 2005; Akerlof et al. 2013), (3) experience of previous disaster was found to be a predictor of an individual's level of preparation for future disaster (Sattler et al. 2000), and (4) individuals who experience a high level of exposure to natural disasters are more likely to engage with the issue and create coping strategies (Spence et al. 2011; Fillmore et al. 2008; Work et al. 1999; Lave and Lave 1991).

There are potentially different reasons that could explain the significance of age. First, older people may be more vulnerable to extreme events (Armas 2006; Granger and Hayne 2001), and therefore, have higher inclination to take protective actions. Alternatively, they have lived through different socio-political eras that may have shaped their cultural perspective and brought a sense of belonging as part of community. As a consequence, they would display a greater interest to flood risk and uptake of resilience measures, which are representative of higher perceptions of SR. This supports previous research which found that greater pro-environmental behaviour was related to increasing age (Lindell and Whitney 2000; Siegel et al. 2003; Takahashi and Selfa 2015) and higher perceived risk

(Whitmarsh and O'Neill 2010; Floyd et al. 2000; Neuwirth et al. 2000). The significance of South Asian ethnic group suggests that participants from the South Asian ethnic group displayed a greater awareness and acceptance of flood risk and were more likely to adopt resilience measures than participants from the White and Black ethnic groups. This finding is in line with previous research which has found that individuals who regard themselves as belonging to the South Asian ethnic group may hold different perceptions of a community's response to and recovery from a flooding event (Tapsell 2000; Tapsell et al. 1999). It is worth noting that although ethnicity differences in the perception of SR as found in Mullins and Soetanto (2013) are confirmed by this finding, the importance of ethnicity to the perception of SR (β = 0.165) is about three times less than those of 'experience of flooding' (β = 0.502) and 'age of respondent' (β = 0.446). This suggests that, in any initiatives to improve community resilience, younger members of community with no experience of flooding should be the priority group whose SR perceptions should be raised. The non-significance of gender is in line with the findings of Harvatt et al. (2011) and Adelekan and Asiyanbi (2016). However, the perception of SR, as a novel concept in community resilience to flooding, is not deemed to be directly comparable to the perception of risk and vulnerability to environmental hazards as the subject of previous research.

6.1 Implications for risk communication strategies

The research underlines the need to develop strategies that can enhance the perception of SR of those community members who are likely to have lower perception of SR. They may be those who have no previous experience of flooding are younger and from non-South Asian ethnic group (in this research, they are majority from White and Black ethnic groups). Approaches for interventions which are aimed to change individual behaviour may be made through informational and structural strategies (Steg and Vlek 2009). Information strategies are aimed to change motivations, perceptions and norms, whereas structural strategies are aimed to change the circumstances under which behaviour choices/decisions are made. Although a combination of strategies may yield the most effective outcomes, informational strategies may be firstly adopted to enhance the perceptions of SR due to its persuasive influence on individual attitude for behavioural change (Lindell and Perry 2004). Further, the finding highlights the need for education and communication of resilience information and disaster responses, which should be targeted to these members of community. Informational strategies can (1) increase individual awareness of flood risks and knowledge of behaviour that can enhance community resilience, (2) persuade individuals to take actions and participate in their community and (3) provide social support and role models of effective behaviours (Steg and Vlek 2009). This strategy could be further enhanced by providing a mechanism to encourage community participation via, for example, community-based organisations (Karanci 2013), or local resilience forums in the UK. This would not only enhance individual's self-confidence and efficacy of their own actions, but also strengthen their sense of belonging as part of community. These strategies would also help to: (1) overcome the awareness barriers, (2) achieve a greater sense of empowerment and involvement for community groups and (3) reduce externalisation of responsibility to other parties (e.g. local governments and agencies).

In communicating risk information to the community, targeted and tailored strategies were found to yield more desirable outcomes (Abrahamse et al. 2007). For example, flood risk information could be disseminated to all, but 'tailored' to be more effective with different ethnic groups in certain geographical areas. Further, to reach younger members of community, specific forms and venues of dissemination of information should carefully be considered, including the use of internet and social media. Community-based organisations (such as local resilience forums) could hold regular meetings and events to foster two-way communication as an active and effective form of community engagement (Olofsson 2007). This engagement allows individual's concerns to be heard and discussed, and thus, helps to nurture a sense of belonging to the community. Continuous communication is required as people tend to forget the need to prepare for flooding after long period of non-flooding (Jha et al. 2012). Kievik and Gutteling (2011) recognised the need to induce high levels of flood risk perception and enhance self-efficacy belief to increase the intention to adopt resilient behaviours. Thus, communication regarding flood risk should not only inform the risks and actions that individuals should take, but also enhance individual's confidence of the efficacy and positive impacts of resilient actions could have on the community resilience.

7 CONCLUSION

This research has made a contribution by enhancing the understanding of factors influencing the perception of SR, which is argued as one of the important motivators behind the willingness to undertake resilient behaviours. The research was conducted within the community, allowing behaviour of individuals to be contextualised within a social group. This has also permitted an exploration of interrelationships between different community groups and locations. The research found the significant effect of experience of flooding, age and ethnicity on the perception of SR. The knowledge generated will allow appropriate strategies to be developed and deployed to enhance the perception of SR. The finding points towards informational strategies for flood risk communication, in which information should be made available to all, but 'tailored' to be more effective for targeted audience who are likely to have lower perception of SR. The finding should also inform engagement with community groups and policy makers to improve institutional policies related to flooding and community resilience measures. The way in which policy makers engage with the community is important, and meetings could be held between all key community groups to specifically identify existing barriers within current policies, which may also be overcome through group meetings and the information exchange process. Age, experience and ethnicity should be given consideration when attempting to understand motivating factors for engagement with policies and agendas.

8 LIMITATIONS AND RECOMMENDATIONS FOR FURTHER RESEARCH

The finding of this research is limited by non-probabilistic sampling of the respondents and unequal sample representation, especially for ethnic groups and age. Future research should apply stricter controls over different variable representation within the community groups. Larger sample sizes of each controlled community group would then provide even greater ability to explore perceptions within and between these groups. Future age research may wish to also include distinct age groups, which should include an over 65 age group to allow for exploration of extremes and the effects displayed within an even greater range of ages. This may also be partly responsible for the unexplained variations of the SR score (55%). These unexplained variations may be explained by another variables not included in this research. Future research should also consider the other potentially influencing variables (such as beliefs and religions). The finding underlines the need for an in-depth investigation on the nature of 'experience of flooding' (such as severity, frequency, duration and impact) and 'ethnicity' with underlying culture (such as individual origin, norms, beliefs, values).

ACKNOWLEDGEMENTS

The research is funded by the Engineering and Physical Science Research Council (EPSRC), UK, and this support is gratefully acknowledged.

REFERENCES

1. Abrahamse W, Steg L, Vleg C, Rothengatter T (2007) The effect of tailored information, goal setting, and tailored feedback on household energy use, energy-related behaviors, and behavioral antecedents. J Environ Psychol 27:265–276
2. Adelekan IO, Asiyanbi AP (2016) Flood risk perception in flood-affected communities in Lagos, Nigeria. Nat Hazards 80:445–469
3. Adger WN (2000) Social and ecological resilience: are they related? Prog Hum Geogr 24(3):347–364
4. Ajzen I (1985) From intentions to actions: a theory of planned behaviour. In: Kuhl J, Beckman J (eds) Action control, from cognition to behaviour. Springer, New York, pp 11–39
5. Ajzen I (1991) The theory of planned behavior. Organ Behav Hum Decis Process 50:179–211
6. Akerlof K, Maibach EW, Fitzgerald D, Cedeno AY, Neuman A (2013) Do people "personally experience" global warming, and if so how, and does it matter? Glob Environ Change 23:81–91
7. Araujo A, Quesada-Aguilar A (2007) Gender equality and adaptation, USA. Women's Environment and Development Organisation (WEDO)/IUCN. http://www.gdnonline.org/resources/ IUCN_FactsheetAdaptation.pdf. Accessed 26 Feb 2016
8. Armas I (2006) Earthquake risk perception in Bucharest, Romania. Risk Anal 26(5):1223–1234
9. Armas I, Damian R, Sandric I, Osaci-Costache G (2003) Vulnerabilitatea versanţilor subcarpatici la alunecari de teren (Valea Prahovei). Fundatiei Romania de Maine, Bucharest
10. Armitage C, Conner M (2001) Efficacy of the theory of planned behaviour: a meta-analytic review. Br J Soc Psychol 40:471–499
11. Balbus JM, Malina C (2009) Identifying vulnerable subpopulations for climate change health effects in the United States. J Occup Environ Med 51(1):33–37
12. Banerjee SB (2007) Corporate social responsibility: the good, the bad and the ugly. Edward Elgar, Cheltenham
13. Bannerman L (2007) Flash floods and torrents of rain hit first exodus of summer. The Times. http://www.timesonline.co.uk/tol/news/weather/article2112848.ece
14. Bartlett S (2008) Climate change and urban children: impacts and implications for adaptation in low and middle income countries. International Institute for Environment and Development (IIED) Human Settlements Discussion Paper—Climate Change 2. IIED, UK
15. Be Birmingham (2011) Birmingham's Climate Portfolio. http://www.bebirmingham.org.uk/uploads/Birmingham%20Climate%20Portfolio(3).pdf. 28 Sept 2016
16. Berkowitz L, Lutterman KG (1968) The traditional socially responsible personality. Pub Opin Quart 32:169–185
17. Bethel JW, Burke SC, Britt AF (2013) Disparity in disaster preparedness between racial/ethnic groups. Disaster Health 1(2):110–116
18. Boehm A (2009) Business social responsibility: perspectives of businesses and social workers. J Soc Serv Res 35(3):262–273
19. Bourque LB, Regan R, Kelley MM, Wood MM, Kano M, Mileti DS (2012) An examination of the effect of perceived risk on preparedness behaviour. Environ Behav 45(5):615–649
20. Bryman A, Cramer D (2011) Quantitative data analysis with IBM SPSS 17, 18 and 19. Routledge, London
21. Buckle P, Marsh G, Smale S (2001) Assessing resilience and vulnerability: principles, strategies and actions. Emergency Management Australia Project
22. Chacon F, Menard M, Sanz M, Vecina ML (1998) Psychosocial factors that influence volunteer work: a pilot study. Psychol Spain 2:108–115

23. Cheah ET, Jamali D, Johnson JEV, Sung MC (2011) Drivers of corporate social responsibility attitudes: the demography of socially responsible investors. Br J Manag 22:305–323
24. City of London Corporation (2014) Responding to flood risk: surface water flooding case study report. https://www.cityoflondon.gov.uk/services/environment-and-planning/sustainability/Documents/surface-water-flooding-case-study-report.pdf
25. Clark CE (2000) Differences between public relations and corporate social responsibility: an analysis. Pub Relat Rev 26(3):363–380
26. Clayton J (2008) Report and recommendations relating to the Severe Flood Event in Selly Park South on 6th September 2008. http://sellyparksouth.org.uk/files/SPS%20FLOOD%20REPORT%201.pdf 26 Sept 2016
27. Collins C, Chambers S (2005) Psychological and situational influences on commuter-transport mode choice. Environ Behav 37:640–661
28. Colten CE, Kates RW, Laska SB (2008) Community resilience: lessons from New Orleans and Hurricane Katrina. CARRI Research Report 3. Community and Regional Resilience Initiative. Oak Ridge National Laboratory
29. Corraliza J, Berenguer J (2000) Environmental values, beliefs and actions, a situational approach. Environ Behav 32:832–848
30. Croydon SFRA (2009) Croydon strategic flood risk assessment: level 2 final report. Prepared by Scott Wilson for Croydon Council. https://www.croydon.gov.uk/sites/default/files/articles/downloads/sfralevel2final.pdf. 23 Sept 2016
31. Croydon Council (2011) Surface water management plan. https://www.croydon.gov.uk/sites/default/files/articles/downloads/swplan.pdf. 23 Sept 2016
32. Cutter SL, Barnes L, Berry M, Burton C, Evans E, Tate E, Webb J (2008) A place-based model for understanding community resilience to natural disasters. Glob Environ Change 18:598–605
33. Dayani A (2007) More rain on the way for Midlands. Birmingham Mail, June 18th 2007. http://www.birminghammail.net/news/tm_method=full&objectid=19313463&siteid=50002-name_page.html. 3 May 2012
34. DEFRA (2005) Making space for water: taking forward a new government strategy for flood and coastal erosion risk management in England. First Government response to the Autumn 2004 making space for water consultation exercise. Department for the Environment, Food and Rural Affairs, London
35. DEFRA (2009) National rank order of settlements susceptible to surface water flooding. Department for Environment, Food & Rural Affairs, UK
36. Delanty G (2003) Community. Routledge, London
37. Douglas M, Wildavsky A (1982) Risk and culture: an essay on the selection of technical and environmental dangers. University of California Press, Berkeley
38. Dunlap RE, Mertig A (1995) Global concern for the environment: is affluence a prerequisite? J Soc Issues 51(4):121–138
39. Ellis RJ, Thompson F (1997) Seeing green: cultural biases and environmental preferences. In: Ellis RJ, Thompson M (eds) Culture matters : essays in honor of Aaron Wildavsky. Westview Press, Boulder
40. Environment Agency (2014) Perry Barr and Witton flood risk management scheme: Environmental statement version 1, July. Environment Agency, Bristol, UK
41. Fillmore EP, Ramirez M, Roth L, Robertson M, Atchison CG, Peek-Asa C (2008) After the waters receded: a qualitative study of university officials' disaster experiences during the Great Iowa Flood of 2008. J Community Health 36(2):307–315
42. Floyd DL, Prentice-Dunn S, Rogers RW (2000) A meta-analysis of research on protection motivation theory. J Appl Soc Psychol 30(2):407–429
43. Fordham MH (1998) Making women visible in disasters: problematising the private domain. Disasters 22(2):126–143
44. Fowler HJ, Kilsby CG (2003) A regional frequency analysis of United Kingdom extreme rainfall from 1961 to 2000. Int J Climatol 23(11):1313–1331
45. Frieden JB, Downs PE (1986) Testing the social involvement model in an energy conservation context. J Acad Mark Sci 14(3):13–20

46. Gorte J (2005) Corporate social responsibility: close to victory. J Invest 14(3):140–141
47. Granger K, Hayne M (2001) Natural hazards and the risk they pose to South-East Queensland. Technical Report. Geoscience Australia, Commonwealth Government of Australia, Canberra
48. Gyawali D (1999) Institutional forces behind water conflict in the Ganga Plains. GeoJournal 47:443–452
49. Hahn R (2012) Standardizing social responsibility? New perspectives on guidance documents and management system standards for sustainable development. IEEE Trans Eng Manag 99:1–11
50. Hakes JK, Viscusi WK (2004) Dead reckoning: demographic determinants of the accuracy of mortality risk perception. Risk Anal 24(3):651–663
51. Hardy MA (1993) Regression with dummy variables. In: Lewis-Beck MS (ed) Regression analysis, International handbooks of quantitative applications in the social sciences. Sage Publications Ltd, London
52. Harries T (2013) Responding to flood risk in the UK. In: Joffe H, Rossetto T, Adams J (eds) Cities at risk: living with perils in the 21st century, Advances in Natural and Technological Hazards Research 22. Springer, Dordrecht
53. Harris P (1996) Sufficient grounds for optimism? The relationship between perceived controllability and optimistic bias. J Soc Clin Psychol 15(1):9–52
54. Harvatt J, Petts J, Chilvers J (2011) Understanding householder responses to natural hazards: flooding and sea-level rise comparisons. J Risk Res 14(1):63–83
55. Horgan S (2005) Kids as stakeholders in business. Young Consum 6(4):72–81
56. ISO (2010) ISO 26000: guidance on social responsibility. International Standard Organization, Geneva
57. Jha AK, Bloch R, Lamond J (2012) Cities and flooding: a guide to integrated urban flood risk management for the 21st century. Global Facilities for Disaster Reduction and Recovery, the World Bank, Washington DC
58. Johnston D (1999) Volcanic hazard perceptions: comparative shifts in knowledge and risk. Disaster Prev Manag 8(2):118–126
59. Joseph R, Lamond J, Proverbs DG (2015) Homeowners' perceptions of property-level flood risk adaptations (PLFRA) measures: the case of the summer 2007 flood event in England. Int J Safe Secur Eng 5(3):251–265
60. Kahan DM, Braman D, Gastil J, Slovic P, Mertz CK (2007) Culture and identity-protective cognition: explaining the white-male effect in risk perception. J Empir Legal Stud 4:465–505
61. Kahan DM, Jenkins-Smith H, Braman D (2010) Cultural cognition of scientific consensus. J Risk Res 14:147–174
62. Karanci N (2013) Facilitating community participation in disaster risk management: Risk perception and preparedness behaviours in Turkey. In: Joffe H, Rossetto T, Adams J (eds) Cities at risk: living with perils in the 21st century, Advances in Natural and Technological Hazards Research 22. Springer, Dordrecht
63. Kievik M, Gutteling JM (2011) Yes, we can: motivate Dutch citizens to engage in self-protective behaviour with regard to flood risks. Nat Hazards 59:1475–1490
64. Kotecha R, Thornes J, Chapman L (2008) Birmingham's local climate impacts profile (LCLIP). http://www.bebirmingham.org.uk/uploads/LCLIP.pdf. 26 Sept 2016
65. Krasovskaia I (2005) Perception of flood hazard in countries of the North Sea region of Europe. Norwegian Water Resources and Energy Directorate, Oslo
66. Lamond JE, Joseph RD, Proverbs DG (2015) An exploration of factors affecting the long term psychological impact and deterioration of mental health in flooded households. Environ Res 140:325–334
67. Larson L, Whiting J, Green G (2011) Exploring the influence of outdoor recreation participation on pro-environmental behaviour in a demographically diverse population. Local Environ 16(1):67–86
68. Lave TR, Lave LB (1991) Public perception of the risks of floods: Implications for communication. Risk Anal 11(2):255–267

69. Leiserowitz A (2006) Climate change risk perception and policy preferences: the role of affect, imagery, and values. Clim Change 77:45–72
70. Lewis-Beck MS (1993) Applied regression: An introduction. In: Lewis-Beck MS (ed) Regression analysis, International handbooks of quantitative applications in the social sciences. Sage Publications Ltd, London
71. Li W, Wright PM, Rukavlna PB, Pickering M (2008) Measuring students perceptions of personal and social responsibility and the relationship to intrinsic motivation in urban physical education. J Teach Phys Educ 27(2):167–178
72. Lindell MK (2013) North American cities at risk: household responses to environmental hazards. In: Joffe H, Rossetto T, Adams J (eds) Cities at risk: living with perils in the 21st century, Advances in Natural and Technological Hazards Research 22. Springer, Dordrecht
73. Lindell MK, Perry RW (2000) Household adjustment to earthquake hazard: a review of research. Environ Behav 32(4):461–501
74. Lindell MK, Perry RW (2004) Communicating environmental risk in multiethnic communities. Sage Publications, Thousand Oaks
75. Lindell MK, Whitney DJ (2000) Correlates of household seismic hazard adjustment adoption. Risk Anal 20:13–25
76. Megicks P, Memery J, Williams J (2005) Ethics and social responsibility issues in grocery shopping: a preliminary typology. Qual Mark Res 8(4):399–412
77. Mullins A, Soetanto R (2013) Ethnic differences in perceptions of social responsibility: informing risk communication strategies for enhancing community resilience to flooding. Disaster Prevent Manag 22(2):119–131
78. Myers G (1994) Words in ads. Routledge, New York
79. Nagel J (1994) Constructing ethnicity: creating and recreating ethnic identity and culture. Soc Probl 41(1):152–176
80. Neuwirth K, Dunwoody S, Griffin RJ (2000) Protection motivation and risk communication. Risk Anal 20(5):721–734
81. Nicholson-Cole SA (2005) Representing climate change futures: a critique on the use of images for visual communication. Comput Environ Urban Syst 29:255–273
82. Norris FH, Stevens SP, Pfefferbaum B, Wyche KF, Pfefferbaum RL (2008) Community resilience as a metaphor, theory, set of capacities, and strategy for disaster readiness. Am J Community Psychol 41(1–2):127–150
83. Norusis MJ (1995) SPSS 6.1: guide to data analysis. Prentice Hall, Englewood Cliffs
84. Olofsson A (2007) The preparedness of local authorities for crisis communication with people who have foreign backgrounds—the case of Sweden. Int J Mass Emerg Disaster 25(2):145–173
85. ONS (2001) 2001 census profile: Birmingham. http://www.birmingham.gov.uk/cs/Satellite?c=BCC_Standard_C&childpagename=Planning-and-Regeneration%2FBCC_Standard_C%2FStandardContentSummary&cid=1223091749672&pagename=BCC%2FCommon%2FWrapper%2FWrapper. Accessed 18 Jan 2016
86. Ouellette JA, Wood W (1998) Habit and intention in everyday life: the multiple processes by which past behaviour predicts future behaviour. Psychol Bull 124(1):54–74
87. Pain R, Barke M, Fuller D, Gough R, MacFarlane R, Mowl G (2001) Introducing social geographies. Arnold, London
88. Panwar R, Han X, Hansen E (2010) A demographic examination of societal views regarding corporate social responsibility in the US forest products industry. For Policy Econ 12:121–128
89. Peterson RT, Jun M (2007) Perceptions on social responsibility: the entrepreneurial vision. Business Society. http://bas.sagepub.com/cgi/rapidpdf/0007650307305758v1, Accessed 2 March 2016
90. Pitt M (2008) The Pitt review—learning lessons from the 2007 floods. Cabinet Office, London
91. Poortinga W, Steg L, Vlek C (2002) Environmental risk concern and preferences for energy-saving measures. Environ Behav 34(4):455–478

92. Quimby CC, Angelique H (2011) Identifying barriers and catalysts to fostering pro-environmental behaviour: opportunities and challenges for community psychology. Am J Community Psychol 47(3):388–396
93. Rea Catchment Partnership (2016) Planning applications submitted—two flood alleviation schemes in the Rea catchment. http://www.reacatchmentpartnership.co.uk/latest-news/item/67-planning-applications. 23 Sept 2016
94. Reed VA, Jernstedt GC, Hawley JK, Reber ES, DuBois C (2005) Effects of a small-scale, very short-term service learning experience on college students. J Adolesc 28:359–368
95. Rushton JP (2004) Genetic and environmental contributions to pro-social attitudes: a twin study of social responsibility. Proc R Soc B Biol Sci 271(1557):2583–2585
96. Safi AS, Smith WJ Jr, Liu Z (2012) Rural Nevada and climate change: vulnerability, belief, and risk perception. Risk Anal 32(6):1041–1069
97. Sattler DN, Kaiser CF, Hittner JB (2000) Disaster preparedness: relationships among prior experience, personal characteristics, and distress. J Appl Soc Psychol 30(7):1396–1420
98. Semenza JC, Hall DE, Wilson DJ, Bontempo BD, Sailor D, George LA (2008) Public perception of climate change voluntary mitigation and barriers to behavior change. Am J Prev Med 35(5):479–487
99. Siegel JM, Shoaf KI, Afifi AA, Bourque LB (2003) Surviving two disasters: Does reaction to the first predict response to the second? Environ Behav 35:637–654
100. Siegrist M, Gutscher H (2008) Natural hazards and motivation for mitigation behaviour: people cannot predict the affect evoked by a severe flood. Risk Anal 28:771–778
101. Sinatra GM, Kardash CM, Taasoobshirazi G, Lombardi D (2012) Promoting attitude change and expressed willingness to take action toward climate change in college students. Instr Sci 40:1–17
102. Slovic P (1999) Trust, emotion, sex, politics, and science: surveying the risk-assessment battlefield. Risk Anal 19(4):689–701
103. Smit B, Wandel J (2006) Adaptation, adaptive capacity and vulnerability. Glob Environ Change 16:282–292
104. Spence A, Poortinga W, Butler C, Pidgeon NF (2011) Perceptions of climate change and willingness to save energy related to flood experience. Nat Clim Change 1(1):46–49
105. Spittal MJ, McClure J, Siegert RJ, Walkey FH (2005) Optimistic bias in relation to preparedness for earthquakes. Australas J Disaster Trauma Stud 1:1–10
106. Steg L, Sievers I (2000) Cultural theory and individual perceptions of environmental risks. Environ Behav 32(2):248–267
107. Steg L, Vlek C (2009) Encouraging pro-environmental behaviour: an integrative review and research agenda. J Environ Psychol 29:309–317
108. Sundblad EL, Biel A, Garling T (2007) Cognitive and affective risk judgments related to climate change. J Environ Psychol 27:97–106
109. Takahashi B, Selfa T (2015) Predictors of pro-environmental behaviour in rural American communities. Environ Behav 47(8):856–876
110. Tanida N (1996) What happened to elderly people in the great Hanshin earthquake. Br Med J 313(7065):1133–1135
111. Tapsell SM (2000) Follow-up study of the health effects of the 1998 Easter flooding in Banbury and Kidlington. Final report to the Environment Agency. Flood Hazard Research Centre, Middlesex University, Enfield
112. Tapsell SM, Tunstall SM, Penning-Rowsell EC, Handmer JW (1999) The health effects of the 1998 Easter flooding in Banbury and Kidlington. Report to the Environment Agency, Thames Region. Flood Hazard Research Centre, Middlesex University, Enfield
113. Tapsell SM, Tunstall SM, Wilson T (2003) Banbury and Kidlington four years after the flood: An examination of the long-term health effects of flooding. Report to the Environment Agency. Flood Hazard Research Centre, Middlesex University, Enfield
114. Tapsell S, McCarthy S, Faulkner H, Alexander M (2010) Social vulnerability to natural hazards. CapHaz-Net WP4 Report. Flood Hazard Research Centre, Middlesex University. http://caphaz-net.org/outcomes-results/CapHaz-Net_WP4_Social-Vulnerability2.pdf. Accessed 14 Dec 2015

115. Terpstra T, Gutteling JM (2008) Households' perceived responsibility in flood risk management in the Netherlands. Int J Water Resour Dev 24:555–565
116. Tierney K, Bruneau M (2007) Conceptualizing and measuring resilience: a key to disaster loss reduction. TR News 250:14–17
117. Valentine G (2001) Social geographies: space and society. Prentice Hall, Harlow
118. van der Linden S (2015) The social-psychological determinants of climate change risk perceptions: Towards a comprehensive model. J Environ Psychol 41:112–124
119. Waddock S (2008) Corporate responsibility/corporate citizenship: the development of a construct. In: Scherer AG, Palazzo G (eds) Handbook of research on global corporate citizenship. Edward Elgar, Cheltenham
120. Wells VK, Ponting C, Peattie K (2011) Behaviour and climate change: consumer perceptions of responsibility. J Mark Manag 27(7–8):808–833
121. Whitmarsh L (2008) Are flood victims more concerned about climate change than other people? The role of direct experience in risk perception and behavioral response. J Risk Res 11(3):351–374
122. Whitmarsh L (2011) Scepticism and uncertainty about climate change: dimensions, determinants and change over time. Glob Environ Change 21:690–700
123. Whitmarsh L, O'Neill S (2010) Green identity, green living? The role of pro-environmental self-identity in determining consistency across diverse pro-environmental behaviours. J Environ Psychol 30:305–314
124. Whittaker M, Segura GM, Bowler S (2005) Racial/ethnic group attitudes toward environmental protection in California: is 'environmentalism' still a white phenomenon? Polit Res Q 58(3):435–447
125. Work PA, Spencer MRJ, Osborne R (1999) Flood retrofit of coastal residential structures: outer banks, North Carolina. J Water Resour Plan Manag 125(2):88–93

CHAPTER 7

How to increase earthquake and home fire preparedness: the *fix-it* intervention

Helene Joffe, Gabriela Perez-Fuentes, Henry W. W. Potts, Tiziana Rossetto

[1] *Department of Clinical, Educational and Health Psychology, University College London (UCL), 26*

Bedford Way, London WC1H 0AP, UK

[2] *Institute for Health Informatics, University College London (UCL), 222 Euston Road, London NW1 2DA, UK*

[3] *Department of Civil, Environmental and Geomatic Engineering, University College London (UCL), Gower Street, London WC1E 6BT, UK*

ABSTRACT

Published, evaluated community intervention studies concerning natural hazard preparedness are rare. Most lack a rigorous methodology, thereby hampering the development of evidence-based interventions. This paper describes the rationale and methodology of a cross-cultural, longitudinal intervention study on earthquake and home fire preparedness, termed *fix-it*. The aim is to evaluate whether and how the intervention brings about behaviour change in the targeted communities in two coastal cities with high seismic risk: Seattle, USA and Izmir, Turkey. Participants are adult residents of these cities. The intervention group attends a 6-h workshop, which focuses on securing items in the household. The control group does not attend the workshop. All participants complete baseline and post-intervention, as well as 3- and 12-month follow-up assessments. The primary outcome measure is an observational measure of nine preparedness items for earthquake and fire in participants' homes. This is evaluated alongside participants' self-reports concerning their preparedness levels. Secondary outcomes are changes in levels of self-efficacy, perceived outcome, trust, corruption, empowerment, anxiety and social cohesion. Results from the first of the studies, conducted in Seattle in September 2015, indicate that while the *fix-it* intervention is effective, in the longer term, multi-hazard preparedness is increased by the mere act of going into people's homes to observe their preparedness levels along with assessing self-reported preparedness and sociopsychological orientation towards natural hazards. This protocol and study aim to augment the empirical literature on natural hazard preparedness, informing national and international policy on delivery of evidence-based community interventions to promote multi-hazard preparedness in households.

Keywords: Community preparedness, intervention, Multi-hazard preparedness, Earthquake preparedness, Fire preparedness, Behaviour change

1. INTRODUCTION

Globally, natural disasters have claimed an average of 68,000 lives per year over the past 20 years, with earthquakes and tsunamis killing more people than all other types of disasters together: they claimed approximately 750,000 lives between 1994 and 2013 (CRED and UNISDR 2015). Furthermore, natural disasters cost an average of US$250–300 billion globally each year (Maskrey and Safaie 2015). Community preparedness for natural hazards is critical in reducing the risk of injury, death and damage to homes; it also facilitates coping with the temporary disruption caused by the hazard (Paton 2000). Cities and communities that are prepared for several hazards are more resilient than those that prepare for a single hazard (Scawthorn et al. 2006b). Consequently, the multi-hazard approach is gaining popularity in the field of natural hazard preparedness (Scawthorn et al. 2006a). Disaster preparedness measures can range from securing heavy objects, structural retrofitting and storing food and water, to having insurance, an emergency kit and communication and evacuation plans.

While significant efforts have been made in recent years to increase public awareness and to train communities in hazard preparedness, such work has not proven sufficient in increasing the adoption of mitigating measures. Community preparedness for natural hazards remains low across cultures (Ballantyne et al. 2000; Lindell and Whitney 2000; Paton 2000), even among those where awareness of the natural hazard risk is high (Joffe et al. 2013; Karanci et al. 2005; Lindell 2013; Solberg et al. 2010)

Despite the importance of natural disaster preparedness interventions in reducing personal, social and economic losses, the literature on such interventions is scarce and its methodology is often not well documented (Perez-Fuentes and Joffe 2015). There are a wide range of mass media and internet-based natural disaster preparedness engagement methods, yet their documentation and evaluation are rare (Verrucci et al. 2016). When preparedness campaigns and interventions are evaluated, most are vaguely described. Furthermore, those interventions that are evaluated tend to be based on self-report measures alone, where social desirability threatens the validity of the results (Eisenman et al. 2009; Glik et al. 2014; Karanci et al. 2005). Lastly, few of the existing published interventions on earthquake and home fire preparedness have been proven to demonstrate increased disaster preparedness behaviours at the household level. Therefore, more explicit and better designed natural hazard preparedness interventions are needed so that they can be replicated and improved. The long-term objective is to expose publics only to evidence-based interventions. This paper aims to describe one such intervention, which was devised on the basis of the existing psychological literature concerning successful natural hazard preparedness and health behaviour change interventions.

1.1 Effective intervention studies on earthquake and fire preparedness

Despite their scarcity, there are a number of examples of successful interventions and campaigns regarding community preparedness that provide a transparent methodology. Findings from these studies inform the present intervention. Overall, disaster preparedness interventions have proven successful in affecting adjustment measures when including *hands-on training* (Miller et al. 2014), *drills* (Simpson 2002), *face-to-face interactions* as opposed to sending a pamphlet or internet-based information, targeting *social empowerment,* a process of increasing personal, interpersonal or political power so that individuals can take action to improve their life situations (Paton et al. 2003) and *community cohesion* (Eisenman et al. 2009; Glik et al. 2014), defined as a "set of attitudes and norms that include trust, a sense of belonging, and the willingness to participate and help, as well as their behavioural manifestations" (Chan et al. 2006).

Based on additional evidence, other factors must also be built into interventions to make them effective. Proven predictors of preparedness are: *perceived self-efficacy*, that is, beliefs regarding personal capacity to act effectively, which influences the number and quality of plans, effort and perseverance in risk reduction behaviours (Paton et al. 2003); *collective efficacy,* defined as people's shared beliefs in their collective power (Paton and Johnston 2015); *perceived outcome expectancy*, the perception of whether ones' actions will reduce the problem or threat (Paton et al. 2003); *behavioural intention*, which refers to the motivation to act and precedes adjustment adoption (Paton et al. 2000, 2010); *critical awareness*, defined as the frequency of talking and thinking about a specific hazard (Paton and Johnston 2015; Paton et al. 2005); *sense of community*, that is, feelings of belonging and attachment to people and places (Bishop et al. 2000; Paton 2000); *community participation*, which refers to individuals who form groups to collaborate with management agencies (Karanci and Aksit 2000; McGee and Russell 2003; Paton and Johnston 2015; Paton et al. 2005; Tierney et al. 2001); and *trust in authorities*, defined as trust in various sources of information (Dillon and Phillips 2001; Paton and Johnston 2015). In addition to targeting these predictors in this intervention, sociocultural and emotive variables, such as *anxiety* and *fatalism*, were added as mediators of the relationship between awareness and actual preparedness behaviours, as well as that between awareness and personal responsibility (Joffe et al. 2013; Johnston et al. 2013; Paton 2008; Paton et al. 2010).

1.2 Theoretical models and preparedness intervention studies

In order to be effective, behaviour change interventions should be based on theoretical models that explain the behaviour change process (Michie et al. 2008). Various theoretical models have been used in natural hazard intervention studies, such as the person-relative-to-event model (Duval and Mulilis 1999), the precaution adoption process model (Weinstein and Sandman 1992) and the emotion-focused coping model (Lazarus and Folkman 1984). In the health field, the prediction of preparedness behaviours has been modelled according to the theory of planned behaviour and protection motivation theory (Abraham et al. 1998; Bennett and Murphy 1997). The inclusion of largely cognitive factors, as found in many of these models, is not sufficient for understanding the relationship between risk awareness and hazard preparedness. More holistic studies have attempted to develop models to predict the adoption of natural hazard preparedness with good success (Paton et al. 2003, 2005).

Thus, the study reported in this paper utilised a modified version of Paton et al.'s model of natural hazard risk reduction preparedness (Paton et al. 2003, 2005), which proposes that motivation, outcome expectancy and self-efficacy beliefs are the main predictors of preparedness. Paton et al. add a fourth variable to their model: intentions to prepare. In addition, drawing from Paton et al.'s model, empowerment and social cohesion are added to the intervention as potential mediators of preparedness. In addition to drawing on Paton et al.'s model of preparedness, this study is also rooted in and draws on previous work on social representations of earthquakes in Seattle and Izmir (Joffe et al. 2013). Like the Paton et al. model, this work showed self-efficacy to be important, but added anxiety, trust, sense of responsibility and sociocultural variables such as fatalism versus sense of control as potential determinants of preparedness. Thus, two key models are drawn on in shaping this intervention with the former giving more emphasis to cognitive and empowerment variables and the latter to emotive and sociocultural factors.

This paper is divided into two parts. The first provides a detailed description of the methodology and design of the intervention study so that it can serve as a basis for the development of future community intervention studies. The second part describes results of the intervention conducted in Seattle in September 2015.

2. METHOD AND DESIGN

2.1 Study design and assessment periods

This is a cross-cultural, quasi-experimental, non-randomised, controlled, longitudinal, face-to-face intervention study, with a pretest–posttest design. Study objectives are (1) to increase household preparedness for earthquakes and home fires in Seattle (USA) and Izmir (Turkey) by way of an intervention, (2) to evaluate changes in levels of critical awareness, self-efficacy, perceived outcome, trust, empowerment, anxiety and social cohesion, as well as levels of adjustment following the intervention, and (3) to compare results of the intervention across the different cultures. Participants in the intervention group each participate in two 3-h workshops (intervention), while controls do not receive the intervention. All participants complete assessments 1 week before and 1 week after the intervention, and they are followed up 3 and 12 months after the intervention to measure preparedness behaviour change (Table 1).

Table 1 Study design chart

Groups (*n* = 160)	**Pre-survey and checklist (1 week before workshop)**	**Work shops**	**Post-survey and checklist (1 week after the end of workshops)**	**Follow-up survey and checklist (3 months after the end of workshops)**	**Follow-up survey and checklist (12 months after the end of workshops)**
Intervention (*n* = 80)	X	X	X ($250)	X ($40)	X ($40)
Control (*n* = 80)	X	--	X ($100)	X ($40)	X ($40)

Monetary incentives are given for participation in the study. In the study conducted in Seattle,[1] participants in the intervention group were paid $250 for completing baseline and post-intervention assessments, as well as a 6-h workshop divided over 2 days (3 h each day). Participants in the control group were given $100 for completing the baseline and post-intervention assessments. To ensure completion of the pretest and posttest assessments, participants in both groups were paid only after completion of the second assessment. Participants are paid $40 per completed follow-up assessment, at 3 and 12 months after the intervention. Study incentives for Izmir are equivalent to the monetary incentives in Seattle.

2.2 Sample

In order to define sample size, an a priori power analysis using G*Power statistical software, version 3.1 (Faul et al. 2009) was conducted. Equal-sized sample groups were planned, meaning that the allocation ratio of N1 to N2 is 1 (Bredenkamp 1969; Orme et al. 1990). To perform a two-group comparison of a primary outcome with a standard alpha of 0.05, 64 individuals per group, or a total of 128 individuals, provide 80 % power to detect a significant difference of 5 % between the intervention group and the control group in a two-tailed analysis. Consequently, we planned to recruit 100 participants per group to account for dropouts, in order to have a minimum of 64 participants per group left in the final longitudinal follow-up. The post-intervention sample in Seattle consisted of 157 people (85 intervention group and 72 control group), which decreased at 3 months to 145 (73 intervention group and 72 control group).[2]

The sample in Seattle consisted of adult residents (aged 18–80) of North Seattle, with a mean age of 50 (SD = 13). Sixty per cent of residents were female, 77 % identified as Caucasian, 47 % identified as Christian, while 36 % reported not having any religion; 59 % were married or living with their partner and 75 % were home owners. More than half of the sample reported being employed (52 %).

2.3 Area selection

Since sense of community and community participation have been found to be variables that affect preparedness and adoption of mitigation measures (Bishop et al. 2000; Paton 2000), intervening in an existing community seemed appropriate. Consequently, for the Seattle study, one area was selected for the intervention group and the intervention was conducted there. A geographically separate, equally large area served as the field for those in the control condition. In February 2015, the team of researchers travelled to Seattle to explore potential areas to conduct the study and ultimately selected two of them.[3] Additional aims for the scoping trip included establishing contact with a recruitment agency and arranging meetings with Seattle emergency management experts, such as the Red Cross, the Seattle Fire Department and the Seattle Office for Emergency Management for information and collaboration in the intervention. This first trip to Seattle proved to be highly important in the design of the intervention and in making connections with experts and survivors of earthquakes, who agreed to serve as motivational speakers in the intervention. A similar trip was conducted for Izmir to explore and select the intervention areas as well as to establish contact with the respective emergency management authorities and trusted informants.

2.4 Participant recruitment and allocation

The sample constitutes adults (aged 18–80), one representative per household, who are residents of Seattle and Izmir. In each of these cities, the study areas constituted two geographically separated areas of equivalent demographic mix assigned to be the intervention and control groups, respectively. A recruitment agency from each city was hired to recruit participants. Area selection in the two cities was based on the same criteria.

In each of the two cities, the agency is instructed to recruit 100 participants from the intervention area and 100 from the control area. Specifically, they are asked, for both groups, to recruit evenly across each age group (18–35, 36–54, 55–80), 50 % male and 50 % female within each age group, two-thirds home owners and one-third renters and 50 %

below $45,000 and 50 % above $45,000 annual household income (and the equivalent respective salary in Izmir). All recruiting in Seattle was completed over the phone by the recruitment agency.

One week before the intervention in Seattle, participants were called by the agency to arrange a specific date and time to visit their homes to complete the baseline assessment. After completion, the intervention group participants were called and reminded of the day, date, time and location for their workshops. One week after completion of the workshops, participants in both groups were called by the agency to schedule a specific date and time to complete the post-workshop assessments and to receive their respective incentives for participation.

2.5 Intervention design and delivery

The intervention is named *fix-it,* as it focuses on fixing and securing items in the home. *Fix-it* is a name that is easy to remember and has positive/neutral connotations, as opposed to threat and fear-inducing names with negative and catastrophic connotations. According to the literature on campaigns, in order to be effective in facilitating action, names and messages of preparedness campaigns should be positive in tone, promote a sense of empowerment, as well as being short, clear, consistent and motivating of action (Jones 2012; Kocher 1993). The intervention name received good feedback in the pilot study, with participants stating that the name was, "easy to remember", "fine", "concise and it implies action".[4] *Fix-it* was also accepted as a good name for the intervention in Izmir, as a similar Turkish term exists.

Fix-it includes the key ingredients of previous preparedness interventions that have proven successful, such as hands-on training, face-to-face interactions, empowerment and community cohesion. In addition, determinants of behaviour such as self-efficacy, outcome expectancy and motivation are targeted through different behavioural techniques (Michie et al. 2008). The intervention design was influenced by the work of Michie et al. (2011) concerning effective behaviour change techniques to challenge specific determinants of behaviour. To address motivation, the use of rewards, incentives (gift cards for group challenge winners, magnets and mugs with acronyms and the name of the intervention), social encouragement and support (engaging with the *fix-it* Facebook group page, taking pictures of their homes) were used, as was persuasive communication. Self-efficacy was addressed using self-monitoring (homework review), rehearsal (videogame play), coping skills learning, social encouragement and support and feedback (homework review); all behaviour change techniques have proven effective in changing self-efficacy (Michie et al. 2008). In order to affect outcome expectancy and increase motivation, persuasive communications were given by experts who were trusted sources, such as the American Red Cross, the Seattle Fire Department and the Seattle Office of Emergency Management (Michie et al. 2008). AFAD, the equivalent trusted source for emergency management in Turkey, and their experts were used as communicators in the Izmir intervention.

The intervention consisted of two 3-h interactive, face-to-face workshops for each participant in the active group and focused on securing items in the household. The workshops were led by two professionals, experts in emergency management training and psychologists.[5] The workshops included hands-on training, as well as interactive tools, such as playing a video game, reviewing photographs of participants' homes and engaging in a Facebook group page. With the purpose of intervening in an existing community, the intervention was delivered in a community centre (as in the case of Seattle) or space where communities gather to participate in activities. The intervention was carefully designed by the psychology researchers of the project who, in addition to reviewing the literature on natural disaster preparedness intervention studies and that on behaviour change, consulted three experts in the field of interventions, epidemiological studies and behaviour change techniques to review the intervention. This, together with the pilot study conducted in June, shaped the *fix-it* intervention.

2.5.1 Fix-it

The 6-h *fix-it* workshop, divided over 2 days (3 h each), was delivered on weekdays after work. Specifically, it focused on securing nine items in the household (Table 2): securing bookcases and large cabinets, securing TVs and computer screens, making sure that sockets were not overcharged (e.g. not having a multi-plug plugged into another multi-plug), having functioning smoke alarms and knowing how to test them, ensuring that no pictures or frames were hanging above sofas and/or beds, having all exits clear of obstruction and having in-date fire extinguishers. These nine items are taken from the 16 earthquake and 13 fire preparedness measures on questions 17 and 30 of the survey (Appendix I of supplementary material), and they constitute the primary outcome measure of preparedness (Table 2).

The literature on behaviour change interventions advocates that one limits interventions to few, simple behaviours to increase effectiveness (Michie et al. 2014). The *fix-it* intervention chose the nine items from a total of 29 preparedness items for earthquake and fire that individuals could carry out at home with little cost and bureaucracy and without necessarily owning their own homes. This intervention did not include structural mitigation actions because the researchers aimed to target those behaviours over which people had control. These mitigative behaviours were also chosen, as opposed to focusing on promoting kits or plans, because of evidence that they are needed. According to the 2013 American Housing Survey (U.S. Census Bureau 2015), which added questions on emergency and disaster preparedness for the first time, 52 % of American citizens reported having an emergency kit, 82 % reported having non-perishable food stored, 62 % stated having stored water and over 30 % claimed having an emergency meeting location and a communication plan (Bureau 2015). Since having a kit and storing food and water had already been adopted by 60–80 % of American residents, the *fix-it* intervention's focus was on improving other adjustment measures that are low cost and easy to adopt, such as securing items in the household, which were not measured in the American Housing Survey. In addition, securing items in the house has been identified as one of the basic categories of preparedness by victims of earthquakes, reducing injuries and allowing them to focus on other categories of preparedness, such as assisting others and following their emergency plans (Paton and Johnston 2015). Despite being considered as a key aspect of preparedness and its adoption being critical for people's outcomes in earthquakes, most preparedness studies focus on survival measures (e.g. storing food and water) and not on such mitigative behaviours (Russell et al. 1995; Spittal et al. 2006).

Table 2 Checklist of the nine *fix-it* preparedness measures

Please check whether each of the following 9 measures has been taken or not by the participant:

	No (None)	**Yes (Some)**	**Yes (All)**	**Not appli cable**
1. TV is secured (with Velcro, pad, straps, or locks)	○	○	○	○
2. Computer is secured (with Velcro, pad, straps, or locks)	○	○	○	○
3. Bookcase is secured to the wall (using nylon strap)	○	○	○	○

Please check whether each of the following 9 measures has been taken or not by the participant:				
	No (None)	**Yes (Some)**	**Yes (All)**	**Not appli cable**
4. Large cabinet is secured to the wall (using nylon strap)	○	○	○	○
5. No multiple plugs plugged in another multiple plug (Ask "Do you have any multiple plug in use?" If so "Let me have a look at them") Check NO if they have a multiple plug plugged into another multiple plug[a]	○	○	○	○
6. No objects (framed pictures, mirrors) are placed above sofas AND beds If posters without frame then OK; if posters WITH frame click, "No"	○	○	○	
7. Exits (doors) are clear of obstruction	○	○	○	
8. Functioning smoke alarms and respondent knows how to test them ("Please, show me how to test it") All alarms have a test button, even hardwired ones. If a person does not know how to test them, then mark "No"	○	○	○	
9. Having an in-date fire extinguisher (if expired, then click "No")	○	○	○	

[a]Item removed from the analyses due to an error in instructions/interpretation and inconsistent scoring

Day 1 of *fix-it* aimed to enhance the awareness and importance of being prepared. The workshop started with a free association task (Joffe and Elsey 2014), where participants were asked to write or draw their associations to "home" on a grid. This gauged how they represented the very homes that the intervention aimed to target and allowed them to share their thoughts and feelings about their homes and to begin to interact with others in their group. The grids were collected for content analysis by the researchers, with feedback provided at the next workshop. Participants were then presented with common cognitive and emotive responses regarding earthquakes that residents in each of these cities had held, as per findings of the previous study (Joffe et al. 2013), such as "denial of fear", or "fascination with earthquakes". A discussion was then held on how these might affect preparedness. Subsequently, a series of videos of previous earthquakes in their own city, survivor stories from each city and a home fire simulation were shown. Each of these videos was followed by a group discussion, where participants shared their experiences of earthquakes and fires and the impact that these had on them. Participants then engaged in playing "Beat the Quake", a videogame concerning earthquake and fire preparedness, in teams of five. To create social encouragement, promote group cohesion and give incentives, participants were then divided into groups where they participated in a group challenge based on a set of questions on earthquake and home fire preparedness. Before the end of the first day of the workshop, participants were given an assignment, to be reviewed on the second day of the workshop. The assignment involved taking pictures of the following items in their houses and sending them via email to the workshop leader: a smoke alarm, a multi-plug, an important exit, a bookshelf, a bed/and or couch, a fire extinguisher, a television and a computer. These pertained to the nine preparedness measures that *fix-it* focused on. Participants were also invited to join a *fix-it* Facebook group page, with the purpose of creating a sense of community empowerment, feedback and an opportunity for social interaction with the researchers and the rest of the intervention participants, as well as for the sharing of learning. The Facebook group page was designed and managed by one of the researchers in the team.

Day 2 of the workshop focused on "how to" *fix-it*. The evening started with the feedback from participants' associations to home, followed by a discussion on the possibility of their homes being a space of danger in the event of an earthquake or a home fire. A review of their assignments followed, where photographs of the participants' homes were screened for discussion. This was followed by an examination of the nine preparedness measures with videos that explained to participants how to carry out the preparations. Trusted experts (In Seattle, the Red Cross for, the Seattle Fire Department and the Office of Emergency Management in the case of Seattle, AFAD in Izmir) then gave presentations and demonstrated live with tools how to secure items. Participants were invited to touch the earthquake preparedness and fire safety tools and interact with the experts by asking them questions at the "hands-on stations". At the end of the workshop, the prize for the winners of the group challenge was given, along with a fridge magnet with an acronym for the nine *fix-it* measures, a mug with the name of the intervention and a certificate of completion. These giveaways, as well as being a reward for having participated in the workshops, aimed to serve as reminders after the intervention was over, ensuring that the *fix-it* measures were adopted.

2.5.2 Beat the Quake

In order to ensure that participants learned, practiced and mastered the nine *fix-it* preparedness measures, they played a videogame (on individual tablets) at the end of the first day of the workshop. *Beat the Quake* was originally designed by Earthquake Country Alliance in 2006 as part of the *Great Shake Out* earthquake preparedness campaign in California. The game is free and easily accessible from its website (http://www.dropcoverholdon.org/beatthequake/game/). In the original version of the game, the player is asked to click on items in a virtual living room and correctly identify how to secure 14 of these items (including *fix-it* measures) with appropriate methods and tools. The player has to choose an answer from a list of options. The more items the player correctly secures, the higher the score received. The player has to act quickly before the big earthquake hits. The player receives audio, visual and textual feedback throughout the game and additional visual feedback at the end of the game. For the purpose of the *fix-it* intervention, *Beat the Quake* was modified with the help of UCL's EPICentre, by adding four fire preparedness items. A Turkish version also existed and was also modified to include the fire items. *Beat the Quake* was improved for the purpose of this study and is now a multi-hazard preparedness educational tool. *Beat the Quake* has proven to generate significant preparedness learning outcomes (Tanes and Cho 2013).

2.6 Consent

Ethical permission was obtained to conduct the study (UCL Ethics Project ID Number: 1392/001). Informed written consent was obtained from each participant before completing the baseline survey. A member of the recruitment agency collected all signed consent forms from participants in both groups. Both participants and recruitment agency members were blind to group allocation.

2.7 Measures

A 25-min computer-assisted survey and a checklist comprised the assessment that participants in both groups completed at the following time points in their homes[6]: (1) at baseline, 1 week before the intervention, (2) 1 week after the intervention, (3) 3 months after and (4) 12 months after the intervention. At the first home visit, a member of the recruitment agency met with each participant and asked them to complete the baseline survey online. Once finished, the agency member proceeded with the second part of the assessment, which was to conduct a tour around the house and examine with a checklist whether the participant had or had not adopted each of the nine preparedness measures. This constituted the observational measure of the study. Home visits to observe and document participants' preparedness behaviours, rather than sole reliance on self-reports, offer a novel approach to this field.

2.7.1 Survey

A self-administered computer-assisted survey (see Appendix I of supplementary material) using Qualtrics software was used to assess the following variables:

- *Demographics*: Sociodemographic characteristics including gender, age, ethnicity, marital status, number of children, employment status, education level and annual income level.
- *Building characteristics* including information such as type of residence, year and material of construction of the residence and earthquake retrofitting.
- *Barriers to preparedness for earthquake and for fire* including a set of items that the person had to tick (all that apply) indicating barriers for not preparing, such as the lack of time, the financial cost, the need for cooperation with others, the skill required, information overload, none and other.

Precursor variables:

- *Critical awareness*: assessed by four questions that asked how often one thinks and talks about earthquakes and how often one thinks and talks about home fires, with responses on a Likert-type scale from 1 (never) to 5 (once a week or more). This is a modified version of Dalton's measure of critical awareness (Dalton et al. 2001) that has been used in other studies with good success (Paton et al. 2003)

- *Earthquake anxiety*: measured by a set of items (e.g. "I avoid thinking about earthquakes") on a Likert-type scale ranging from 1 (not at all) to 4 (a great deal). It was developed by Johnston (Paton et al. 2003) and used in several earthquake preparedness studies by Paton.
- *Self-efficacy*: used a modified version of Riggs and Knight's scale (Riggs and Knight 1994).
- *Intention to prepare* was assessed by the question "Do you intend to do it in the next week or two?" for each of the nine preparedness measures.
- *Outcome expectancy*: The perception of whether personal actions would effectively mitigate or reduce the threat was assessed by a set of five questions for each of the nine preparedness measures, to which respondents could answer yes or no.

Mediators:

- Trust: measured by asking the participant how much trust they had in several institutions including education, armed forces, government, scientists and the church, rated from 1 (none at all) to 4 (a great deal) (Joffe et al. 2013).
- *Fatalism*: evaluated by a set of questions used in a previous study (Joffe et al. 2013)
- *Empowerment:* used a modified version of Speer and Peterson's empowerment scale (Speer and Peterson 2000).
- *Social cohesion*: used a reduced version of the scale used to measure social cohesion in European countries (Dickes and Valentova 2013).
- *Collective efficacy*: used Riggs and Knight's scale version in modified form (Riggs and Knight 1994).
- *Sense of responsibility:* assessed by the question "how much responsibility do the following people and institutions have for reducing your risk from earthquakes?" The same question was asked for home fires (Joffe et al. 2013).
- *Perceived level of corruption* of the government and the construction industry was assessed by asking: "On a scale from 1 (None at all) to 4 (A great deal) how would you rate the level of corruption of the following: the local government, the state government, the federal government, the construction industry?" (Joffe et al. 2013)
- *Religiousness*: was assessed using a single question ("How religious would you say you are?") rated on a five-point scale, ranging from "not at all" to "very much".

Outcome variables:

- *Earthquake preparedness* was assessed by a set of 16 earthquake preparedness items, which included the five earthquake-related *fix-it* measures; participants had to indicate whether they had adopted them or not. This set of items used to assess preparedness is a modified version of the set of items used in a previous study (Joffe et al. 2013).
- Fire preparedness was measured by a set of 13 fire preparedness items, including the four *fix-it* measures that gauged fire preparedness.

2.7.2 Checklist

The nine-item checklist constitutes the primary outcome measure (Table 2). As described above, a member of the recruitment agency examined the participant's household and checked whether each of the nine preparedness measures had been adopted by the participant or not. This is a purely observational measure, and it was evaluated alongside the self-report measure.

2.7.3 Follow-ups

Follow-up assessments included a modified version of the survey (with additional questions about the workshop) and the checklist. Follow-up assessments were conducted 3 and 12 months after the end of the workshops.

2.8 Interrater reliability

Interrater reliability on 5 % of the checklists in all countries, using two independent raters, produced a satisfactory result (kappa = 0.65).

2.9 Data analysis and management

The SPSS statistical software package, version 20 (Nie et al. 2011), was used for conducting the statistical analysis. Behaviour change was measured in both groups, and t tests were conducted to evaluate differences between means of behaviour change between groups. Logistic regression analysis of the outcome measure on the active group was conducted, with adjusted confidence intervals and *p* values. In addition, cross-tabulation and χ^2 statistics were carried out to evaluate correlations between self-report and observational preparedness measures to determine the extent of agreement between them. Moderated mediation analysis among relevant variables was carried out to explore mediators of preparedness. Levels of anxiety in relation to preparedness behaviours were also examined. Although preparedness is measured as an overall score, including earthquake and fire measures, separate analyses for earthquake and home fire preparedness were conducted in order to explore differences in preparedness for the more every day risk (i.e. fires) versus the more long return period risk (i.e. earthquakes). It must be noted that it is the difference in scores between assessment times that is the key outcome: that is, the study measures behaviour change.[7]

3. RESULTS: SEATTLE STUDY

3.1 Baseline analyses (Tables 3, 4, 5, 6)

The sample in Seattle was adult residents (aged 18–80) of North Seattle, with a mean age of 50 (SD = 13). Sixty per cent of residents were female, 77 % identified as Caucasian, 47 % as Christian, while 36 % reported not having any religion; 59 % were married or living with their partner; and 75 % were home owners. More than half of the sample reported being employed (52 %). Baseline differences in sociodemographic, psychosocial and preparedness characteristics of participants were assessed before the intervention in order to rule out selection bias due to the non-randomised sampling design of the study. There were no differences in sociodemographic features or psychological variables between the intervention and control groups at baseline. In addition, no significant differences were found between the intervention and control groups in terms of their preparedness levels (Table 3).

Both the intervention and control groups reported barriers to earthquake and fire preparedness. The most common barrier reported was "other things/problems to think about", followed by "the time required to do it" and "the skill or knowledge required". The least reported barrier for preparing for earthquakes was "the financial cost" (Table 4).

Predictors of preparedness before the intervention are listed in Table 5. It is important to note that all results are given first for overall preparedness, which refers to an average of the five earthquake and the three fire preparedness items.[8] Results show that *empowerment* ($F(1,154) = 9.1$; $p = 0.003$, $R^2 = 0.05$), *trust* ($F(1,154) = 4.0$; $p = 0.046$, $R^2 = 0.02$), *corruption* ($F(1,155) = 6.2$; $p = 0.01$, $R^2 = 0.03$) and *social cohesion* ($F(1,155) = 1.6$; $p = 0.04$, $R^2 = 0.23$) predicted overall preparedness in a univariate regression analysis at baseline.

Table 3 Preparedness levels at baseline

	Intervention (*n* = 85)		**Control** (*n* = 72)		**Total** (*n* = 157)	
	M	**SD**	**M**	**SD**	**M**	**SD**
Overall preparedness	13.32	2.76	13.32	2.51	13.32	2.64
Earthquake preparedness	6.11	2.22	5.68	1.89	5.91	2.08
Fire preparedness	7.21	1.51	7.64	1.46	7.41	1.50

Table 4 Barriers for preparedness at baseline

Barriers for earthquake preparedness	**Mean**	**SD**	**Barriers for fire preparedness**	**Mean**	**SD**
Other things/problems to think about	0.49	0.50	Other things/problems to think about	0.49	0.50
Time required to do it	0.40	0.49	Skill or knowledge required	0.28	0.45
Skill or knowledge required	0.30	0.45	Time required to do it	0.24	0.43
The financial cost	0.26	0.44	Need for cooperation with others	0.16	0.36
Need for cooperation with others	0.20	0.40	The financial cost	0.11	0.31
Information overload	0.14	0.34	Information overload	0.10	0.29

Table 5 Significant predictors of overall preparedness, earthquake preparedness and fire preparedness at baseline[1]

Predictors	Overall preparedness					Earthquake preparedness					Fire preparedness				
	B	SE	β	R^2	p-level	B	SE	β	R^2	p-level	B	SE	β	R^2	p-level
Empowerment	1.49	0.49	0.23	0.053	0.003*	0.84	0.39	0.16	0.029	0.034	0.65	0.28	0.18	0.034	0.022
Social cohesion	0.33	0.11	0.22	0.053	0.004	0.28	0.09	0.24	0.060	0.002*	0.05	0.06	0.06	0.004	0.416
Trust	0.84	0.41	0.16	0.026	0.046	0.65	0.32	0.15	0.025	0.049	0.18	0.23	0.06	0.004	0.431
Corruption	−0.97	0.39	−0.19	0.039	0.014	−0.73	0.30	−0.18	0.035	0.019	−0.24	0.22	−0.08	0.007	0.287

[1] Significant predictors shown using univariate analysis

* Significant predictors also using multivariate analysis

Table 6 Correlations between self-report and observational measures of earthquake and fire preparedness

	TV secured	Computer secured	Bookcase secured	Large cabinet secured	No objects above sofas/beds	Exits clear	Fire extinguisher	Smoke alarm
TV secured	0.333**							
Computer secured		0.010						
Bookcase secured			0.601**					
Large cabinet secured				0.438**				
No objects above sofas/beds					–0.122			
Exits clear						0.034		
Fire extinguisher							0.557**	
Smoke alarm								0.198*

* Correlation significant at the 0.05 level (2-tailed)

** Correlation is significant at the 0.01 level (2-tailed)

Predictors of earthquake preparedness, in particular, at baseline were *empowerment* ($F(1,154) = 4.5$; $p = 0.034$, $R^2 = 0.02$), *trust* ($F(1,154) = 3.9$; $p = 0.04$, $R^2 = 0.02$), *corruption* ($F(1,155) = 5.6$; $p = 0.01$, $R^2 = 0.03$) and *social cohesion* ($F(1,155) = 9.8$; $p = 0.00$; $R^2 = 0.05$). *Empowerment* was the only predictor for fire preparedness at baseline.

Correlations between self-reported preparedness in the survey and observational preparedness measures in the checklist are described in Table 6. All items correlated significantly with each other except "computer is secured", "all exits are clear of obstruction" and "no objects are hanging above sofas/beds".

3.2 Changes in preparedness levels after the intervention (Table 7; Figs. 1, 2, 3)

One week post-intervention, the intervention group had significantly improved its overall preparedness ($t(72) = 3.8$, $p < 0.001$), as well as its earthquake ($t(72) = 2.4$, $p = 0.018$) and fire preparedness ($t(72) = 3.7$, $p < 0.001$) compared to baseline. Those in the control group improved their overall preparedness compared to baseline, though not significantly. The control group's levels of earthquake preparedness decreased slightly though not significantly from baseline, but they had significantly increased their fire preparedness ($t(71) = 2.4$, $p = 0.019$). Between-group differences of preparedness change means 1 week after the intervention were significant for overall preparedness change ($p = 0.001$) and for earthquake preparedness change ($p = 0.019$), with the intervention group showing significantly more change (Table 7).

Three months after the intervention, both the intervention ($t(66) = 3.3$, $p = 0.001$) and control groups ($t(67) = 2.2$, $p = 0.032$) had significantly improved their overall preparedness levels compared to baseline. Although both groups had improved their earthquake preparedness 3 months after the intervention, the scores did not reach statistical significance. Fire preparedness improved for both groups at 3-month follow-up compared to baseline, but only the intervention group's fire preparedness reached statistical significance ($t(66) = 5.6$, $p < 0.001$) (Table 7).

Multivariate analysis of variance was conducted to explore levels of preparedness change between groups across the two points in time, 1 week after the intervention and at the 3-month follow-up. Results show a significant interaction of time and group ($p = 0.013$), with the intervention group significantly decreasing their overall preparedness from time 1 (post-intervention) to time 2 (3-month follow-up) compared to the control group (Fig. 1).

Table 7 Changes in preparedness levels 1 week after the intervention and at 3-month follow-up

Post-intervention[a]	Intervention ($n = 73$)				Control ($n = 72$)				Between groups	
	M	SD	t test	p-level	M	SD	t test	p-level	t test	p-level
Overall preparedness change	1.48	3.30	3.83	<0.001	0.18	2.90	0.52	0.60	2.51	0.013
Earthquake preparedness change	0.85	2.99	2.42	0.018	−0.22	2.42	−0.77	0.40	2.36	0.019
Fire preparedness change	0.62	1.44	3.65	<0.001	0.38	1.32	2.39	0.019	1.05	0.296
3-month follow-up										
Overall preparedness change	0.93	2.28	3.31	0.001	0.91	3.44	2.18	0.032	0.27	0.978
Earthquake preparedness change	0.07	1.98	0.30	0.80	0.47	2.82	1.37	0.20	−0.94	0.348
Fire preparedness change	0.88	1.29	5.55	<0.001	0.38	1.65	1.90	0.061	1.94	0.054

[a] One week after the end of the intervention

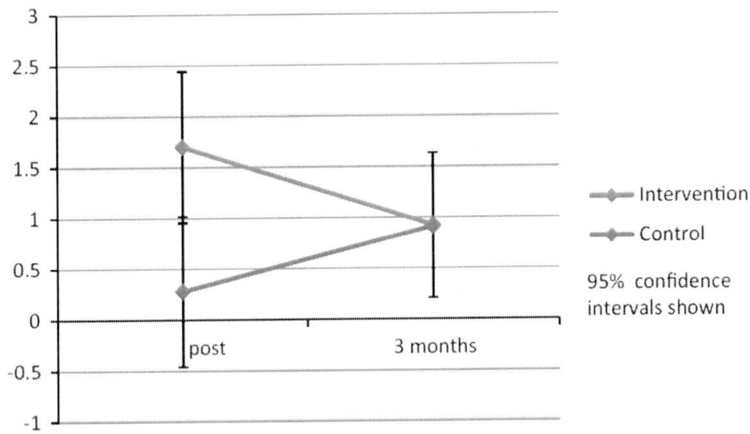

Figure. 1 Overall preparedness change 1 week and 3 months after the intervention

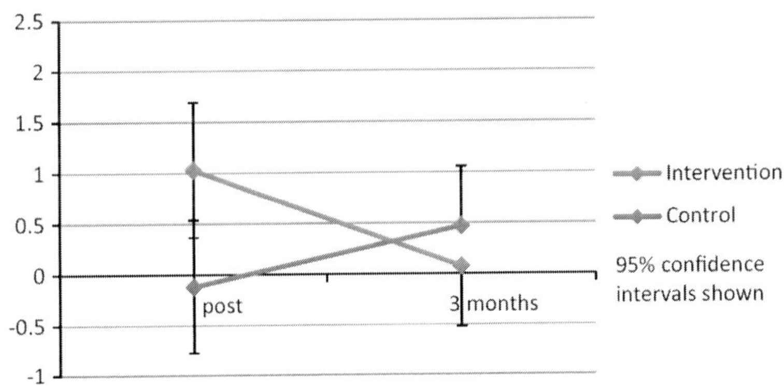

Figure. 2 Earthquake preparedness change 1 week and 3 months after the intervention

A similar pattern can be found regarding earthquake preparedness (p = 0.002) (Fig. 2). There was no significant interaction for group or time in fire preparedness between the two time points (Fig. 3).

3.3 Predictors of preparedness change (Table 8)

No significant predictors of changes in overall preparedness, earthquake or fire preparedness were found for the intervention group at 3-month follow-up using both univariate and multivariate analysis. However, for the control group, anxiety ($F(9,58) = 3.5$; $p = 0.011$, $R^2 = 0.355$) and trust in different systems (e.g. education, armed forces, government, scientists and church) ($F(9,58) = 3.5$ $p = 0.030$, $R^2 = 0.135$) were found to be predictors of overall preparedness at 3 months, both using univariate and multivariate analysis (multivariate results are presented in Table 8). In addition, using multivariate analysis (including the variables of self-efficacy, outcome expectancy, fatalism, collective efficacy, anxiety, empowerment, trust, corruption and social cohesion), self-efficacy ($p = 0.03$) and collective efficacy ($p = 0.04$) were significant predictors of overall preparedness for the control group (Table 8).

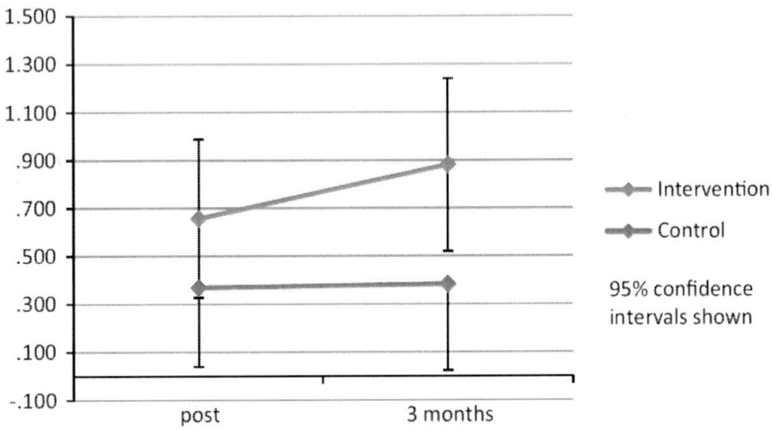

Figure. 3 Fire preparedness change 1 week and 3 months after the intervention

In terms of the specific predictors of earthquake and fire preparedness change at the 3-month follow-up, only anxiety ($F(9,58) = 2.7$; $p = 0.043$, R^2 - 0.297) and trust ($F(9,58)$ - 2.7; p - 0.009, $R^2 = 0.297$) were significant predictors of earthquake preparedness, for the control group only (within a multivariate analysis of self-efficacy, outcome expectancy, fatalism, collective efficacy, anxiety, trust, corruption, empowerment and social cohesion). The only predictor for fire preparedness change at 2 months was self-efficacy ($F(9,58) = 1.49$; $p = 0.045$, $R^2 = 0.188$), and this was only for the control group.

Table 8 Significant predictors of preparedness change at 3-month follow-up[1]

	Intervention (n = 73)				Control (n = 72)				
	B	SE	β	R^2	B	SE	β	R^2	p-level
Predictors overall preparedness									
Self-efficacy	0.14	0.71	0.29	0.097	1.78	0.84	0.25	0.355	0.039
Collective efficacy	−0.23	0.70	−0.04	0.097	−1.93	0.92	−0.23	0.355	0.041
Anxiety (earthquake)	1.01	0.93	0.16	0.097	2.35	0.90	0.30	0.355	0.011*
Trust	0.10	0.64	0.02	0.097	−1.35	0.60	−0.25	0.355	0.030*
Predictors of earthquake preparedness									
Anxiety (earthquake)	0.65	0.77	0.12	0.166	1.60	0.77	0.24	0.297	0.043
Trust	−0.12	0.54	−0.03	0.166	−1.40	0.52	−0.32	0.297	0.009
Predictors of fire preparedness									
Self-efficacy	−0.09	0.41	−0.03	0.069	0.93	0.45	0.27	0.188	0.045

[1]Results of multivariate analysis reported

* Significant predictors with univariate analysis

4. DISCUSSION

This cross-cultural, longitudinal, community-based intervention aims to improve earthquake and home fire preparedness at the household level. The intervention's theoretical background, methodology, content and assessment process are thoroughly described above. In addition, results from the Seattle study and the 3-month follow-up are presented. Such detailed methodological descriptions, as well as the combination of self-report and observational measures to assess preparedness, are extremely rare in the natural disaster intervention literature. Therefore, this protocol seeks to provide a guideline for developing future, empirically based natural disaster preparedness interventions.

One week after the intervention, the intervention group had significantly improved its overall preparedness levels, as well as its earthquake and fire preparedness levels, showing the effect of the *fix-it* workshop on behaviour change in the intervention group. The control group, on the other hand, did not increase overall preparedness significantly: earthquake preparedness decreased slightly but fire preparedness increased significantly.

However, 3 months after the intervention, both groups had significantly improved their overall preparedness levels compared to baseline. Earthquake preparedness improved for both groups at the 3-month follow-up compared to baseline levels, though not significantly, while fire preparedness improved for both groups at the 3-month follow-up compared to baseline, but only significantly in the intervention group. Overall, results show the effect of the *fix-it* workshop on behaviour change in the intervention group, significantly improving their preparedness levels both 1 week after and 3 months after the intervention. More surprisingly, preparedness levels for the control group also increased significantly at 3 months. The control group caught up with the intervention group, who showed an unsurprising drop off compared to 1 week after the intervention. This leaves both groups significantly above baseline at 3 months, but not different from each other. Since controls did not receive the *fix-it* intervention, the home visits, which were conducted face-to-face and included a survey and observation of home earthquake and fire preparedness, seem to have played a role in subsequently increasing preparedness behaviour, particularly regarding earthquakes, in what had originally been designed as a control group.

The novel approach of combining self-report and observational measures of preparedness also proved to be of value. Correlation between self-report and observational data was good but not excellent, suggesting that there was variation in how items were observed and showing that results from self-report and observation differ. This study cannot determine which one is the more ecologically valid measure. Future studies would need to refine this.

Corroborating both the Paton et al. (2003, 2005) and Joffe et al. (2013) models, baseline results showed that overall preparedness is predicted by feeling empowered and by living in a socially cohesive community, as well as by having higher levels of trust in the systems that surround one. Perhaps more surprising is that representations of one's government being somewhat corrupt predict preparedness. Overall, the findings suggest that empowered, cohesive communities, who have trust in their social systems (e.g. education, scientists, the church and health) and who come from an individualistic culture, such as the USA, might put more effort into self-preparation, rather than relying on "corrupt" governmental authorities to protect them.

Another important finding is the significant role that emotions play in this study. Consistent with the existing literature, anxiety was found to have a direct effect on preparedness behaviours (Lee and Lemyre 2009; Miceli et al. 2008; Paton et al. 2003; Rüstemli and Karanci 1999) but only for the control group. Despite some research suggesting that high anxiety can hinder preparedness, this study suggests that the home visits may have produced an optimum level of anxiety, since the literature indicates that optimum levels are needed to bring about preparedness behaviours. Anxiety levels for the control group increased over time, compared to the intervention group, causing participants to increase their preparedness behaviour. It is likely that their anxiety increased due to having home assessments and yet not being given information on preparedness, as opposed to the intervention group, who received a 6-h workshop over two days on exactly what to do to be better prepared and whose levels of anxiety decreased after the intervention.

Trust in various systems (e.g. education, police, government, church and scientists) was another emotion that directly predicted preparedness only for the control group. This is consistent with previous research that shows that trust is a determinant of community action (Dillon and Phillips 2001). Other studies have found trust to moderate the relationship between intention to prepare and actual preparedness (Paton et al. 2003) (Johnson-George and Swap 1982; Paton 2007; Sjoberg 1999). In this study, since the control group had no information about how to prepare as they did not attend the workshops, trust in authorities may have increased over time as a way of coping with their worry about not knowing what to do to become better prepared.

The possibility that the home visits acted as an intervention could have been determined by the inclusion of a second control group, who would not have undergone the baseline assessment. This possibility was considered, but the very assessment process in this study, done at whatever time point, potentially creates change and so this option was not seen as useful. Another potential limitation was that participants were not randomly assigned to the intervention versus control groups, as the researchers purposively sought to run the intervention in particular communities and to draw the control group from equivalent but separate communities. Despite non-random assignment, results showed no differences between groups on sociodemographic, psychosocial and preparedness variables at baseline.

Overall, it seems that the very act of going into people's houses, interacting with them face-to-face and having someone check their preparedness levels, as well as their emotions and cognitions, produced behaviour change, in this case an increase in overall, earthquake and fire preparedness over time. When people are left with anxiety, concerning what they should be doing to improve their preparedness and are having their preparedness levels gauged at predetermined intervals by an observer as well as via self-report, they may seek to engage in preparedness measures and may thereby improve their preparedness.

This study constitutes one of the first to evaluate both earthquake and home fire preparedness and that aims to compare results across two different cultures: Seattle (USA) and Izmir (Turkey).[9] In addition, this multi-hazard preparedness intervention study uses both self-report and observational measures. Furthermore, the intervention was carefully designed by a team of multidisciplinary researchers, from the fields of psychology and engineering, who had previously evaluated the social representations of earthquakes in the people of Seattle and Izmir (Joffe et al. 2013). Results of these thorough interviews have allowed this team to develop detailed interventions tailored to match the social representations of each location (Seattle and Izmir). The researchers approached and designed both the survey and the intervention with depth knowledge of how people in the two cities conceptualise earthquakes. That study had highlighted the potential role played by emotive and sociocultural factors, rather than purely cognitive factors, in shaping earthquake-related conceptualisations and behaviours. This sensitised the researchers to ensure that the current intervention study should carefully assess emotions such as anxiety and trust. They were, indeed, found to be critical in increasing the adoption of preparedness behaviours, though only in the control group. The sociocultural elements identified in the previous study can only be fully assessed once the Seattle component of the intervention is compared with the results of the Izmir intervention.

This protocol aims to improve the field of hazard preparedness, being one of the few papers that fully describe a rigorous methodology and rationale for a natural hazards intervention. This study has significant implications for the field of natural disaster preparedness at an international level, for earthquake and fire preparedness more specifically and, in particular, for the two cities studied, since it allows for replication and therefore improvement of intervention studies. Furthermore, in terms of impact, the present study suggests that earthquake and fire preparedness can be increased significantly by going into people's homes to observe whether they have prepared for earthquakes and fires at multiple, prearranged time points, as well as surveying their psychosocial characteristics relevant to preparedness, without any further intervention. This face-to-face element may well contain an ingredient that precipitates change in a way that digital methods do not.

5. GOVERNANCE AND ETHICAL CONSIDERATIONS

The research was funded by the Engineering and Physical Science Research Council, UK (Grant ref. EP/F012179/1). There are no known conflicts of interest. Ethical permission was obtained to conduct the study: UCL Ethics Project ID Number: 1392/001. Participants all signed an informed consent form before the study commenced.

FOOTNOTES

1. The intervention in Seattle was conducted in September 2015, and the three-month follow-up assessment was conducted in January 2016. The intervention in Izmir was conducted between May and June of 2016.

2. Eleven participants in the intervention group were living in subsidised housing, and thus, they were not able to make the required preparedness fixes at home. For this reason, their data were not included in the post-intervention and 3-month analysis. These data call for a separate analysis, to explore barriers to preparedness and the effect of the intervention on a vulnerable population.

3. The radius from which participants were chosen was enlarged at the recruitment stage because the agency could not find sufficient numbers of people willing to partake, who fitted the criteria in the original, smaller communities sampled.

4. A pilot of this study was conducted with 11 residents of North-east Seattle in June 2015. The intervention proved to be effective in increasing preparedness behaviours in participants, with participants having

adopted significantly more preparedness measures one week after the intervention, compared to their baseline levels of preparedness (p = 0.001).

5. The workshops were conducted with approximately 20 people in each. Therefore, four identical workshop groups were run on consecutive nights the first week, and then, people attended their second workshop one week later than their first.

6. The assessments were done on paper in Izmir as this was deemed more appropriate by the Turkish researchers employed on the project.

7. If an item is not applicable at both time points, then there is no change. However, if a participant goes from not applicable to showing the behaviour, then there is behaviour change because some act has taken place. For example, if a person did not have a TV at time 1, but had a TV at time 2, which they had secured, then they had carried out a positive act (securing the TV) (see Appendix II of supplementary material).

8. The item of safe multi-plugs was removed from the analyses due to an error in instructions and inconsistent scoring. Thus, analyses included eight preparedness items, five checklist items for earthquake and three for fire preparedness.

9. Results from Izmir and longitudinal (1-year follow-up) results from Seattle will be presented in subsequent papers.

REFERENCES

1. Abraham C, Sheeran P, Johnston M (1998) From health beliefs to self-regulation: theoretical advances in the psychology of action control. Psychol Health 13:569–591
2. American Housing Survey for the United States: 2013 (2015) US census bureau. http://www.census.gov/programs-surveys/ahs.html
3. Ballantyne M, Paton D, Johnston D, Kozuch M, Daly M (2000) Information on volcanic and earthquake hazards: the impact on awareness and preparation. Institute of Geological and Nuclear Sciences Limited Science Report Wellington
4. Bennett P, Murphy S (1997) Psychology and health promotion. Open University Press, Buckingham
5. Bishop B, Paton D, Syme G, Nancarrow B (2000) Coping with environmental degradation: salination as a community stressor. Network 12:1–15
6. Bredenkamp J (1969) The application of significance tests in theory-testing experiments. Psychol Beitr 11(3):275–285
7. Chan J, To H, Chan E (2006) Reconsidering social cohesion: developing a definition and analytical framework for empirical research. Soc Indic Res 75:273–302
8. CRED, UNISDR (2015) The human cost of weather-related disasters, 1995–2015. United Nations, Geneva
9. Dalton JH, Elias MJ, Wandersman A (2001) Community psychology. Wadsworth, Belmont

10. Dickes P, Valentova M (2013) Construction, validation and application of the measurement of social cohesion in 47 European countries and regions. Soc Indic Res 113:827–846
11. Dillon J, Phillips M (2001) Social capital discussion paper. Unpublished manuscript, Curtin University, Perth, Western Australia
12. Duval T, Mulilis JP (1999) A person-relative-to-event (PrE) approach to negative threat appeals and earthquake preparedness: a field study 1. J Appl Soc Psychol 29:495–516
13. Eisenman DP, Glik D, Gonzalez L, Maranon R, Zhou Q, Tseng CH, Asch SM (2009) Improving Latino disaster preparedness using social networks. Am J Prev Med 37:512–517
14. Faul F, Erdfelder E, Buchner A, Lang AG (2009) Statistical power analyses using G* power 3.1: tests for correlation and regression analyses. Behav Res Methods 41:1149–1160
15. Glik DC, Eisenman DP, Zhou Q, Tseng CH, Asch SM (2014) Using the precaution adoption process model to describe a disaster preparedness intervention among low-income Latinos. Health Educ Res 29:272–283
16. Joffe H, Elsey J (2014) Free association in psychology and the grid elaboration method. Rev Gen Psychol 18:173
17. Joffe H, Rossetto T, Solberg C, O'Connor C (2013) Social representations of earthquakes: a study of people living in three highly seismic areas. Earthq Spectra 29:367–397
18. Johnson-George C, Swap WC (1982) Measurement of specific interpersonal trust: construction and validation of a scale to assess trust in a specific other. J Pers Soc Psychol 43:1306
19. Johnston D et al (2013) Community understanding of, and preparedness for, earthquake and Tsunami Risk in Wellington, New Zealand. In: Joffe H, Rossetto T, Adams J (eds) Cities at risk, vol 33. Advances in natural and technological hazards research. Springer Netherlands, pp 131–148. doi:10.1007/978-94-007-6184-1_8
20. Jones AM (2012) Use of fear and threat-based messages to motivate preparedness: costs, consequences and other choices: part one. J Bus Contin Emerg Plan 6:180–191
21. Karanci N, Aksit B (2000) Building disaster-resistant communities: lessons learned from past earthquakes in Turkey and suggestions for the future. Int J Mass Emerg Disasters 18:403–416
22. Karanci AN, Aksit B, Dirik G (2005) Impact of a community disaster awareness training program in Turkey: does it influence hazard-related cognitions and preparedness behaviors. Soc Behav Personal Int J 33:243–258
23. Kocher KW (1993) The stop AIDS story, 1987–1992. Swiss Aids Foundation and the Federal Office of Public Health, Switzerland
24. Lazarus RS, Folkman S (1984) Stress, appraisal, and coping. Springer, New York
25. Lee JE, Lemyre L (2009) A social-cognitive perspective of terrorism risk perception and individual response in Canada. Risk Anal 29:1265–1280
26. Lindell M (2013) North American cities at risk: household responses to environmental hazards. In: Rossetto T, Joffe H, Adams J (eds) Cities at risk. Springer, Netherlands, pp 109–130
27. Lindell MK, Whitney DJ (2000) Correlates of household seismic hazard adjustment adoption. Risk Anal 20:13–26. doi:10.1111/0272-4332.00002
28. Maskrey A, Safaie S (2015) GAR global risk assessment. In: EGU general assembly conference abstracts, p 6494
29. McGee TK, Russell S (2003) "It's just a natural way of life…" an investigation of wildfire preparedness in rural Australia Global Environmental Change Part B. Environ Hazards 5:1–12
30. Miceli R, Sotgiu I, Settanni M (2008) Disaster preparedness and perception of flood risk: a study in an alpine valley in Italy. J Environ Psychol 28:164–173. doi:10.1016/j.jenvp.2007.10.006
31. Michie S, Johnston M, Francis J, Hardeman W, Eccles M (2008) From theory to intervention: mapping theoretically derived behavioural determinants to behaviour change techniques. Appl Psychol 57:660–680

32. Michie S, Van Stralen MM, West R (2011) The behaviour change wheel: a new method for characterising and designing behaviour change interventions. Implement Sci 6:42
33. Michie S, Atkins L, West R (2014) The behaviour change wheel: a guide to designing Interventions. Silverback, UK
34. Miller TR, Bergen G, Ballesteros MF, Bhattacharya S, Gielen AC, Sheppard MS (2014) Increasing smoke alarm operability through theory-based health education: a randomised trial. J Epidemiol Community Health 68:1168–1174
35. Nie N, Hull C, Bent D (2011) IBM statistical package for the social sciences (SPSS version 20) computer software. Chicago, SPSS
36. Orme JG, Borenstein M, Cohen J (1990) Statistical power analysis: a computer program. Soc Serv Rev 64(3):518–520
37. Paton D (2000) Emergency planning: integrating community development, community resilience and hazard mitigation. J Am Soc Prof Emerg Manag 7:109–118
38. Paton D (2007) Preparing for natural hazards: the role of community trust. Disaster Prev Manag Int J 16:370–379
39. Paton D (2008) Risk communication and natural hazard mitigation: how trust influences its effectiveness. Int J Glob Environ Issues 8:2–16
40. Paton D, Johnston DM (2015) The Christchurch earthquake: integrating perspectives from diverse disciplines. Int J Disaster Risk Reduc 14:1–5
41. Paton D, Smith LM, Johnston DM (2000) Volcanic hazards: risk perception and preparedness. N Z J Psychol 29:86–91
42. Paton D, Smith L, Johnston D, Johnston M, Ronan K (2003) Developing a model to predict the adoption of natural hazard risk reduction and preparatory adjustments. Earthquake Commission, New Zealand
43. Paton D, Smith L, Johnston DM (2005) When good intentions turn bad: promoting natural hazard preparedness. Aust J Emerg Manag 20(1):25–30
44. Paton D, Bajek R, Okada N, McIvor D (2010) Predicting community earthquake preparedness: a cross-cultural comparison of Japan and New Zealand. Nat Hazards 54:765–781
45. Perez-Fuentes G, Joffe H (2015) An intervention to increase earthquake and fire preparedness. Paper presented at the sustainable development conference proceedings
46. Riggs ML, Knight PA (1994) The impact of perceived group success-failure on motivational beliefs and attitudes: a causal model. J Appl Psychol 79:755
47. Russell LA, Goltz JD, Bourque LB (1995) Preparedness and Hazard Mitigation Actions before and after 2 earthquakes. Environ Behav 27:744–770. doi:10.1177/0013916595276002
48. Rüstemli A, Karanci AN (1999) Correlates of earthquake cognitions and preparedness behavior in a victimized population. J Soc Psychol 139:91–101
49. Scawthorn C, O'Rourke T, Blackburn F (2006a) The 1906 San Francisco earthquake and fire—enduring lessons for fire protection and water supply. Earthq Spectra 22:135–158
50. Scawthorn C, Schneider PJ, Schauer BA (2006b) Natural hazards—the multihazard approach. Nat Hazards Rev 7:39–39
51. Simpson DM (2002) Earthquake drills and simulations in community-based training and preparedness programmes. Disasters 26:55–69
52. Sjoberg L (1999) Consequences of perceived risk: demand for mitigation. J Risk Res 2:129–149
53. Solberg C, Rossetto T, Joffe H (2010) The social psychology of seismic hazard adjustment: re-evaluating the international literature. Nat Hazards Earth Syst Sci 10:1663–1677
54. Speer PW, Peterson NA (2000) Psychometric properties of an empowerment scale: testing cognitive, emotional, and behavioral domains. Soc Work Res 24:109–118
55. Spittal MJ, Walkey FH, McClure J, Siegert RJ, Ballantyne KE (2006) The Earthquake Readiness Scale: the development of a valid and reliable unifactorial measure. Nat Hazards 39:15–29. doi:10.1007/s11069-005-2369-9
56. Tanes Z, Cho H (2013) Goal setting outcomes: examining the role of goal interaction in influencing the experience and learning outcomes of video game play for earthquake preparedness. Comput Hum Behav 29:858–869

57. Tierney KJ, Lindell MK, Perry RW (2001) Facing the unexpected: disaster preparedness and response in the United States. Joseph Henry Press, Washington
58. Verrucci E et al (2016) Digital engagement methods for earthquake and fire preparedness—a review. Nat Hazards 83:1583
59. Weinstein ND, Sandman PM (1992) A model of the precaution adoption process: evidence from home radon testing. Health Psychol 11:170

CHAPTER 8

Natural Disaster and International Development

Luke Strongman

Social Sciences, Open Polytechnic of New Zealand, Lower Hutt, New Zealand

ABSTRACT

Recovery from natural disaster has for many years been seen in objective terms as simply the time taken to replace damaged infrastructure. Increasingly, however, social scientists are describing the large part that human capital plays in the recovery from natural disaster in the form of 'resilience'. The purpose of the chapter is to delineate, from a social scientific perspective, the main factors involved in disaster rehabilitation from a necessarily superficial but nevertheless accurate and useful viewpoint. The main areas to be considered are infrastructural impacts, psychological impacts and communication factors. The chapter concludes by defining various perspectives that contribute to the quality of resilience that underscores the investment in human capital in post-disaster zones.

Keywords: natural disaster, infrastructure, psychology, social networks, psychological effects, communication, resilience

1. INTRODUCTION

A natural disaster is a major adverse event resulting from natural processes of the earth. Natural disasters may have the properties of being environmentally disruptive (war, earthquake, flood), a hazard (major but temporary disruption to environment) and/or socially focused (no major impact on land or infrastructure but on a human populous). The classic definition of a disaster is of an 'occurrence causing widespread disruption or distress'. As Lindell points out, a disaster may be characterised by three temporal periods—pre-impact, trans-impact and postimpact; however, some disasters have multiple or secondary impacts [1]. Furthermore, disasters are sometimes identified by a series of impact zones which display irregularities in effect. In general, society tends to have more knowledge about what has gone wrong in natural disasters than what goes right in recovery [2]. Even today, relatively little is known about how quickly an area affected by natural disaster may be rehabilitated and in what circumstances and under what conditions. As the World Bank reported in 2014:

During the 1980–2012 period, estimated total reported losses due to disasters amounted is US$3.8 trillion.... Hydro-meteorological disasters accounted for 74% (US$2.6 trillion) of total reported losses, 78% (18,200) of total disasters and 61% (1.4 million) of total lives lost. Looking ahead, climate change will have major implications on global ecosystems, agriculture and water supply, sea level rise and storm surges [3].

It can be argued that natural and human-made disasters that effect human populations are becoming more common worldwide due to a myriad of factors related to population growth and urbanisation. But in fact natural disasters are not becoming more prevalent; rather, there is increased reporting of them. In fact, as Lowrey et al. point out, of the 15,833 reported disasters worldwide from 1900 to 2006, over a third of these occurred between 2000 and 2006 [4]. Nevertheless, the prevalence of natural disasters themselves is unlikely to be much increasing, though the reporting of them is, but as Strömberg states, '... the average magnitude of the reported disasters has fallen' [5]. While more than half the globe becomes urbanised, the prevalence of the effect of natural disasters is greater than previously experienced, although due to increased technical know-how in building and in healthcare, the effects of natural disasters may be decreasing.

It is possible that prior to modernisation, urbanisation and globalisation, many events that could have been assigned the status of natural disaster remained unreported due to their geographical remoteness or the lack of impact on large populated areas. It is therefore not strange that the incidence of natural disaster causing events is increasing, though the impact of them has become more measured. In this context rehabilitation from natural disaster is a topic that increasingly necessitates a multidisciplinary approach to social scientific analysis.

Traditionally it was assumed that disaster recovery would be predicated on the extent of damage within a natural disaster zone. As Lindell points out, disaster effects may be determined by three pre-impact conditions: 'hazard exposure, physical vulnerability and social vulnerability—as well as three event-specific conditions—hazard event characteristics, improvised disaster responses and improvised disaster recovery' [6]. It is no longer necessarily the case that as Dacy and Kunreuther state that '[i]t just seems reasonable to assume that the speed of recovery following a disaster will be determined primarily by the magnitude of the physical damage' [7]. Rather, increasingly social networks and community resilience are shown to play a larger part in recovery from catastrophe than simply that measured by the restoration of infrastructure. The overcoming of catastrophe does not produce replacement communities; it produces reinvented, renewed and differently evolved communities. As Dynes has suggested, based on experiences of survivors in a number of disasters, social capital may be the basis for resilience as it provides information and resources at a critical moment [8]. However, as Aldrich relates, population density (the population per unit area) determines the rate of recovery, the greater the density the population conversely the more slowly it may recover due to the difficulty in providing permanent and temporary housing in the disaster period [9]. Consequently, rather than area damage, some social scientists have argued that population density—the population per unit area—alters the rate of recovery. Areas with greater population densities may recover more slowly because of the difficulty in providing temporary and permanent housing for displaced people during the post-disaster period. Nevertheless, there are behavioural typologies which do produce resilient and rehabilitated communities.

According to Aldrich, four factors are thought to influence the rate of recovery from disaster zones resulting from earthquakes. These are the amount of damage, the population density of the affected area, the efficacy of human capital and the amount of available economic capital [2].As Ross and Carter explain, '[t]he spontaneous activation of "social capital"—bonds within and between social groups—has been impressive, a solace to those affected, as well as a huge practical and economic benefit. Organised volunteers, such as from the Red Cross, were the foundation of evacuation centres' [10]. However, Aldrich also notes that 'social capital', namely, the 'networks, norms and social trust that facilitate coordination and cooperation for mutual benefit' [11], best predicted a population recovery. 'Human capital' is the ability of people to remain cooperative and resilient in communities and to work in a coordinated way together on sustainable relief. This is a requisite of community rehabilitation from earthquakes. Finally, Lindell suggests that '[p]hysical impacts can be reduced by hazard mitigation practices and emergency preparedness practices, whereas social impacts can be reduced by recovery preparedness practices' [6]. This requires extensive pre-impact planning (see Table 1).

Table 1. Factors of vulnerability in disaster impacts (after Lindell [1, 6, 42]).

Structural vulnerability	Extreme stresses	High wind, seismic shaking
Physical vulnerability	Susceptibility to biological changes—casualties	Impacts on anatomical structures and physiological functioning
Social vulnerability	Limitations in physical assets	Buildings, contents
	Psychological	Knowledge, skills and abilities
	Social	Community integration
	Economic	Financial savings
	Political	Public policy resources

2. FACTORS IN SOCIETY FOR CONSIDERATION FOR INTERNATIONAL DEVELOPMENT

2.1. Building and infrastructure

Arguably, permanence may be sacrificed for flexibility in earthquake environments when infrastructure is considered. Since at least 1751 and the devastation of the earthquake at Port-au-Prince in the Caribbean Haiti, it has been known that timber structures (houses and buildings) are able to withstand earthquakes more easily due to their flexibility of construction design [12]. Even in contemporary cities which are earthquake prone, buildings of mid-height (ten levels or under) built on base isolators which separate the motion of the building from the land underneath, or that have counterweights which control the extent of 'sway', perform best in earthquakes. So while a sense of relative permanence in built materials is sacrificed, nevertheless, in domestic architecture timber constructions are able to withstand many of the violent movements associated with earthquakes. The domestic built environment of colonial settlements was largely accomplished in timber constructions—that being the readily available material in comparison to stone used in public structures. Such domestic infrastructure could sustain rebuilding. But choice and style of rebuilding materials are not the only considerations in earthquake-zone rehabilitation.

2.2. Social networks

Motivation, empathy and communication play a tangible role in disaster rehabilitation as do resources required to rebuild infrastructure. Furthermore, social ties serve as informal insurance; people also network for financial, physical and logistical guidance. Networked communities are more politically active and better connected to overcome barriers to demand. Information sharing also increases with trust; however, embedded networks raise the cost of an 'exit' for individuals from a community (in terms of benefits forfeited). Furthermore, closely interlinked communities increase the probability that demands are articulated and obstacles are overcome. However, in communities with a comparatively higher level of social capital, individuals and groups can work towards a solution [13].

As the English Parliamentary Office of Science and Technology (2012) state:

Bangladesh is an example of a country where the spread of mobile phones and improved scientific understanding of natural hazards has helped reduce the number of deaths caused by annual flooding and more extreme hazards, such as cyclones. These advances have been exploited effectively through enhanced contingency planning and early warning systems. The mobile phone network does not reach everyone, but once the message is out, other solutions, like cyclists with megaphones, can be used [14].

Sebag-Montefiore noted that after the quake and tsunami in eastern Japan in 2012, the disaster provoked an 'unprecedented mobilisation of volunteers, public outcry following the Fukushima cover-up...the stirrings of new construction and architectural ambitions [and a] prevalent feeling among the young [to] pave the way to discard the stifling traditions of the past, including the patriarchal customs of provincial cities' [15]. This includes artistic and artisan responses such as mural paintings and the remaking of 22.5 million tons of tsunami debris into furniture (an enterprise also featured in the response to the Christchurch earthquakes of 2010 and 2011). Thus, recovery from loss is a key motivational factor, along with self and other community rehabilitations and new business opportunities that evolve from adapting dislocated environments into restructured ones. However, natural disaster environments can also serve as nodal points for learning.

Olshansky outlines seven lessons learned from reconstructions following natural disasters. These are [16]:

- Economic and social networks are more resilient than buildings—recovery begins with people first.
- Taking time to plan to ensure full participation of stakeholders is necessary in reconstruction and is proportionate to speed of recovery.
- Bureaucracies tend to lack the flexibility for community-based organisations emerge.
- Areas with fewest resources get minimal attention from aid organisations.
- Relocation is resisted by citizens, but without citizen support, relocations often fail.
- Cities see some improvement after disasters but never as much as planners intend; improvement occurs over a longer-than-expected timescale.
- Setting funding priorities is a difficult part of recovery—money is always insufficient.

Consequently, recovery from natural disasters always takes longer than most initial planning indicates. Disasters radically alter community and societal equilibrium and create huge differentials in experiences which take a long time to readjust from.

2.3. Building and infrastructure: historical sources

Olshansky argues that lessons learned from the Kobe earthquake in Japan, in 1995, involve three main factors. These are, firstly, the development of a clear vision in post-disaster response; secondly, an emphasis on people (community, welfare, health and habitat) and, thirdly, thinking creatively about property—temporary and permanent housing facilities and new sustainable building designs and materials [17]. The maintenance of social networks is considered necessary. Relatedly, an historical example shows what can be accomplished by modernisation. A memorandum (termed the Dissertação) by Manuel da Maia who was responsible for the elaboration of proposals for the renovation of the lower area of Lisbon in the wake of the 1755 earthquake (and Reconstruction Plan of 1758) revealed a method for fair transfer of old properties into new ones with appropriate equivalent values. This was achieved by 'dividing the total new surface area created within the Plan by the total value of the old properties'. The result was uniformity of architectural facades and increased modularity [18]. One of the central features of the rebuild was a Property Register established after the earthquake and fire, in which a record of the ownership and state of buildings was maintained—surveying the dimensions of 'streets, squares, public and private building' [18]. Thus, modernism was progressed in large part on quantification of infrastructure relative to value.

2.4. Building and infrastructure: modern examples

In Christchurch, Canterbury, New Zealand, following the devastating earthquakes of 2010 and 2011, the Canterbury Earthquake Recovery Authority (CERA) divided its response plan into four strategic areas of social, economic, natural and built but also leadership and integration applying to the other areas; the CERA Act defines 'rebuilding' as pertaining to physical resources and communities as well as 'improvement' and 'enhancement' of the latter in the recovery [19]. In the post-disaster environment, opportunities become apparent to incorporate new sustainable building practices into redevelopments and to adapt and reinvent environments from relocated materials. Building innovations might include 'green' building (or building in sympathy with the natural environment) and designing for maximum efficiency, for example, by practices which involve renewable energy sources or insulation and adoption of technological innovations such as building information modelling. Eversely has argued that evidence from Haiti after the devastating 2010 quake near Port-au-Prince, '... the single most important resource in the hours and days immediately following such events, relevant knowhow and practical help will become vital too' [20]. According to Gill cited in Eversely, '[i]t also involves replacing or upgrading basic water and electricity infrastructure, providing immediate areas for markets and other informal trading and providing materials and on-site advice for early self-help reconstruction by Haiti: replacement housing needs to be resilient to future disaster risks families and small communities. Finally it calls for a focus for continuing governance' [20]. Clearly, for developing communities, commerce, trade, durable physical resources and human capital and organisation are critically important. However, time is another all important factor—and the recognition of a cityscape that is both pre- and post-quake. As Jonathan Ling, chief executive of Fletcher Building Ltd., the company appointed by the government as project manager for the reconstruction of homes damaged in the September 4, 2010 and February 22, 2011, quake, much of the Christchurch rebuild said, 'There will not be a finite day on which this is finished'. 'There will be new buildings going up' in the central business district, Mr. Ling said, 'on empty blocks probably for 10, 15 or 20 years' [21]. Some of the uninhabited blocks will change owners numerous times before they are eventually built on.

Three years following the Christchurch quakes, the inner-city landscape is denuded of building—over 1000 demolitions have taken place (1611 full or partial demolitions were scheduled in total). The atmosphere is one of floating shells of building amid the dissolving remnants of built space. Familiar landmarks remain, but they are sighted from unfamiliar angles. Thus, the imagined pace of a rebuilt Christchurch assumes its shape progressively, incrementally amid the barren spaces of urban renewal. Amid the hiatus in construction plans stands the Spartan and surprisingly uniform and unornamented remnants of a built cityscape. The predominant remaining buildings are 1970s concrete multistoreys (under ten storeys), indicating there must have been an engineering practice in reinforced concrete introduced during this era that made these buildings more resilient to the movement that causes earthquake damage. While the vast majority of community aid following disaster is based on infrastructure rehabilitation, social networks are also an important aspect of rebuilding and the nurturing of social networks develops trust and interconnectedness [22]. However, it is the social effect of the trauma of earthquake, or constant demolition, or the large rents in the fabric of the built form that has characterised Christchurch's citizen's everyday lives that also demands the most attention. Trauma is partly characterised as 'loss of communality' [23]. The effect of trauma on the psychology of the earthquake survivors cannot be underestimated and ameliorative processes are regarded to be 'restoring connectedness, social support and a sense of collective efficacy' [24]. If there is one concept that summarises the requisites for survival during and after natural disaster, it is resilience. As Smith et al. explain, '[r]esilience is the ability of an individual or group to carry on and solve problems so that survival of hard times is more likely' [25]. Furthermore, resilience is a protective factor that describes an ability to rebound, make improvements in physical or psychosocial conditions and to recover from loss or illness [26]. Resilient behaviours include motivational contributions to others' lives, strengthening of coping skills, improving knowledge about caring activities and encouragement with nurturing activities [27]. People with higher health ratings report lower depressive symptom scores but higher resilience scores [27]. This is consistent with people enjoying good health having a higher band of tolerance for minor negative health problems.

2.5. Psychological effects

As Somasundaram and Sivayokan observe, both natural and man-made disasters are recognised as causing a variety of 'psychological and psychiatric sequelae' [23]. Sometimes, people at risk from natural disaster don't prepare for its eventuality because to acknowledge the need for preparation creates anxiety. Montgomery and Morris (n.d.) point out that people at risk have two choices—either to behave in ways that reduce risk or to discount the risk [28]. Thus, taking precautions to prepare for a disaster also needs people to manage their anxiety. Three orders of response are possible. These include adaptive and resilient coping responses, nonpathological distress and maladaptive behavioural patterns to diagnosable psychiatric disorders [23]. Collective traumatisation may result for large groups of effected people and 'loss of community' but is also best addressed by community interventions along the dimensions—economic development, social capital, information and communication and community competence [29]. However, after large-scale natural disasters, the effects of community disturbance may be prolonged. As Gordon and Fleming state, 'In the third year after a disaster the immediate crisis has passed. International experience shows that in most cases people have confirmation of their personal situations—problems and opportunities—and can begin to heal' [30]. In any context health is seen as '... complete physical, mental, familial, social, cultural, spiritual and ecological well-being and not merely an absence of disease or infirmity' (WHO) (see Figure 1) [31] [24].

Norris et al. identify higher rates of anxiety following disaster and depression. They identify five main effects in the post-disaster scenario. These are non-specific distress, health problems or concerns (somatic complaints and medical conditions, quality of sleep), chronic problems in living (interpersonal relationships, occupational stress and financial stress, ecological stress), psychosocial resource loss (perceived social support, social embeddedness, self-efficacy, optimism) and problems specific to youth (clinginess, dependence, loss of sleep, aggressive behaviour, separation anxiety) [32]. As Weine points out, it is only comparatively recently that scholars have integrated 'social capital measures into a quantitative study of rehabilitation' [33]. Social capital is defined as the 'networks, norms and social trust that facilitate coordination and cooperation for mutual benefit' [11]. Quantitative study of rehabilitation has tended to focus on the restoration of infrastructural facilities and the capital input in so doing as opposed to the human causal linkages that bring such restoration into effect. Relatedly, areas with larger population density recover after disasters more slowly because of the difficulty in providing temporary housing. However, the traditional formula for the recovery rate is provided by Dacy and Kunreuther who argue that 'the speed of recovery following a disaster will be determined primarily by the magnitude of physical damage' [7]. It was thus thought that the amount of damage would determine the rate of recovery, but more recent research has turned instead to community regeneration through the resilience of social capital in postcrisis recovery [34].

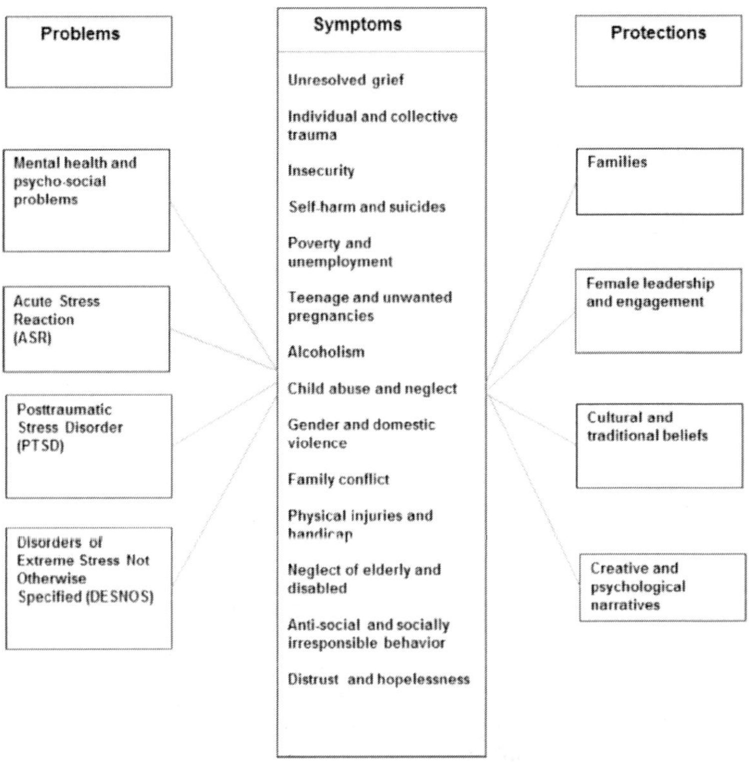

Figure 1. Problems, symptoms and protections of mental health issues in disaster zones.

As Weine also suggests, social ties form a kind of 'informal insurance' of ready-made support networks—providing information, financial and administrative support and guidance. Politically active communities can present demands and extract resources better. Trustworthy neighbours share information, prevent duping and looting and maintain the relative integrity of community relations [35]. Furthermore, embedded networks raise the costs of 'exit' for individuals—as networks carry latent effects which benefit individuals in a shared community. Such networked neighbourhood (rich in human capital) is more likely to articulate community needs to authorities and work together to overcome obstacles [36]. Thus, it stands to reason, if, following catastrophe efforts and made to strengthen social networks, leading to a better chance of recovery. There is a second reason why social capital is a constituent factor for community resilience and urban renewal and that is because it enables the mobilisation of information and resources at critical moments [37].

2.6. Communication factors

Lien et al. state that survival in post-disaster zones is most likely if people (trapped by building collapses in earthquakes) are rescued within 72 h [38]. Survival tolerance probably diminishes after 3 days. The question then becomes what are effective communication practices for post-disaster zones. As Quintanilla states, 'It is critically important that affected communities know how, when and where aid services can be accessed, what's going on around them and how they can connect with aid providers' [39]. The need for information increases and the availability and clarity of information decreases. In terms of hardware requirements for communication after natural disasters, in developed societies where information and computer technology (ICT) is prevalent, the environmental constraints for a disaster area may include temporary Internet links, unavailability of servers and limited or no Internet access. However, where broadband computing is available through relief or remaining facilities, communication is improved when devices with simple interfaces are used, Wi-Fi notebook formats are popular and where power generation is also available [40]. New technologies—mobile, SMS, social media—increase the ability to access quality information [41]. Reliability and trustworthiness of information is deemed critical. Information-sharing networks can quickly disseminate news, for example, via mobile phone texting or the Twitter service. In this context, peoples' responses to disaster are not always that of panicked confusion; rather, there is an adaptive response termed normalcy bias that provokes 'confirmation' prior to remediation or protective action [42]. As Aldrich relates, '[i]f social networks prove to be an important part of the rebuilding process, then aid givers should allocate resources to ensure that social structures are not damaged in the evacuation and resettlement processes' [43]. In post-disaster environments, most people respond by texting or tweeting about uncertain or precarious conditions if they are in densely populated areas, while those in remote areas will communicate whether or not they are safe; as Longstaff and Yang state in an immediate post-disaster situation such as an earthquake, individuals and organisations require an ability to receive (and transmit) information in secure trustworthy routes (to report damage and rehabilitation needs) and in the absence of such systems, panic may ensue [44]. Therefore, information sharing through effective communication systems is a form of investment in social capital, or 'safety valve', which helps to strengthen community resilience. As Quintanilla suggests, '... communication itself is a form of aid' [45]. Secondary communication issues include mitigation of real threats companies might suffer to their operational abilities; these might include collapse of the share price, serious legal claims, loss of credit, possible bankruptcy, damage to the company's image or reputation, loss of employees and possible closure [46]. Mitigation is best achieved through planning. However, communication in a crisis needs clarity and conciseness; it needs logical progression, a focus on the facts and the avoidance of jargon [47].

3. CONCLUSION: RESILIENCE IN DISASTERS

History has shown that while natural disasters are by definition, devastating, survivors can rebuild their lives given the opportunity. In combination with the complex rebuilding of infrastructure in post-disaster zones, social scientists and human resource practitioners are increasingly advocating that a psychology of resilience is necessary for community rebuilding. Resilience is defined by Norris as 'A process linking a set of adaptive capacities to a positive trajectory of functioning and adaptation after a disturbance' [48]. Longstaff and Yang observe that resilience is the 'capacity of a system to absorb disturbance, undergo change' yet still functions in the same way [49]. Thus, resilience is the ability of a system to recover from setback with no major change in its form or functioning. Norris also defines five dimensions along which resilience may be enhanced by 'networked adaptive capacities'. These are, firstly, developing economic resources—reducing inequities; secondly, enhancing social capital through minimising harm through damage mitigation; thirdly, adaption and utilisation of pre-existing organisational networks; fourthly, protecting naturally occurring social supports and, fifthly, community planning [48]. Consequently, central planners engaged in community rehabilitation strategies should look to strengthen these community functions in post-catastrophe environments. Hence, resilience may refer to adaptive governance 'across spatial scales (local to national) and between sectors (government, community)'; it may refer to knowledge, communication and social learning; maintaining a positive outlook and drawing on social networks for support [10]. Surviving a natural catastrophe and rebuilding a community after it require the investment in human capital which has the quality of being able to be self-sustaining by exhibiting resilience.

REFERENCES

1. M. K. Lindell, Disaster studies. *Current Sociological Review*. 61, 5–6, 797 (2013).
2. D. P. Aldrich, Social, not physical, infrastructure: the critical role of civil society after the 1923 Tokyo earthquake. *Disasters*. 36, 3, 398 (2012).
3. The World Bank, Managing Disaster Risks for Resilient Development. Retrieved from: http://www.worldbank.org/en/results/2013/04/12/managing-disaster-risks-resilientd evelopment (2014, April 10).
4. W. Lowrey, W. Evans, K. K. Gower, J. A. Robinson, P. M. Ginter, L. C. McCormick and M. Abdolrasulina, Effective media communication of disasters: pressing problems and recommendations. *BMC Public Health*. 7, 97, 2 (2007).
5. D. Strömberg, Natural disasters, economic development and humanitarian aid. *Journal of Economic Perspectives*. 21, 3, 202 (2007).
6. M. K. Lindell, Disaster studies. *Current Sociological Review*. 61, 5–6, 799 (2013).
7. D. C. Dacy and H. Kunreuther, *The Economics of Natural Disasters: Implications for Federal Policy*. (New York, NY: The Free Press, 1969). p. 72.
8. R. R. Dynes, The dialogue between Voltaire and Rousseau on the Lisbon earthquake: the emergence of a social science view. *University of Delaware Disaster Research Center Preliminary Paper*. 293, 399–400 (1999).
9. D. P. Aldrich, Social, not physical, infrastructure: the critical role of civil society after the 1923 Tokyo earthquake. *Disasters*. 36, 3, 399–400 (2012).
10. H. Ross and R. W. Carter, Natural disasters and community resilience. *Australasian Journal of Environmental Management*. 18, 1, 2 (2011).
11. R. Putnam, Bowling alone: America's declining social capital. *Journal of Democracy*. 6, 1, 67 (1995).
12. J-F. Mouhot, History matters. The tragic annals of Haiti. *History Today*. 60, 4 ,3 (2010, April).
13. D. P. Aldrich, Social, not physical, infrastructure: the critical role of civil society after the 1923 Tokyo earthquake. *Disasters*. 36, 3, 401 (2012).
14. The Parliamentary Office of Science and Technology, Resilience to Natural Hazards in Developing Nations. *Postnote* 402. Retrieved from: www.parliament.uk/briefing-papers/ POST-PN-402.pdf (2012).
15. C. Sebag-Montefiore, Tradition, washed away. *New Statesman*. 141, 5095, 14 (2012, 5th March).
16. R. Olshansky, San Francisco, New Orleans: lessons for rebuilding. *Social Policy*. 36, 2, 17 (2005–2006).
17. R. Olshansky, San Francisco, New Orleans: lessons for rebuilding. *Social Policy*. 36, 2, 18 (2005–2006).
18. [R. Dos Santos, M. Helena, S. I. Trias and T. S. I. Ferran, Trading properties after the earthquake: the rebuilding of eighteenth-century Lisbon, *Planning Perspectives*. 26, 2, 302 (2011).
19. V. Smith, C. Hart and S. Oliver, What CERA did. *Planning Quarterly*. Issue 183, 27 (2011).
20. J. Eversely, RTPI News. Post-earthquake rebuild underlines need for professions' joint approach. *Planning Issue*. 1854, 24 (2010).
21. J. Ling cited Chris Bourke, Christchurch Faces 20-Year Post-Quake Rebuild, Fletcher CEO Says. Bloombergo.com (April 26, 2011)

22. S. M. Weine, Developing preventive mental health interventions for refugee families in resettlement. *Family Process*. 50, 3, 412 (2011).
23. D. Somasundaram and S. Sivayokan, Rebuilding community resilience in a post-war context: developing insight and recommendations – a qualitative study in Northern Sri Lanka. *International Journal of Mental Health Systems*. 7, 3, 2 (2013).
24. D. Somasundaram and S. Sivayokan, Rebuilding community resilience in a post-war context: developing insight and recommendations – a qualitative study in Northern Sri Lanka. *International Journal of Mental Health Systems*. 7, 3, 3 (2013).
25. P. R. Smith, Resilience: resistance factor for depressive symptom. *Journal of Psychiatric and Mental Health Nursing*. 16, 829 (2009).
26. P. R. Smith, Resilience: resistance factor for depressive symptom. *Journal of Psychiatric and Mental Health Nursing*. 16, 831 (2009).
27. P. R. Smith, Resilience: resistance factor for depressive symptom. *Journal of Psychiatric and Mental Health Nursing*. 16, 835 (2009).
28. B. Montgomery and L. Morris, (n.d), Psychological Responses to Disasters – Facilitating Preparedness and Recovery. Retrieved from: http://csrp.hku.hk/files/ 1010_3818_941.pdf
29. D. Somasundaram and S. Sivayokan, Rebuilding community resilience in a post-war context: developing insight and recommendations – a qualitative study in Northern Sri Lanka. *International Journal of Mental Health Systems*. 7, 3, 4 (2013).
30. R. Gordon and P. Fleming, Red Cross. News Release: Red Cross Continues Its Support for the People of Christchurch. Retrieved from: https://www.redcross.org.nz/yk-files/836f6024aca36c04a2e6368e42534223/Feb%2022%20media%20release%20.pdf (20 February 2013).
31. World Health Organization, *The Constitution of the WHO*. (Geneva: World Health Organization, 1948).
32. F. H. Norris, M. J. Friedman, P. J. Watson, 60,000 disaster victims speak: part 1. An empirical review of the empirical literature, 1981–2001. *Psychiatry*. 65, 3, 216–217 (2002).
33. S. M. Weine, Developing preventive mental health interventions for refugee families in resettlement. *Family Process*. 50, 3, 410–430 (2011).
34. S. M. Weine, Developing preventive mental health interventions for refugee families in resettlement. *Family Process*. 50, 3, 400 (2011).
35. S. M. Weine, Developing preventive mental health interventions for refugee families in resettlement. *Family Process*. 50, 3, 401 (2011).
36. A. O. Hirschman, *Exit, Voice and Loyalty: Responses to Decline in Firms, Organizations and States*. (Cambridge, MA: Harvard University Press, 1970).
37. R. Dynes, The dialogue between Voltaire and Rousseau on the Lisbon earthquake: the emergence of a social science view. *University of Delaware Disaster Research Center Preliminary Paper*. 293 (1999).
38. Y-N. Lien, H-C. Jang and T-C. Tsai, A MANET based emergency communication and information system for catastrophic natural disasters. In *2009 29th IEEE International Conference on Distributed Computing Systems Workshops*. (2009). p. 412.
39. J. Quintanilla, Disaster communications: covering the last mile. *Guardian Professional*. Retrieved from: http://www.theguardian.com/global-development-professionals-network/2013/apr/03/disaster-communications-ict-internews. (2013, April 3rd). Para. 2.
40. Y-N. Lien, H-C. Jang and T-C. Tsai, A MANET based emergency communication and information system for catastrophic natural disasters. In *2009 29th IEEE International Conference on Distributed Computing Systems Workshops*. (2009). p. 414.
41. J. Quintanilla, Disaster communications: covering the last mile. *Guardian Professional*. Retrieved from: http://www.theguardian.com/global-development-professionals-network/2013/apr/03/disaster-communications-ict-internews. (2013, April 3rd). Para. 5.
42. M. K. Lindell, Disaster studies. *Current Sociological Review*. 61, 5–6, 806 (2013).

43. D. P. Aldrich, Social, not physical, infrastructure: the critical role of civil society after the 1923 Tokyo earthquake. *Disasters*. 36, 3, 412 (2012).
44. P. H. Longstaff and S-U. Yang, Communication management and trust: their role in building resilience to "surprises" such as natural disasters, pandemic flu and terrorism. *Ecology and Society*. 13, 1. Retrieved from: http://www.ecologyandsociety.org/vol13/iss1/art3/ (2008). p. 3.

CHAPTER 9

Analysis of the Twitter Response to Superstorm Sandy: Public Perceptions, Misconceptions, and Reconceptions of an Extreme Atmospheric Hazard

John A. Knox[1*], Brendan Mazanec[1], Emily Sullivan[1], Spencer Hall[1] and Jared A. Rackley[2]

ABSTRACT

Superstorm Sandy was the second-costliest hurricane in U.S. history, causing catastrophic flooding and prolonged power outages in New Jersey and New York. The public's response to this extreme event on the social network Twitter is examined using statistical analysis and manual inspection of 185,000 "tweets" relating to Sandy. Sentiment analysis of tweets from Manhattan Island reveals a statistically significant trend toward negative perceptions, especially on the southern half of the island, as Sandy made landfall. Inspection of all tweets uncovered scientific misconceptions regarding hurricanes, and a surprising and disquieting anthropomorphic *reconception* of Sandy. This reconception, divorced from factual information about the storm, dominated the "Twittersphere" compared to official scientific information. The implications of such reconceptions for social media communication during future extreme events, and the utility of the methodology employed for analysis of other events, are discussed.

Keywords: hurricane, social media, scientific misconceptions, Superstorm Sandy, Twitter

1. INTRODUCTION

At the end of October 2012, Superstorm Sandy—a hurricane with the size of an extratropical cyclone—battered the mid-Atlantic coast of the United States. This chapter examines the response of the public on the social media venue Twitter [1] to the approach, landfall, and immediate aftermath of the storm.

Twitter is one of the most popular social networks in the world, with approximately 180 million users at the time of Sandy and over 300 million users at the end of 2015 [2]. A novel feature of Twitter is the restriction of posts to 140 characters or fewer. This restriction encourages brevity and contractions among users, and limits in-depth discussion, unlike other social networks such as Facebook.

The social media reaction to Sandy has been examined from a variety of angles. Edwards et al. analyzed millions of geolocated "tweets" (i.e., posts on Twitter) and focused on the utility of Twitter to meet needs often provided heretofore by first responders and relief agencies [3]. Similarly, Chatfield et al., in a conference presentation, studied the ability of Twitter users to convey time-critical information during this disaster [4]. Lachlan et al. noted, however, that Twitter communications during the storm were used more for emotional release than for dissemination of information, and that messages from official organizations were largely absent [5]. In addition, a large automated effort to examine Twitter messages during Sandy is underway at the National Center for Atmospheric Research [6].

Our research presented in this chapter attempts to combine the best aspects of statistical analysis of a large dataset with qualitative insights gained by manual, not automated, examination of individual tweets. We have accomplished the latter via the creation of original software which makes visual inspection of Twitter messages easy and efficient. Our initial objective was to characterize misconceptions and their propagation in Twitter posts; however, our research revealed not only misconceptions but an intriguing, and disquieting, *reconception* of Sandy that threatened to drown out factual messages regarding the storm.

Below, we provide an overview of Superstorm Sandy, discuss in detail the two methodologies used to study the dataset of tweets, examine the results from both methodologies, and provide conclusions based on our quantitative and qualitative results.

2. PHYSICAL SCIENCE OVERVIEW

This overview of Sandy relies on the authoritative post-storm analysis of [7]. Sandy began as a tropical wave off the west coast of Africa on 11 October 2012. After traversing the tropical Atlantic and moving westward into the Caribbean without much growth, it reached tropical storm intensity (34 kt or 39 mph or 17 ms^{-1}) on 22 October south of Jamaica (**Figure 1**). At this point the storm received the name "Sandy," from the list created by the World Meteorological Organization for Atlantic hurricanes which alternates between male and female names familiar to the cultures that border the tropical Atlantic. "Sandy" was selected a female name, given to the 18th named storm of the 2012 Atlantic hurricane season, following "Rafael."

Two days later, after executing a loop south of Jamaica, Sandy reached Category 1 hurricane strength (64 kt or 74 mph or 33 ms^{-1}) on the Saffir-Simpson Hurricane Intensity Scale at 1200 UTC 24 October just off the southeast coast of Jamaica. It then crossed Jamaica, moving northward, and became a major hurricane (100 kt or 115 mph or 51 ms^{-1}), i.e. a Category 3 on the Saffir-Simpson scale, just prior to landfall in Cuba. Jamaica, Cuba, Haiti, and the Dominican Republic were all impacted by Sandy, with 69 deaths and hundreds of thousands of homes destroyed, particularly in Cuba. Sandy expanded greatly in size after crossing Cuba, with the radius of tropical-storm-force winds doubling by the time it passed the Bahamas. However, its winds dropped in intensity, below the threshold for a hurricane.

As it moved northward in the western Atlantic, Sandy regained strength and became a menace to the Atlantic coast of the United States. Due to interactions with an approaching trough over North America and the influence of warm Gulf Stream waters underneath it, Sandy intensified as it moved north and reached a secondary maximum of 85 kt (98 mph or 44 ms^{-1}) at 1200 UTC on 29 October about 220 nautical miles (405 km) southeast of Atlantic City, New Jersey. Under the influence of a strong high-pressure system to its north, Sandy executed a highly unusual left turn that brought the storm inland over New Jersey instead of heading eastward out to sea. At 2330 UTC on 29 October, Sandy made landfall just northeast of Atlantic City with an estimated sustained wind of 70 kt (81 mph or 36 ms^{-1}) and a minimum central pressure of 945 mb. It was one of the largest and most intense hurricanes ever observed in the mid-Atlantic, particularly for so late in hurricane season. Due to the increasingly extratropical nature of the storm as it approached the coast, hurricane experts reclassified Sandy as a "post-tropical" storm shortly before landfall. However, in terms of hazards such as wind, rain, and storm surge, Sandy was virtually indistinguishable from a hurricane. In this paper we adopt the descriptor "Superstorm" for Sandy in order to reflect its dual nature.

Figure 1. The track of Sandy from inception in the Caribbean Sea until its demise after extratropical transition (ET) over the northeastern United States. Saffir-Simpson categories are indicated by number and shading. Image adapted from https://coast.noaa.gov/hurricanes/?redirect=301ocm.

Sandy's impacts in the United States were unusually severe and widespread, owing to its nearly 1000-mile swath of gale-force winds near the time of landfall. It was the deadliest tropical cyclone to hit outside of the southern U.S. in 40 years, with 72 U.S. deaths directly attributed to Sandy. A majority of these deaths (41 out of 72, or 57%) were due to storm surge along the Atlantic coast, but 20 of the deaths (28%) were due to falling trees. In addition, 87 indirect deaths were caused by Sandy, mostly related to loss of power during cold weather. At least 650,000 homes were damaged or destroyed in the U.S., approximately half in the state of New York near or along the coastline. So many homes perished because of the exceptional storm surge, aided by a full-moon tide, which reached a record 9.40 feet (2.87 m) above normal at the Battery on the southern tip of Manhattan Island (**Figure 2**). Similar high tides occurred just to the right of Sandy's eye at landfall; before it failed, the Sandy Hook tide gauge recorded a surge of 8.57 feet (2.61 m) above normal. Throughout the region, from New Jersey to Connecticut, barrier islands were inundated and in some cases breached; coastal areas were flooded to depths of several feet.

Figure 2. The Battery Park underpass at the southern tip of Manhattan Island immediately after Sandy (top), and what it normally looks like (bottom). Photo courtesy K.C.Wilsey/FEMA.

Superstorm Sandy caused approximately $50 billion in damage and was the second-costliest hurricane in U.S. history, topped only by Hurricane Katrina in 2005. About 8.5 million people lost electrical power during the storm, most of them in the hard-hit regions of New Jersey and New York; some customers were without power for months. It was the worst disaster in the history of the New York subway system. The flooding and power outages due to Sandy also closed the New York Stock Exchange for 2 days, the NYSE's longest closure in 124 years. It is in the context of this extreme event that we now turn to the public's response on Twitter.

3. DATA AND METHODOLOGY

As a reminder, data for this study are Twitter posts, which are uncensored public utterances on a social media platform. Readers are advised of more-than-occasional strong language that is inevitably included in this narrative.

According to Pew Research, there were over 20 million tweets about Superstorm Sandy from October 27, 2012 through November 1, 2012 [8]. An analysis of that volume of data was beyond the scope of this project. Instead, we chose to isolate subsets of this immense trove of tweets and eventually created our own software to examine, both qualitatively and quantitatively, a sizable subset of the full trove. The intent was to create an intermediate level of breadth of tweets and depth of analysis of the tweets, rather than either to crunch statistics on a huge dataset or else to scrutinize in fine detail a small number of tweets. Our approach merges both statistical analysis with informed qualitative impressions based on the personal reading of thousands of tweets, made efficient via our software.

To this end, a third-party Twitter export service, GNIP Company, was used to acquire the necessary Twitter data needed for this study. The service used the two keywords "sandy" and "superstorm" to sift through all of the tweets posted between October 25, 2012, and November 2, 2012, tag the tweets that contained one or both of those keywords, and export the tagged tweets to two JavaScript Object Notation (JSON) files. A JSON file is a representation of JavaScript Objects in text form. These files are referred to below as "the full dataset" or "the complete dataset," and contain approximately 185,000 individual tweets. To our knowledge this wealth of data makes our research one of the more comprehensive analyses to date of social media during the Sandy event.

Two methodologies were pursued in the course of our research, and are discussed in detail below.

3.1. Sentiment analysis methodology

3.1.1. Data

Initial analysis of the dataset was begun using the open source software OpinionFinder [9], which can identify subjectivity and positive or negative sentiment in phrases. This software is used widely in multiple disciplines [10]. Tweets are classified on a numeric scale with positive tweets set to greater than zero, negative tweets set to less than zero and neutral tweets set to zero. Problems inherent in the OpinionFinder classifying system are described in Ref. [11]. This study assumes that OpinionFinder correctly identifies tweets as positive, negative, or neutral, but also notes the tendency of OpinionFinder to over-classify tweets as neutral. Fortunately, a neutral zero does not skew the data. Numerous other studies have used OpinionFinder analysis of tweets to conduct research, including some work on Superstorm Sandy [12].

A small subset of the full dataset, tweets on Manhattan Island, was examined. The point of this component of the research was to track the evolution of sentiment over time; therefore, classified tweets were divided into equal time intervals. The first analysis uses 17 time intervals of 12 hours each, while the second compares a before-event interval of to a during/after event period. Twelve-hour intervals were chosen for simplicity while maintaining high enough time resolution to denote change. The two 12-hour periods from 1200 UTC 29 October to 1200 UTC 30 October also encompass an appropriate time period of the direct impacts from Sandy on Manhattan. The time of 1200 UTC 29 October was chosen as the mark between pre-event and event based on Manhattan weather observations.

Classified tweets for each time period were then aggregated using ArcMap to census tracts in Manhattan, with each tract taking the sum of the classified tweets (i.e., tracts that have a higher proportion of positive tweets to negative tweets obtain a more positive rank). Aggregating the data points to census tracts made it more manageable and help to smooth out pockets of many tweets versus areas with fewer tweets. Census tracts also help to distribute the data by population, because tracts are roughly similar in population count. The census tracts in Manhattan are small enough in size that it can be assumed residents within the area experienced similar effects from Sandy.

Bias is inherent in using data such as tweets. Here, the database is obviously biased towards those with access to the internet via desktop (and especially access to smartphones, because power outages occurred), and is presumably also biased toward a younger demographic that utilizes Twitter. For this study, we assume that the opinions reflected in the tweets adequately reflect the opinion of others in the area that did not have access to Twitter.

3.1.2. Statistical analysis

The first step in analyzing the evolution of the opinions over time was to determine if there is a statistically significant change. To accomplish this, an analysis of variance (ANOVA) model was run using the RStudio statistical IDE [13]. ANOVA can effectively determine if one or more of the time intervals has a statistically different positive/negative opinion. In our work, the null hypothesis was that no change in opinion occurred over the 9 days.

To further assess an evolution in opinion, a second test was used to determine if there was a difference in opinion between two time periods: before Sandy's impacts were felt in Manhattan and after the effects began to be felt. In order to determine which test is appropriate for measuring the difference between these time pairs, that data were tested for normality using the Kolmogorov-Smirnov (K-S) Goodness-of-Fit test. If the data were determined to be non-normal by the K-S test, then the Wilcoxon Matched Pairs Signed-Ranks test was then used to determine a difference. (The Wilcoxon Matched Pairs Signed-Ranks test is appropriate for determining whether there are significant differences in a pair of non-parametric data such as this.) Again, the null hypothesis was that there was no change in opinion before and after Sandy's impacts. Finally, sums of the classified opinions for each of the 12 time intervals were plotted using R's barplot command. This allowed for visualization of the data and further determines directionality of the alternate hypothesis.

3.2. Tweetreviewer methodology

3.2.1. Data

After our initial OpinionFinder effort, we decided to explore the full dataset in more breadth and depth. The JSON files were then imported to a MySQL database using an import program written in PHP to accomplish this more in-depth analysis.

3.2.2. Quantitative and qualitative analysis

Once the data were in the MySQL database, a tool was needed to aid in reviewing the tweets, sorting them as relevant (pertaining to Superstorm Sandy in some manner) or irrelevant (not referring to Superstorm Sandy) to the project, and bookmarking the tweets of interest. The program TweetReviewer was built by the second author on the .Net 4.0 Framework and written in C# for this purpose (**Figure 3**).

A set of filter words were created and plugged into the program to help determine the relevance of the numerous tweets. The filter words are listed in the Appendix. These filters were used to aid the researchers as they went through the tweet dataset by hand to determine the relevance of the tweets the filters did not tag. The tweets deemed relevant either through the filter tagging process or by hand, were sorted together. The non-relevant tweets were also marked accordingly and separated from the relevant tweets. However, none of the non-relevant tweets were deleted. The entire database of tweets was archived and retained.

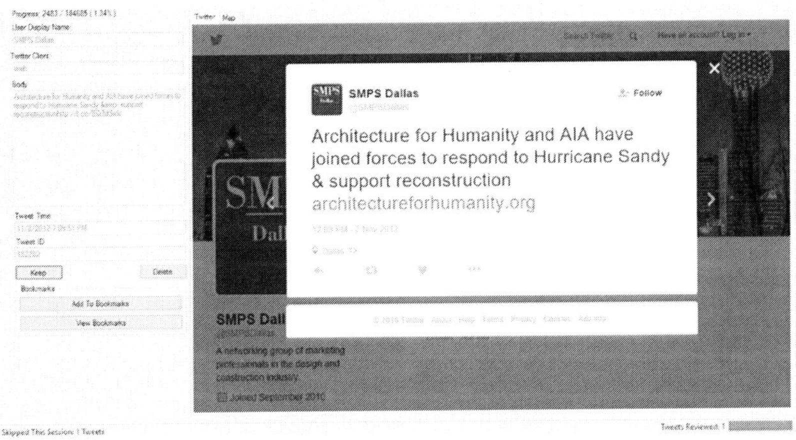

Figure 3. TweetReviewer software graphical user interface.

Once the relevant and non-relevant tweets were separated, new keywords were searched using MySQL's search capabilities to identify the number of times that particular keyword was used. This process helped provide a clearer picture of the perception of Superstorm Sandy through the lens of Twitter.

The posts have been analyzed spatially as well as temporally, with word counts and word clouds as a function of day and time scrutinized as quantitative measures of public responses. We have also examined tweets individually by the thousands using the database, which has permitted the authors to develop qualitative insights into the public response that would be difficult or impossible without simple visualization and bookmarking tools.

4. RESULTS

4.1. Sentiment analysis

Analysis using ANOVA allowed for a rejection of the null hypothesis that no change in opinion occurred over the 9 days examined. ANOVA showed that there was a significant difference between the opinions of the different 12-hour time intervals. As seen in **Table 1**, based on an *F*-value of 3.971 calculated using [14] and a *p*-value of 0.0464, a difference was found between the time intervals that are significant at the 95% confidence level.

The second set of tests included the Kolmogorov-Smirnov (K-S) Goodness-of-Fit test, and Wilcoxon Matched Pairs Signed-Ranks test. Results from the K-S tests, shown in **Table 2**, demonstrate that the data are not normally distributed. *p*-Values for both "before" and "after" data are vanishingly small (<< 0.001), indicating that the null hypothesis can be rejected and a non-parametric test statistic would be more appropriate, such as the Wilcoxon Matched Pairs Signed-Ranks test.

Table 1. ANOVA results for Manhattan Island sentiment analysis. An asterisk indicates that the result is significant at the 95% confidence level.

	Df	**Sum Sq**	**Mean Sq**	***F* value**	**Pr(>*F*)**
Time	1	2.1	2.0868	3.971	0.0464*
Residuals	4792	2518.4	0.5255		

Table 2. Lilliefors (Kolmogorov-Smirnov) Normality test results for the Manhattan Island sentiment analysis.

Before	***After***
$D = 0.2183$, *p*-value < 2.2e−16	$D = 0.1573$, *p*-value < 2.2e−16

The results of the Wilcoxon Matched Pairs Signed-Ranks test give p << 0.001 (**Table 3**). A *p*-value this small indicated that, once again, the null hypothesis can be rejected. We can confidently say that opinions for the time period before Sandy's effects were felt in Manhattan differ from the period during and after effects were felt.

Table 3. Wilcoxon Matched Pairs Signed-Ranks test results for the Manhattan Island sentiment analysis.

Data: *Before* vs. *After*	
V = 18,766	p-value = 9.881e−07

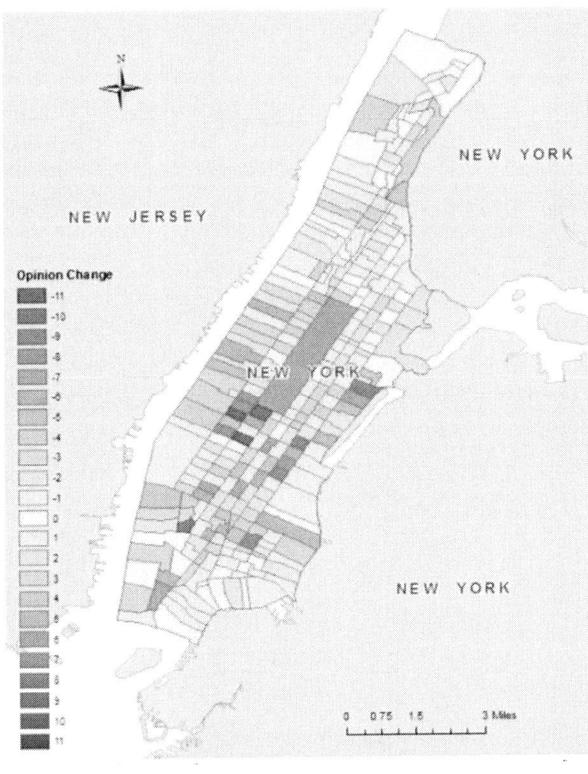

Figure 4. Sentiment analysis of opinion change in tweets on Manhattan Island from before Sandy to during/after Sandy's impact on the area.

Figure 4 demonstrates this change geographically across Manhattan Island. The most negative changes occurred south of Central Park, broadly consistent with the region that experienced a power blackout due to Sandy [15]. Lesser negative trends, and some positive trends, were found most often north of Central Park, further from the blackout and from the Atlantic coast.

In addition, sums of opinions were plotted over time using R's barplot command. The resulting diagram (**Figure 5**) allows a clear visualization of the evolution of opinions over time. The bar graph shows that overall opinions through 1200 UTC on 28 October tended to be positive, with the exception of slightly negative tweets from 1200 UTC to 1200 UTC on 26 October–27 October. After 1200 UTC 28 October, tweets became much more negative, reaching their negative peak at 1200 UTC on 29 October through 1200 UTC 30 October. Tweets remained negative in sentiment until the end of the period.

Overall, the results indicate that public opinion of Sandy did change in the Manhattan area. It appears that public opinion, expressed through tweets, was more lighthearted before effects from Sandy were felt in Manhattan. This correlates with observed skepticism from many in the path of the storm prior to actual impacts (see the following section). Interestingly, tweets become more negatively skewed 24 hours prior to actual impacts to the local Manhattan area. This could be related to non-meteorological impacts from Sandy that occurred prior to the storm's direct impacts. As stores, subways, and other services closed, public opinion may have begun to shift. These early negative tweets could also be the result of evacuation orders, or news of destruction as Sandy swept up the East Coast. Further tests would be needed to determine the impacts of these and many other variables. It is clear, however, that the ratio of positive to negative opinions became most negative at the time when Sandy's impacts were being felt in Manhattan (1200 UTC–1200 UTC 29–30 October).

4.2. Tweetreviewer analysis

The results uncovered by sentiment analysis failed to capture other insights that the researchers felt were important but were outside the scope of OpinionFinder. This led to the creation and use of TweetReviewer software to allow us to examine, both quantitatively and qualitatively, other aspects of the full database.

While analyzing the full database, it was determined there were two major common categories of interest with regard to the public's perceptions of Superstorm Sandy: scientific misconceptions, and an anthropomorphic reconception of Sandy that, in the parlance, "went viral."

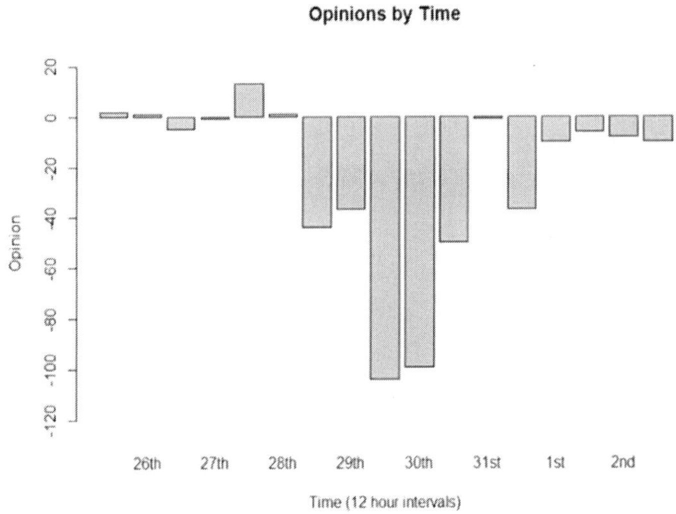

Figure 5. Temporal analysis of sentiment (positive or negative) on Manhattan Island in 12-hour increments from 25 October 2012 through 2 November 2012.

4.2.1. Misconceptions

There were many misconceptions regarding Hurricane Sandy and hurricanes in general in the dataset. One quotidian misconception was the common misspelling of the word "hurricane." In fact, the word "hurricane" was misspelled so often that the list of filtered words had to be adapted to accommodate the many misspellings (e.g., see #51 in Appendix). More substantively, there was a general lack of understanding of what defines a hurricane. These misconceptions can be separated into four categories: *strength, category, size*, and *duration*.

The strength of a hurricane is most commonly associated with the Saffir-Simpson Hurricane Wind Scale categories. Overall, the Twitter users had a fairly decent grasp of the numerical scale as far as the numbers were concerned, but the understanding of what those numbers stood for was almost entirely absent. Throughout the dataset (**Table 4**) were tweets, many from Florida, that downplayed Hurricane Sandy claiming there was little to worry about because it was "only cat 1." Instead of acknowledging the force of the hurricane, users discriminated against the lower numbers even though the Saffir-Simpson Scale is based on wind only. It does not take into account the damage from other impacts like the storm surge and rainfall, and neither did the Twitter users. A few tweets mentioned Hurricane Sandy as having a storm surge more often associated with a

Category 3 or Category 4 hurricane, such as this (tweet #76357 in the full database):

"Irene was a category 3. Sandy is a 1 but with the storm surge it's supposed to act as a 4 at most."

Table 4. Tweet word counts related to hurricane strength.

	Category 1	**Aftermath**	**Apocalypse**	**Worse**	**Power**
25 Oct. 2012	2	0	1	6	10
26 Oct. 2012	13	1	4	28	28
27 Oct. 2012	5	1	4	21	55
28 Oct. 2012	17	7	343	61	182
29 Oct. 2012	26	8	1136	213	1326
30 Oct. 2012	14	212	398	220	1748
Total	**77**	**229**	**1886**	**549**	**3349**

The size and duration of Hurricane Sandy was also frequently misunderstood (**Tables 5** and **6**). Very few tweets implied that the users fully grasped the sheer size of any hurricane, let alone the immense size of Sandy. They appeared to assume that a hurricane was a small storm that would be bad, much like a supercell or "derecho-sandy thing" (tweet #4885) that would "touchdown" or "touch land" (tweet #100800), wreak havoc, close schools, and leave, all within a matter of hours. There were several tweets that asked if the storm was "over yet?" Hardly any of the Twitter users seemed to grasp the fact that a hurricane is actually a huge storm spanning hundreds of miles which can last for days. Some tweets even called Hurricane Sandy a tornado (tweets #38455 and #49032) or, as one CEO put it, a "tornadocaine" (tweet #5267).

Table 5. Tweet word counts related to hurricane size.

	Tornado	**Touchdown**	**Landfall**	**How Long**
25 Oct. 2012	13	0	10	0
26 Oct. 2012	21	0	11	1
27 Oct. 2012	33	1	10	1
28 Oct. 2012	126	7	22	2
29 Oct. 2012	213	4	142	13
30 Oct. 2012	170	2	43	10
Total	**576**	**14**	**238**	**27**

Table 6. Tweet word counts related to hurricane duration.

	Over/ov er yet	**Fini sh**	**Final ly**	**Bring it**	**Passed/ past**
25 Oct. 2012	28	1	0	2	2
26 Oct. 2012	69	2	3	13	3
27 Oct. 2012	76	2	6	35	6
28 Oct. 2012	244	17	17	112	22

	Over/ov er yet	Fini sh	Final ly	Bring it	Passed/ past
29 Oct. 2012	784	38	60	191	60
30 Oct. 2012	785	30	67	23	82
Total	**1986**	**90**	**153**	**376**	**175**

Among many visual misconceptions propagated during Superstorm Sandy, one of the most prominent and widely shared was a Photoshopped image of a supercell thunderstorm over the Statue of Liberty in New York City (e.g., tweet #60774). The supercell thunderstorm was indicated to be Sandy—another sign of confusion regarding the vast differences in size, strength, and duration between hurricanes and tornadic thunderstorms. This misconception became so popular on social media that a story about it appeared in the Philadelphia media [16].

4.2.2. The reconception of sandy

As Superstorm Sandy approached the mid-Atlantic coast, a number of Twitter users began to do something peculiar, at least from the perspective of scientists or emergency managers. Instead of relaying factual information regarding the storm, accounts pretending to be the personification of Hurricane Sandy started appearing, as venues for posting jokes about the hurricane from a first-person perspective. With the help of these accounts, the Twitter community took the idea of personifying Sandy and expanded upon it. Without any visible evidence of premeditation regarding the nature of this anthropomorphic Sandy, the Twitter community banded together and simply accepted a persona of their creation without question. They also hurled insults at the hurricane based on its fabricated persona and its perceived status as a female.

Sandy was suddenly no longer an impersonal, inanimate hurricane; she was an "independent sassy black hurricane who don't need no man" (tweets #13365 and 29 subsequent tweets), who grew up in the ghetto, went to school with (Hurricane) Irene, cursed up a storm, voted Democratic, knew how to work the pole, had an avid and colorful sex life, and was the brunt of many cruel taunts related specifically to female anatomy. **Table 7** provides a sampling of word counts related to this persona in the days leading up to landfall.

Table 7. Tweet word counts related to the invented persona of Sandy.

	Bitch	**Whore**	**Twerk**	**Pole**	**Shark**	**Black**	**Nigga**
25 Oct. 2012	8	0	0	0	2	0	0
26 Oct. 2012	35	2	0	1	0	1	7
27 Oct. 2012	61	3	4	0	0	8	3
28 Oct. 2012	253	17	25	2	6	22	29
29 Oct. 2012	614	41	42	16	32	96	68
30 Oct. 2012	382	31	14	21	74	168	39
Total	**1353**	**94**	**85**	**40**	**114**	**295**	**146**

The various Twitter accounts created to portray this user-crafted persona encouraged its incredibly fast proliferation and popularity across the website. They posted tweets which consisted mainly of rather dirty jokes, inappropriate suggestions, racial slurs, female-specific insults, and Republican-specific insults. One of the aforementioned tweets talked about the Sandy persona tossing a trailer at a woman in a minivan simply because she had a bumper sticker of a Republican presidential candidate on her car (tweet #21587).

It is possible that the timing of the hurricane's landfall, within a week of the 2012 presidential elections when tensions between the Democratic and Republican parties were already high, amplified the politically slanted comments. But the outright cruelty and twisted content went beyond simple political reasons. To make matters worse, these offensive tweets were shared hundreds of times by users who were apparently from a variety of ages, races, and political beliefs.

The tweets referring to the Hurricane Sandy persona consistently referred to her as a "bitch" and joked about her coming to "blow" the entire east coast and make everyone "wet" or simply told her to "fuck off." This was such an exceedingly common theme that the words "bitch," "blow," "wet," and "fuck" became keywords for both the filters during the process of going through the tweets as well as for searching their word counts. Of these four words, "wet" was used the least common, only garnering a word count of 544 times used. The word "blow" was used 1129 times, "fuck" was used 3655 times, and "bitch" led all epithets in the full database with a word count of 4335 times used with regard to Hurricane Sandy.

These word counts are, of course, small compared to the total number of words used in the full database. To give a sense of how the persona of Sandy dominated the "Twittersphere," **Table 8** presents the fractional representation of scientific/hurricane-related terms in the full database versus the top three persona terms. As shown in the table, the persona was many times more popular on Twitter than were factual reports about Sandy.

Table 8. Ratio of word count of scientific terms to the word counts of the three most common Sandy-persona terms.

NWS	MPH	Category/cat	Landfall	Windspeed
0.022	0.133	0.387	0.079	0.0001

5. CONCLUSIONS

Our results shed light on public perceptions, misconceptions, and reconceptions of an extreme atmospheric hazard. Superstorm Sandy was a virtually unprecedented event along the Atlantic coast of the United States, in terms of intensity, size, and path. What did the public make of this event on Twitter?

From our analysis of Manhattan Island data, it is plausible that residents underrated the storm's ferocity until the last 24 hours before landfall. The largest sentiment swings occurred, quite naturally, in and near the regions most affected by the storm: the coastline and the south Manhattan neighborhoods blacked out due to a power plant failure.

Our more fine-grained analysis using the TweetReviewer software revealed additional aspects of the public's reaction to Sandy. Profound confusion exists regarding the size, strength, and duration of hurricanes. The public seems to confuse hurricanes with tornadoes; this confusion during Sandy extended to the Photoshopped image of a supercell thunderstorm over the Statue of Liberty. Expectations that Sandy would be as brief and intense as a tornado were not met; in particular, the unusually large extent of Sandy (the largest storm in terms of diameter of gale-force winds since records began in 1988; see [7]) was not well understood by those on Twitter. The more subtle point that Sandy could be "only a Category 1" and still do extensive damage due to storm surge was also not grasped.

What the "Twittersphere" did seem to eagerly grasp was a user-generated anthropomorphic "Sandy" who dealt out death and destruction like a villain in a superhero comic or movie. This personified "Sandy" was then made the object of race-based and gender-based slurs that were widely perceived to be amusing, rather than offensive if they had been made in public. This created an increased level of noise that tended to drown out the true signal: reports of important developments in Hurricane Sandy's category changes, the watches, advisories, and warnings, and tweets with legitimate scientific information that could have better informed the public. Our results thus align more with those of [5] than with the more positive, life-saving impacts of Twitter found by other researchers. Our analysis of geo-located tweets may bias our results somewhat in this respect, however (R. Morss, pers. comm., January 2016).

When much of New York City lost electricity, the theme of attacking the hurricane based on its manufactured persona quickly fell by the wayside in favor of complaining about the power loss and realizing Sandy was in fact a hurricane and not a fictitious creation. The persona continued to play a substantial role in the dataset, but the newest point of interest was the loss of power. In fact, the word "power" was found 8649 times in our database, making it more popular than "bitch."

We conclude, with some surprise, that until the hurricane interfered directly with people's personal lives, Twitter users seemed content with obsessing over the invented persona of Hurricane Sandy. Rather than bemoaning this flight from reality, however, we encourage a more proactive response among emergency management personnel, meteorologists, and others who communicate directly with the public. Perhaps it is possible that this behavior can be utilized for the benefit of the public. If the public could quickly create and propagate a persona for a hurricane, it stands to reason that official outlets could do the same. They could create a Twitter account for a new hurricane, create their own persona for it, and use this to disseminate the important relevant information to the public in a format that is more easily digestible than more esoteric scientific criteria.

For example, a large swath of the U.S. population is familiar with superheroes on some level, whether DC, Marvel (or for even younger audiences, Pokémon). What if scientists began using popular superheroes to help describe the strength of a storm? As outlandish as it sounds, this is something the general public would grasp fairly easily, simply because they are already familiar with the characters. For instance, intense heat could be described as being on par with an attack by Marvel's Human Torch, a member of the Fantastic 4 superhero group. A powerful electrical storm could be compared to Marvel superhero Thor wielding Mjolnir in battle. The incredible storm surge of a hurricane such as Sandy could be compared to the DC supervillain Ocean Master or New Wave in combat. Similar translations of specific hazards into personae would be possible in Pokémon, reaching even younger audiences.

These examples may come across as childish, but the public would easily grasp a general idea of the intensity of the heat, the electrical storm, and the storm surge from them. More importantly, children would easily comprehend these examples. A parent can disregard weather alerts, but not if their children continually bother them. If this method of using pop-culture references to explain something as confusing as weather can bridge the communication gap to the next generation, then they could learn to listen to and obey weather alerts. This "if you can't beat 'em, join 'em" approach to anthropomorphizing atmospheric hazards could capture some of the social media energy that might otherwise propel a completely non-factual (and offensive) personification to prominence, as occurred with Sandy.

Finally, we advocate the use and/or development of software such as our TweetReviewer as a means for visually inspecting thousands of tweets easily and efficiently. As our research indicates, actually reading the tweets provides insights that are unlikely to be gained by mere statistical crunching on datasets. The filtering capabilities of TweetReviewer enable the user to focus on relevant tweets and screen out non-relevant tweets, significantly accelerating the process and permitting human analysis of relatively large datasets.

ACKNOWLEDGEMENTS

The first author gratefully acknowledges the support of an M.C. Michael Award from the University of Georgia, which funded much of the research presented here.

APPENDIX

Filter words used in our TweetReviewer software to narrow the database to the most relevant posts are:

1) "red cross" or "redcross"
2) "#sandy" if either "springs" or "beach" is not present.
3) "#hurricanesandy" or "#hurricainesandy"
4) "hurricane sandy"
5) "jersey shore"
6) "#njsandy"
7) "hurricane death megatron"
8) "sandy aftermathpocalypse"
9) "halloweenpocalypse"
10) "#postsandy"
11) "#aftersandy"
12) "#fucksandy"
13) "#fuckyousandy"
14) "#fusandy"
15) "#fyousandy"
16) "fuck sandy"
17) "fuck you sandy"
18) "#ctsandy"
19) "#bptsandy"
20) "#nhsandy"
21) "#dcsandy"

22) "#wvsandy"
23) "#nysandy"
24) "#ny1sandy"
25) "#nycsandy"
26) "#masandy"
27) "#vasandy"
28) "#mdsandy"
29) "#vtsandy"
30) "#tosandy"
31) "#6abcsandy"
32) "#lifeaftersandy"
33) "storm sandy"
34) If the tweet contains both "Sandy" and "Katrina"
35) If the tweet contains both "Sandy" and "Irene"
36) "#thankshurricanesandy"
37) "#thankssandy"
38) "#thankyousandy"
39) "#damnyousandy"
40) "#damnitsandy"
41) "post-sandy"
42) "post sandy"
43) "#frankenstorm"
44) "superstorm"
45) If the tweet contains both "Obama" and "Fema"
46) "fema"
47) "#damnsandy"
48) "#darnsandy"
49) "#ihatesandy"
50) If the tweet contains both "prayers" and "affected by sandy"
51) "@hurricannesandy"

52) "@ahurricanesandy"

53) "@sandydahurricane"

54) "@sandyshurricane"

55) "#survivingsandy" or "#survingsandy"

56) "#stormsandy"

57) "sandy storm"

58) "Frankenstorm"

59) "hurricane"

60) "bitch sandy"

61) "#pray"

62) "#hurricane"

63) If the entire tweet is "superstorm sandy xuo"

REFERENCE

1. Twitter. [Internet]. 2016. Available from: https://www.twitter.com [Accessed: 2016-04-03]
2. Statista, The Statistics Portal. [Internet]. 2016. Available from: http://www.statista.com/statistics/282087/number-of-monthly-active-twitter-users/ [Accessed: 2016-04-03]
3. Edwards JF, Mohanty S, Fitzpatrick P, Martin D. Assessment of social media usage during Hurricane Sandy and the development of a Twitter-based, web application for real-time communication of storm-related information. 96th American Meteorological Society Annual Meeting. New Orleans, LA, 11 January 2016. https://www.hpc.msstate.edu/publications/docs/2016/01/14285social_media_tool_storm_info_ams.pdf [Accessed 2016-04-03]
4. Chatfield AT, Scholl HJ, Brajawidagda U. #Sandy Tweets: Citizens' co-production of time-critical information during an unfolding catastrophe. Proceedings of the 47th Hawaii International Conference on System Science, IEEE Computer Society, Waikoloa, Hawaii. 2014. pp. 1947-1957.
5. Lachlan KA, Spence PR, Lin X, Del Greco M. Screaming into the wind: Twitter use during Hurricane Sandy. Communication Studies. 2014; 65 (5): 500-518.
6. Demuth JL, Morss RE, Palen L, Anderson K, Anderson J, Kogan M, Stowe K, Bica M. Understanding dynamic communication, risk perception, and decisions during Hurricane Sandy through analysis of Twitter data, Society for Risk Analysis annual meeting, Arlington, VA, 9 December 2015.
7. Blake ES, Kimberlain TB, Berg RJ, Cangialosi JP, Beven III JL. Tropical cyclone report: Hurricane Sandy (AL182012) 22-29 October 2012. [Internet]. 2013. Available from: http://www.nhc.noaa.gov/data/tcr/AL182012_Sandy.pdf [Accessed: 2014-04-03]
8. Pew Research Center. [Internet]. 2013. Available from: http://www.pewresearch.org/fact-tank/2013/10/28/twitter-served-as-a-lifeline-of-information-during-hurricane-sandy/ [Accessed: 2016-04-03]

9. OpinionFinder. 2.x Release Page [Internet]. 2011. Available from: http://mpqa.cs.pitt.edu/opinionfinder/opinionfinder_2/ [Accessed: 2016-04-03]
10. Bollen J, Mao H, Zeng X-J. Twitter mood predicts the stock market. Journal of Computational Science. 2010; 2 (1): 1–8. doi:10.1016/j.jocs.2010.12.007
11. Wilson T, Hoffman P, Somasundaran S, Kessler J, Wiebe J, Choi Y, Cardie C, Riloff E, Patwardhan S. OpinionFinder: A system for subjectivity analysis. Proceedings of Human Language Technologies Conference/Conference on Empirical Methods in Natural Language Processing (HLT/EMNLP 2005), Vancouver, Canada, 2005. pp. 35-36.
12. George Mason University. The GeoSocial Gauge [Internet]. 2016. Available from: http://geosocial.gmu.edu/projects/superstorm-sandy/ [Accessed: 2016-04-03]
13. R Core Team. R: A language and environment for statistical computing. R Foundation for Statistical Computing, Vienna, Austria, 2015. URL https://www.R-project.org/.
14. Free Statistics Calculators. Version 4.0 [Internet]. 2016. Available from: http://www.danielsoper.com/statcalc3/calc.aspx?id=4 [Accessed: 2016-04-03]
15. New York Magazine. From The Editors [Internet]. 2012. Available from: http://nymag.com/nymag/letters/hurricane-sandy-editors-letter-2012-11/ [Accessed: 2016-04-03]
16. Mathis J. That Awesome Hurricane Sandy Photo of the Statue of Liberty? Fake [Internet]. 2012. Available from: http://www.phillymag.com/news/2012/10/29/awesome-hurricane-sandy-photo-statue-liberty-fake/ [Accessed: 2016-04-03]

CHAPTER 10

New estimates of potential impacts of sea level rise and coastal floods in Poland

Dominik Paprotny[1] and Paweł Terefenko[2]

1 Department of Hydraulic Engineering, Faculty of Civil Engineering and Geosciences, Delft University of Technology, Stevinweg 1, 2628 CN Delft, The Netherland

2 Remote Sensing and Marine Cartography Unit, Faculty of Geosciences, University of Szczecin, Mickiewicza 18, 70-383 Szczecin, Poland

ABSTRACT

Polish coastal zone is thought to be one of the most exposed to sea level rise in Europe. With mean sea levels expected to increase between 28 and 98 cm by the end of the century, and storms increasing in severity, accurate estimates of the consequences of those phenomena are needed. Recent advances in quality and availability of spatial data in Poland made possible the reassessment of previous estimates of inundation caused by sea level rise. Up-to-date, detailed information on land use, population and buildings was used here to calculate their exposure to floods at a broad range of scenarios. Inclusion of a high-resolution digital elevation model contributed to a further improvement in estimates. The results revealed that even by using a static "bathtub fill" approach, the amount of exposed land, population or assets is significantly smaller than indicated in previous assessments. In the perspective of the twenty-first century, direct damages caused by sea level rise will be small and adaptation costs will not be significant. However, the increase in the frequency of storm surges could elevate the risk to the population and economy, but cost-effective flood protection measures would be able to mitigate the risk. The exposure of different kinds of assets and sectors of the economy varies to a large extent, though the structural breakdown of potential losses is remarkably stable between scenarios

Keywords: Sea level rise, Coastal floods, Poland, Flood risk, GIS

1. INTRODUCTION

Storms surges are an important factor shaping the Polish coast. Short-term water level variations caused by them significantly alter the coast in the non-tidal Baltic Sea, where extreme water levels depend largely on the volume of water flowing in from the North Sea. Several large coastal floods occurred in late nineteenth and early twentieth century, with a few calmer decades thereafter. More recently, however, storms have been on the rise: the number of surges exceeding the warning level of approximately 70 cm above mean sea level (MSL) soared from about two per year in the 1950s to six per year in 2000s (Wiśniewski and Wolski 2009). Usually these events are too short to cause damages inland. However, during a storm surge in January 1983 water levels exceeded 1.3 m above MSL along the most of the coast, affecting many locations, particularly in the eastern part. In the Żuławy region, a total of 50 km^2 of land was flooded and almost 1300 people were evacuated (Bednarczyk et al. 2006). Wiśniewski and Wolski (2009) estimated that a 100-year coastal flood could be as high as 1.95 m above average in Kołobrzeg, while a 1000-year flood might reach 2.5 m; the highest value actually recorded was 2.22 m in 1872.

Long-lasting storm surges, even though relatively insignificant at the coast, can cause a flood dozens of kilometres inland, on the low-lying parts of Odra and Vistula fluvial plains. In those estuaries, water levels could even be higher than at the coast; this happens when large volumes of water flow in from the sea due to the northerly winds (Wiśniewski and Kowalewska-Kalkowska 2007). The highest recorded water level in Szczecin, located almost 70 km inland, was 1.54 m above average during a storm surge in 1913 (Richter et al. 2012). The most recent storm surge resulting in a large inflow of seawater into the river network occurred in 2009, causing losses of several million euro in locations all over the coast, mainly in the eastern part (KZGW 2011). Across the whole region, flood defences protect the lowest areas, particularly around river mouths and coastal lakes, while the rest of the coast is protected by nature.

At a longer timescale, the contemporary sea level rise (SLR) triggered by the changing climate is also a threat to coastal areas. It is generally considered (Church et al. 2013; Hunter 2010; Wróblewski 1994) that the increase in mean sea level will directly translate into higher storm surge maxima. It is therefore necessary to assess the number of persons and value of assets which could be at stake. In Poland, comprehensive research on coastal flood risk considering sea level rise was originally spawned by IPCC's first assessment report released in 1990. Rotnicki and Borówka (1991) used 1:25,000 topographic maps to calculate the area threatened by rising waters along with a quantification of assets affected, such as land use, infrastructure, settlements and their population. According to an expanded version of that study (Rotnicki et al. 1995), a 2.5-m SLR would impact 2700 km^2 of land, including 93 km^2 of settlements, 374 km of railways, 906 km of roads and 297,000 people. Based on the results of Rotnicki's and Borówka's (1991) study, Pluijm et al. (1992) calculated the monetary value of land and structures exposed to sea level rise. The results indicated that a 1-m rise of mean sea levels would result in a loss of assets worth 24 % of 1990 gross domestic product (GDP). This study was later updated with newer economic data by Zeidler (1997), whose results were in turn reused by Pruszak and Zawadzka (2005, 2008) and cited in several global assessments of sea level rise (e.g. Nicholls and Mimura 1998; Nicholls and Klein 2005). In effect, published assessments of the effects of sea level rise are limited to delimitations of hazard zones made in late 1980s (the earliest version of the study by Rotnicki and Borówka was published in 1989) and socio-economic data from the early 1990s. Poland was included in studies on global and European flood risk based on DIVA model—most recently by Hinkel et al. (2014)—though only Hinkel et al. (2010) and Bosello et al. (2012) included a breakdown of impacts by country. All of these studies used only low-resolution data and were characterized by a high degree of generalization. For instance, only a most recent analysis (Mokrech et al. 2015) included flood defences in the calculation of flood hazard in Europe and only using assumptions of protection levels. Still, no information on country-level impacts was provided.

More recently, due to the obligations imposed by the European Union's "flood directive" (EU 2007), the national government ordered preparation of flood risk maps for selected regions of Poland. They were published in 2013 by the National Water Management Authority (KZGW 2015). Their main advantage is the use of up-to-date, detailed and high-resolution data, including lidar-derived digital elevation models (DEMs), aerial photography, land surveys and administrative registers. However, the main focus has been put on river floods, while only parts of the coast were investigated. Moreover, no sea level rise impacts were assessed, while the analysis of storm surges was limited only to one or two scenarios, depending on location. Finally, the results of flood risk calculations are provided only as a graphical representation on maps, and the only aggregate statistics provided are the estimates of exposed inhabitants per settlement. The total value of exposed assets was not disclosed.

In the light of the above, there is clearly a need for a high-resolution assessment of possible social and economic impacts of storm surges and inundation of land caused by sea level rise in Poland. This is particularly important as the existing studies all indicated that very severe consequences for the population and economy might occur in the twenty-first century. In this paper, we aim to provide more precise estimates of exposure of land, population and assets at a broad range of scenarios, using the up-to-date and detailed cartographic materials, which have become available only recently. Most importantly, the digital elevation model used here includes the majority of flood defences in the area, which were missing in other studies. The main aim is to provide a dataset, which could be used to various, frequently changing, projections of climate, population and economic growth. Furthermore, we analyse the data in the context of IPCC projections, as well as discuss the adaptation measures.

2. MATERIALS AND METHODS

2.1 Regional setting

The Baltic Sea is a non-tidal, semi-enclosed and shallow body of brackish water. Exchange of water through the Danish Straits is the primary regulator of water levels in the basin, causing dangerous surges primarily at its southern and eastern coasts (Ekman 2009; Wolski et al. 2014). Poland has a 500-km-long, diversified coastline. At various locations along the coast, there are sections of cliffs totalling ca. 18 % of entire coastline length (Tomczak 1995). Apart from an alluvial section in the Puck Lagoon (less than 3 % of the total), the remainder is a spit- and barrier-type coast, with dunes ranging in height from less than 2 up to 49 m. Behind the dunes, coastal plains with occasional depressions (down to −1.8 m MSL) are usually observed.

Tide gauge in Świnoujście, at mouth of Odra Estuary in the Baltic Coast, has one of the longest instrumental records in the world (Permanent Service for Mean Sea Level 2015), and during 1811–2006, a trend of 0.4 mm per year was observed here, accelerating to 1.0 mm per year during 1947–2006 (Wiśniewski et al. 2011). This rate rises moving eastward: Kołobrzeg (central part of the Polish coast) recorded an increase of 0.5 mm during 1901–2006 and 1.4 mm during 1947–2006, while in Gdańsk (eastern coast), 1.6 mm increase was observed during 1886–2006 and 2.5 mm during 1947–2006. The difference is largely due to uneven isostatic movement of the crust. Global mean sea level has risen by 21 cm between 1880 and 2009 (Church and White 2011), or 1.6 mm per year. Also, satellite altimetry reveals that the rise in water levels is mostly uniform in the southern Baltic Sea. It amounted to 3.2 mm per year between 1992 and 2016, a pace very similar to the world ocean average, which was 2.9 mm per year during 1992–2016 (NOAA 2016).

Although as much as 7 % of world population lives within 20 km from the coast (Kummu et al. 2016), the Polish Baltic Sea coast itself is not densely inhabited: only 2.8 % of Poles reside in municipalities with direct access to the sea. Excluding the large Tricity agglomeration (Gdańsk, Sopot, Gdynia), the population density in this area is a mere 90 persons per km^2 which is quite low when compared to the national average of 123 (CSO 2015). However, the coast is by far the most popular vacation destination in Poland. Thirty-five percentage of all overnight stays in accommodation establishments during July–August 2013 occurred in 55 municipalities considered by Eurostat as coastal regions (CSO 2015). That corresponds to an average of 71 tourists per 100 residents during the summer. The economies of many coastal resorts are completely dependent on this short holiday season (Łonyszyn and Terefenko 2014). Any damages to infrastructure and buildings could therefore dampen the economic prospects of those localities.

The study area, as referred to hereafter, includes all municipalities, not necessarily bordering the Baltic Sea, in which there is some land laying no more than 5 m above mean sea level. There are 86 such municipalities, including 22 that further divide into a town and a rural area. That sums up to 108 basic administrative units with a combined population of 2.4 million (at 2011 census according to CSO 2015). Figure 1 outlines the study area within the Polish coastal zone.

2.2 Data sources

Two main sources of data were used in this study: the topographic objects database containing information on land use as well as buildings, and a digital elevation model (DEM). Both were derived from national cartographic repository (CODGiK 2014, 2016), which include recent output of large-scale mapping projects. Statistical data on demographic and economic indicators have also been collected.

2.2.1 Digital elevation models

Quality of flood hazard analyses depends mainly on detailed information on the terrain in question. Here, a digital elevation model created through airborne laser scanning technology (lidar) was used. DEMs obtained using this method are commonly applied to coastal flood hazard analyses because of their accuracy (Webster 2010). In recent years, the national government launched a large measurement campaign specifically to provide data for flood hazard mapping, which was carried out in the coastal areas between 2010 and 2013 (mainly in 2011). The density of scanning was usually 4 points per m^2, except for urban agglomerations, where the density was 12 points per m^2. The final dataset in raster format has a spatial resolution of 1 m, which is fine enough to include flood defences and most of other small topographic features that can have a profound impact on flood hazard analyses. Average vertical error of the dataset is nominally less than 20 cm in areas with scanning density of 4 pts/m^2 and less than 10 cm in areas scanned with 12 pts/m^2 (CODGiK 2014). Yet, the DEM has some other imperfections. A large low-lying area, which in reality is protected by higher terrain or flood defence structures, can be marked as potentially flooded if a minor error in the data creates a "gate" for floodwater. Also, some of the hydraulic structures used to manage river flow and maritime transport, such as locks, sluices and weirs, are mostly not present in the DEM.

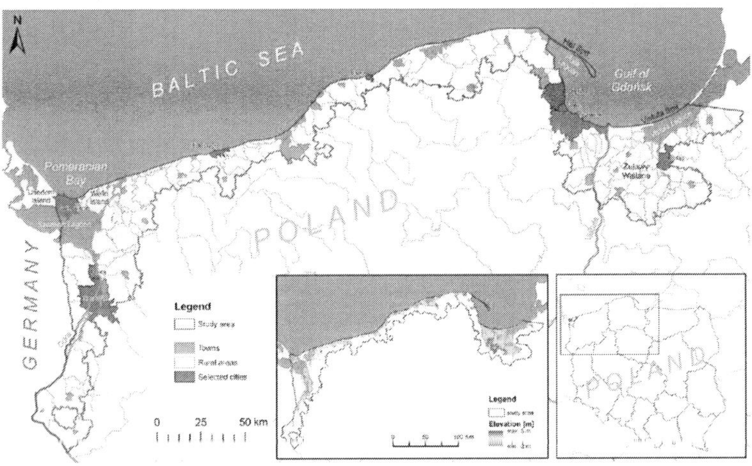

Figure. 1 Study area (all municipalities with some land lying below +5 m MSL) and the Polish coastal zone. The *inner map* presents elevation values

A very small area located in the western part of the coastal zone is not covered by the detailed lidar DEM. The gap was filled with a different DEM, which was created on the basis of aerial photographs. It has a spatial resolution of 40 m and a vertical accuracy of about 1.5 m. These are much worse values than for the lidar DEM, but it had to be used only for less than 0.1 % of the exposed area, so it has a negligible impact on the results.

2.2.2 Topographic objects database

The primary source of information on land use and buildings used here is the topographic objects database, the digital successor of analogue topographic maps, created and maintained by Poland's surveyor general (CODGiK 2016). It consists of almost 300 types of objects with a vectorized representation of their geometry and additional qualitative and quantitative descriptors. Accuracy of the database in terms of location and minimum size of objects approaches a 1:10,000 topographic map. This is enough to represent the natural and socio-economic environment down to a single parcel, road and building. The current structure of the database was implemented in 2011 (Law 2011/279/1642), and its content was updated during 2012–2013. The database used here corresponds to the state of January–July 2013, depending on location. Quantitative information on the objects, such as their area, is expected not to deviate from real values more than ±20 %.

Data on buildings include their area, functional characteristic and, in most cases, the number of storeys. In case the latter is missing, the building was assumed to have only one storey. Roads are represented in the database as a linear object, but for each section, their width is given, so that they could be transformed into a surface. Only paved roads were included in the calculation. Railway and tramlines were transformed into polygons according to the gauge of the tracks. Roads and tracks located on bridges or similar above-ground structures were excluded.

2.3 Calculation of flood hazard

2.3.1 Inundation

Flood hazard analysis involves calculation of the inundation extent and other hydrological parameters of the flooding, such as water depth. To calculate the flood zone, a "bathtub fill" approach was applied for this study (Bates and De Roo 2000; Poulter and Halpin 2008). It is assumed that the sea at a certain scenario will cover all land lying below the assumed water level, as long as there is a direct connection with the source of flood. For the analysis of inundation caused by sea level rise, that is an accurate description. However, for storm surges it is a straightforward simplification. The method does not take into account the kinematics of the flow or sedimentation processes and considers flooding as an instantaneous process. The temporal change in water levels is also an important factor determining the inundation extent. Unfortunately, the data required to include time in such a calculation were not available to the authors. In effect, the storm surge scenarios constitute a description of the worst-case event (Apel et al. 2009; Breilh et al. 2013). The biggest advantage of the chosen method is a possibility of a direct comparison of achieved results with previous studies performed in the Baltic coast. The possibility of directly implementing the method in the geographical information systems is also useful. On the other hand, dynamic effects such as erosion and sedimentation were not included.

The resolution of the DEM is fine enough to include flood defences; therefore, isolated locations, i.e. having no direct connection with the sea, lying below the water level at a given scenario were considered protected by the surrounding high terrain. These areas were discarded to create the final flood zone using a four-side rule, which means that no water flow was allowed in diagonal directions (Poulter and Halpin 2008). Naturally, dikes and other structures could fail and flood the hinterland, but that aspect was not considered in this article. Also, the situation when water floods the land behind structures through culverts was not analysed. Protection from flood defences such as dikes, banks and other earthworks, as well as sluices and weirs is included in the calculation.

The GIS analysis itself was performed using the ArGIS 10 software. Raster datasets with information on elevation (CODGiK 2014) were combined to generate a complete grid of the study area. The grid was then reclassified into two distinct classes: above and below the analysed sea level. The result was a two-coloured map showing the land and the inundation. The latter class was transformed into a vector polygon, and its connection with the sea was investigated by selecting only those polygons that are intersecting with the water at normal level. The resulting polygons were clipped with a mask representing Polish borders in order to delimit the final area at country level. Further analysis included intersections between inundated area for each water level and different layers representing buildings, land types, etc. extracted from the topographic objects database (CODGiK 2016) as well as grouping the results into different systematic classes according to administrative divisions.

2.3.2 SLR and storm surge scenarios

Sea level rise and storm surges were analysed in intervals of 5 cm, providing scenarios of 5, 10, 15, etc., cm up to 5 m above mean sea level. In this way, the exposure of population and assets at different elevations could be assessed in detail. The 5-m value was chosen as the rounded value of the maximum possible SLR indicated in the literature (1.4–2 m in Rahmstorf 2007; Pfeffer et al. 2008) together with a 1000-year coastal flood (about +2.5 m MSL according to Wolski et al. 2014). Mean sea level along the Polish coast varies slightly, with 1947–2006 average in Gdańsk being 7 cm higher than in Świnoujście (Wiśniewski and Wolski 2009). As a minor simplification, the 500 cm "baseline" level of all Polish tide gauges was used here as MSL, which corresponds to about +0.10 m in the European Vertical Reference System EVRS-2007 (Urbański 2012).

The results are also juxtaposed here with future sea level rise scenarios included in the latest IPCC report (Church et al. 2013). Three scenarios are considered: the medium projection for RCP4.5 and the lowest and highest of the projections (lower bound of RCP2.6 and upper bound of RCP8.5). In those scenarios, water levels rise gradually to 28, 53 and 98 cm SLR by 2100 relative to 1986–2005 levels. However, Polish coast is subject to the glacial isostatic adjustment, which causes a yearly uplift by about 0.4–0.5 mm (Peltier et al. 2015). Increase in water levels is also uneven regardless of the movement of the coast; satellite-measured SLR trend during 1992–2015 was between 4.2 and 4.3 mm per year in the Baltic along the Polish coast; only in the far western part, it was slightly higher, up to 4.5 mm per year (NOAA 2016). These two effects translate to a difference of only a few centimetres in a perspective of a century. Bigger spatial differences in SLR can be caused by ground subsidence, which is a very local factor; no large-scale data are available on this matter. Therefore, IPCC scenarios were applied in the analysis without modifications.

For the analysis of coastal floods, the return periods of storm surges were calculated. They vary along the coast and lagoons; therefore, the return periods were calculated from annual maximum water levels recorded at eight gauges during 1948–2007. To each station (5 coastal, 1 in the Szczecin Lagoon and two on Odra river), the nearest basic administrative units were assigned. Water level data from Wiśniewski and Wolski (2009) were used and fitted to Gumbel probability distribution; details on the methodology can be found in Paprotny (2014).

2.4 Calculation of flood risk

Flood risk is a combination of three elements, as follows:

- Hazard, i.e. factor that can cause adverse effects, in our case sea level rise and coastal floods (see previous section)
- Exposure, i.e. the inventory of elements that could potentially be affected by the hazard
- Vulnerability, i.e. the propensity of exposed elements to be adversely affected by the hazard event (Kron 2005; Cardona et al. 2012).

In this section, we present the methodology to calculate exposure and vulnerability in order to obtain an estimate of risk caused by sea level rise and coastal floods in context of SLR.

2.4.1 Calculation of exposure

Following Hallegatte's (2012) framework on investigating impact of SLR on economic growth, we analyse exposure to the following elements:

- Permanent losses of natural capital: market value of land, including an estimate of the area of protected habitats
- Permanent loss of physical capital: gross replacement cost of immovable and movable fixed assets
- Permanent loss of social capital: number of inhabitants

Value of aforementioned assets was calculated here using 2011 data, the latest year for which complete statistics required here were available. The primary source was the online databases of the Central Statistical Office of Poland (CSO 2015). Values of land under different types of use are an estimate of their 2011 market price. Average sale price calculations made by CSO were used for arable land, meadows and pastures. Value of forests was taken from estimates of the State Forests, the manager of the vast majority of Polish woods (Ministry of Treasury 2012). Their calculations include both the value of land parcels and the trees covering them. Value of areas covered by buildings, transport infrastructure or non-built-up areas ready for construction was estimated using the relation between the sale price of those types of land use and the sale price of arable land in Germany in 2011 (Statistisches Bundesamt 2015). For other types of land use, numbers were assigned based on governmental regulations on estimation of real estate value (Law 2004/207/2109). As a result, woodlands or bushes are assumed to be worth equal to poor-quality arable land. Orchards were assigned the same value as arable land, while wastelands and other unutilized land were considered as equal to poor-quality meadows. Loss of inland surface water was not taken into consideration. Summary of land values is presented in Table 1. Agricultural build-up areas do not form a separate category in the topographic database; it was assumed that build-up areas adjacent to arable land and permanent crops fall into this category 4.12 PLN = €1

Information on buildings was compiled from several sources in order to calculate their value relative to their area (Table 2). Value of housing is the average construction cost of new houses as calculated by CSO. Stock of movable assets is difficult to estimate, and no national statistics in this matter are available. However, the ratio between household durable goods' value and GDP is reported to differ only slightly between countries and throughout modern history (Piketty and Zucman 2014). It was therefore assumed that the total value in this category in Poland constitutes 35 % of GDP, which is the average ratio in four countries (Canada, Germany, UK and USA) for which data are provided by Piketty and Zucman (2014). GDP and housing area statistics from CSO were used in order to obtain the amount of movable assets per m^2 in 2011.

Table 1 Estimated commercial value of land in 2011 per hectare, in Zloty (the polish national currency, PLN)

Land use	Value (PLN per ha)
Built-up: services	4,294,000
Built-up: housing (dense urban)	2,802,000
Built-up: housing (scattered urban)	1,896,000
Transport areas, non-built-up areas	559,000
Built-up: agricultural	453,000
Built-up: industrial	444,000
Forests	37,976
Arable land, orchards	20,004
Woodlands and bushes	16,401
Meadows and pastures	14,259
Wastelands and other lands	12,337

Table 2 Value of fixed assets per area in 2011 and damage functions

Type of building or structure	Value of fixed assets		Damage function
	Immovable (PLN per m^2)	Movable (PLN per m^2)	Y—damage (%), D—water depth (m)
Housing	3858	604	$Y_1 = D + 18.77 \times \sqrt{D}$ $Y_2 = 4.83 \times D + 20.69 \times \sqrt{D}$
Industry	2057	1712	$Y = 27 \times \sqrt{D}$
Services	2678	923	Market services: $Y = 27 \times \sqrt{D}$ Non-market services: $Y = 30 \times \sqrt{D}$
Agriculture	393	166	$Y = 27 \times \sqrt{D}$
Transport infrastructure	436	–	$Y = \frac{100}{3}$

For housing, Y_1 is damage to immovable assets and Y_2 is damage to movable assets

Commercial buildings were divided into three branches (industry, services and agriculture), and for each type, its value was calculated using the following formula:

$$V = \frac{F \times L_s}{L \times A_s} \tag{1}$$

where V is the value of a building per area in PLN, F is the value of fixed assets in Poland, L_S is the corresponding land use area in the study area, L is the corresponding land use area in Poland, and A_S is the total building area in the study area. Fixed assets estimates were obtained from Eurostat (2015). Land use data for Poland is from CSO (2015), while for the study area, the corresponding types of land use from the topographic database were used. Finally, building area was extracted from the topographic database, taking into account the multiple storeys several buildings contain.

2.4.2 Calculation of vulnerability

Assumptions on vulnerability of land and assets are presented in Table 3. In case of rising average water levels, land and immovable fixed assets (i.e. buildings and structures) are covered permanently by water and therefore lost completely. Movable fixed assets, which include machine tools, household goods, vehicles or livestock, can be evacuated from the endangered area given the gradual nature of SLR. Therefore, no losses in this category are assumed. In case of coastal floods, no losses are assumed for land, as water covers it only temporarily. Some productivity of land may be lost and consequently its market value could decrease; however, this aspect was not analysed here. Damage to crops is not analysed too, as they are not considered fixed assets, similarly to stocks of produced goods. Losses of fixed assets are calculated using damage functions, which relate losses to water depths (Apel et al. 2009).

Table3 Vulnerability assumptions for sea level rise and storm floods

Asset	Losses caused by...	
	Sea level rise	Coastal flood
Land	100 %	Nil
Immovable fixed assets	100 %	Relative to water depth and type of asset
Movable fixed assets	Nil	Relative to water depth and type of asset

Damage functions, which relate water depth to the relative loss of assets, were applied from Emschergenossenschaft/Hydrotec (2004), who created them using HOWAS database of flood losses (Merz et al. 2004). They were selected for this study, because there are no corresponding functions created from Polish empirical data. Additionally, they are similar to damage curves created as a combination of various European methodologies by Huizinga (2007) and suggested for use in countries without national damage functions. The only exception is the damage function for transportation, which is a constant value rather than a damage function. As can be noticed in Table 2, the equations differ for immovable and movable housing assets, while for commercial activity, they are combined. Finally, the value of transport infrastructure was taken from official risk calculations, according to the Polish government's methodology (Law 2013/104). Total value of losses for a building can be described as:

$$R = A \times V \times Y(D) \tag{2}$$

where R is the total loss for a building in PLN, A is the building area (including all storeys), V is the total value of the building per area, and $Y(D)$ is the appropriate damage function for water depth D (averaged within the building's contour).

Additionally, the number of people affected by sea level rise and coastal floods was estimated. Population data from the 2011 census (CSO 2015) provide information down to settlement level. By combining these data with housing area from the topographic objects database, it was possible to further disaggregate the data and obtain the average number of persons per housing area in settlements and towns. All persons living in a residential building even partially covered by water were assumed as being affected by the event. Similarly, the amount of tourism traffic potentially lost was estimated by disaggregating the number of tourists and nights spent in establishments (at basic administrative unit-level) using the size of appropriate categories of buildings. For the purpose of this study, "tourists" refer only to persons staying overnight in collective accommodation, excluding persons in camping sites.

2.5 Adaptation measures

Currently existing flood defences will be put under pressure from rising sea levels. In this section, we outline the methodology to estimate costs of adaptation to the that phenomena. We look at the most popular means of protection, namely dikes. We estimate the length of flood defences using GIS as follows. Firstly, continuous flood zones that contain any buildings are selected. Then, those zones are intersected with a layer representing water at normal conditions (including channels and streams). The length of the intersection lines between the layers was considered as the dike length required for full protection. Because only flood zones with buildings are considered here, some land and pieces of infrastructure will remain unprotected. However, these flood zones are very numerous, yet very small, and would therefore seriously overestimate of the required investments in flood defences.

Dikes have a high marginal cost, which varies enormously depending on location (Jonkman et al. 2013). Average investment in 1 km of dikes was 1.39 million PLN (€337,000) in 2011, according to official statistics (CSO 2015). This value was used as the "fixed" investment value per km, adding 0.7 % for each centimetre of dike crest level, a value used for cost optimization of flood defences in the Netherlands (Eijgenraam 2006). Using annual sea level rise projections, yearly cost of raising the flood defences could be calculated. It is assumed that an upgrade or new construction will be built overnight with a 50-cm safety margin and will be serve without further upgrades until the year 2100.

A different approach was used for analysing the adaptation to increased storm surge heights. Protection standards of dikes are often chosen arbitrarily. However, it is possible to calculate a flood defence standard in a given location that would be optimal from an economic point of view. In this approach (Van Dantzig 1956; Eijgenraam 2006), investments in flood defences are added to flood damages during the same given period of time. The sum of the two elements reaches a minimum at a certain dike crest elevation; this level constitutes the optimal flood protection. It can be expected that the optimum will shift due to sea level rise. Investments were calculated for each water level as described in the previous paragraph. Risk to assets for each protection standard and SLR scenario was obtained using the following procedure. Firstly, a Monte Carlo simulation was used to estimate the annual losses due to coastal floods D_i:

$$D_i = \frac{1}{85} \sum_{t=2016}^{2100} F_{w(t,i)} \times \frac{P_t}{P_{2011}} + T_{w(t,i)} \tag{3}$$

where t denotes the year (2016–2100), i is the iteration in the Monte Carlo simulation, $F_{w(t,\ i)}$ is the value of buildings, and $T_{w(t,i)}$ is the value of transportation infrastructure affected at water level $w(t, i)$. The latter is a sum of the random storm surge (sampled from the Gumbel distribution) and cumulative sea level rise up to year t at a given scenario. Additionally, the population size P in year t relative to year 2011 was used to approximate the change in value of buildings. The procedure was repeated for each administrative unit 10,000 times and averaged. For each assumed protection level, the values of $F_{w(t,i)}$ and $T_{w(t,i)}$ were set to zero if water level $w(t,i)$ was below the dike height. 2011 prices and GDP were kept constant, therefore assuming that the construction cost of dikes and wealth relative to income will not change in the future, as it is very uncertain how those elements will develop.

Additionally, the number of affected inhabitants per year was obtained similarly to (3)

$$E_i = \frac{1}{85} \sum_{t=2016}^{2100} P_{w(t,i)} \times \frac{P_t}{P_{2011}} \tag{4}$$

where $P_{w(t)}$ is the number of affected inhabitants under water level $w(t,i)$ in our baseline calculations for 2011. Future population numbers were derived using official projections from CSO (2015) up to year 2050, with an extrapolation of the trend to up to 2100. Those projections are at county level, with rural/urban breakdowns, and thus have far lower resolution than the settlement data used in the baseline calculations. However, no other local-scale demographic forecasts are available; it is therefore assumed that the population in the flood hazard zones will change at the same rate as the population outside them.

3. RESULTS

In this section, the consequences of coastal floods and sea level rise are presented. It should be noted that whenever statistics of inundated area are mentioned in the next sections, they do not include land covered by water under normal conditions. Also, all monetary values are in the Polish national currency, Zloty (denoted PLN), in 2011 prices. The exchange rate in 2011 was 4.12 PLN = €1. Firstly, we present the common socio-economic characteristics of the flood zones for both sea level rise and coastal floods (Sect. 3.1). The estimate of economic losses, which differentiates between the two phenomena, is presented in Sects. 3.2 and 3.3. This is followed by an analysis of the temporal dimension of losses (Sect. 3.4) and adaptation measures (Sect. 3.5). Finally, detailed results, as well as data underlying the figures, can be found in the Supplement.

3.1 General results

Water inundates the study area at a different pace depending on the analysed scenario. The flooded area encompasses 342 km^2 at +0.5 m MSL,

1662 km² at 1.5 m up to a maximum of 3323 km² at 5 m (Fig. 2a). The area at risk grows most dynamically between +0.3 and +1.6 m MSL with an average of almost 13 km² for every cm of water level rise. The highest growth of 371 km² is recorded between +1.2 and +1.3 m MSL scenarios (Fig.2b). Besides that, two other peaks at intervals of 0.8–0.9 and 0.9–1.0 m are observed, covering 132 and 212 km², respectively. In the +1.5-m MSL scenario, the flooded area already reaches half the size of the +5-m hazard zone, which indicates a significant slowdown in the growth dynamic of the flood zone at higher water levels. Up until the very last scenario, it increases by 3–5 km² per cm on average.

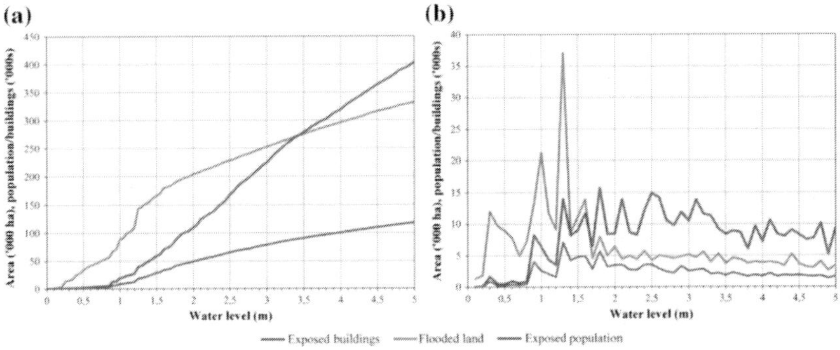

Figure. 2 *Left* Cumulative numbers of exposed land, inhabitants and buildings by water level. *Right* increments of exposed land, inhabitants and buildings by 10-cm intervals ending with the water level given

The number of affected buildings and inhabitants does not follow the aforementioned tendencies. Urban areas as well as other dense inhabited areas are either protected or generally not located at high-flood-risk zones. This results in an insignificant number of endangered buildings and people up to +1 m MSL. Crossing this barrier, in the range from +1.0 to +2.5 m MSL, the number of both inhabitants and buildings in the hazard zones soars (Fig. 2b). For buildings, the rate of increase in hazard peaks in the 1.2–1.3 m interval and significantly slows thereafter. Increase in the number of persons affected reaches maximum in the 1.7- to 1.8-m interval and eases only above +3 m MSL.

As can be seen in Figs. 3 and 4, flood hazard in the Polish coastal zone concentrates in the Vistula and Odra rivers' mouths. These estuarine regions encompass approximately 75 % of the total flooded area at all scenarios. In the area spread between those two regions along 300 km of coastline, the flood zones cluster around numerous coastal lakes. Other locations are generally well protected by dunes and cliffs. It should be noted that some areas can also become surrounded by water without being directly flooded, e.g. the Hel Spit could be severed from mainland at a water level of +2.4 m MSL.

3.1.1 Population

Inhabitants residing in rural municipalities or rural parts of rural–urban municipalities are disproportionately exposed to rising water levels. Those units contain 87 % of land area in the region, but only 18 % of its population. Yet, 51 % of inhabitants within +1 m MSL hazard zone live in rural areas. More persons are at risk in urban municipalities than in the rural type only above +1.7 m MSL. At +5 m MSL, the proportional of rural inhabitants at risk is still higher (26 %) than the regional average.

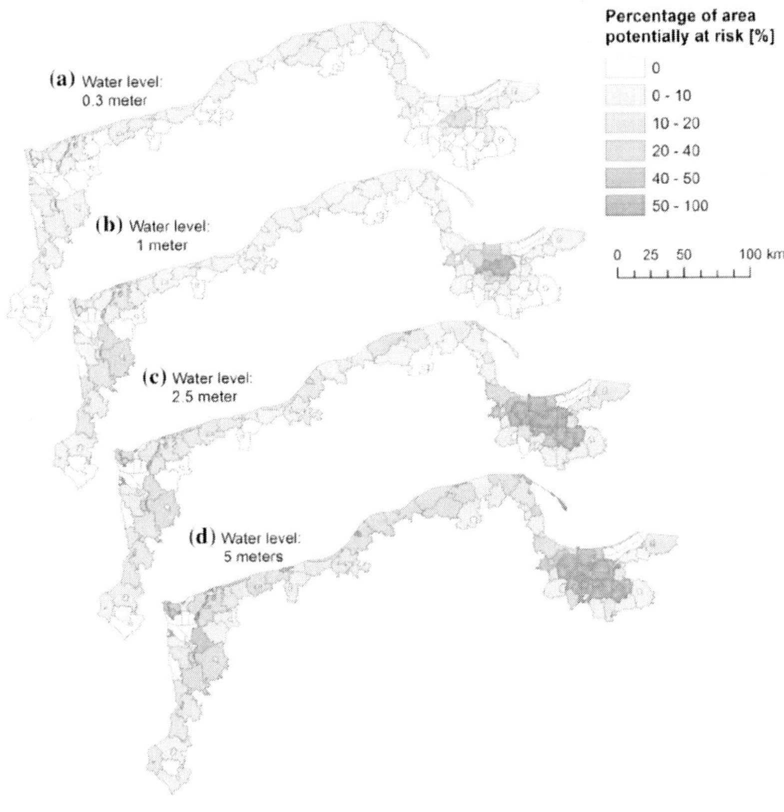

Figure. 3 Percentage of exposed area by water level and town/municipality

The population in main urban centres comes at risk in larger numbers only above +2 m MSL. Gdańsk, located in the Vistula estuary, comprises more than a fifth of inhabitants at risk in scenarios above +2.5 m MSL. Two other urban municipalities in the Tricity are to a vast extent less exposed, with almost no inhabitants in the flood zone below that value. Świnoujście comes second, with exposure soaring above +1.5 m MSL, putting the city's contribution to affected inhabitants at 10 % of regional total.

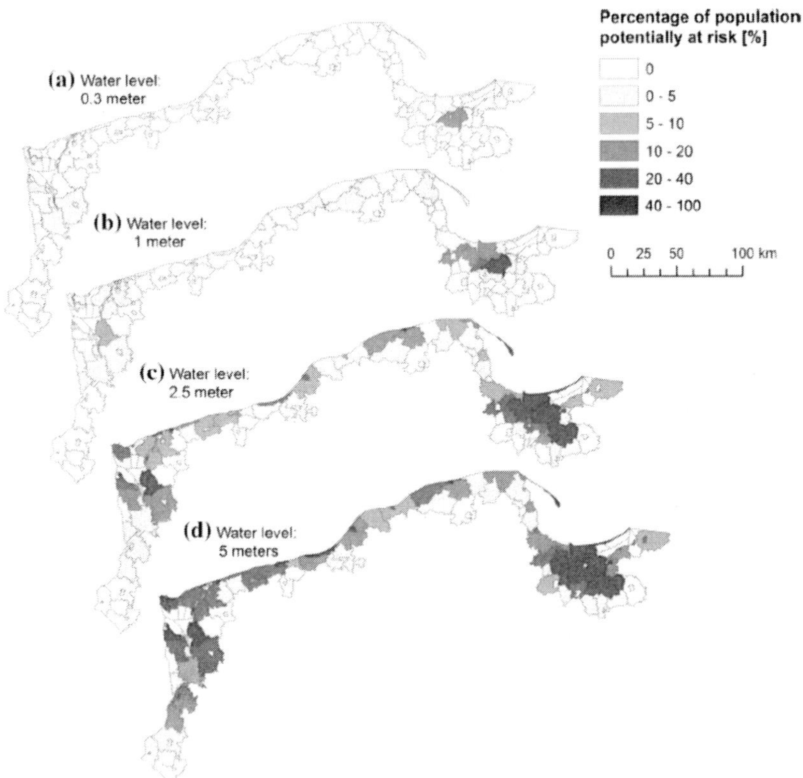

Figure. 4 Percentage of exposed inhabitants by water level and town/municipality

3.1.2 Land

The structure of land in the hazard zones is similar to the general composition of land use both in the region and country. Figure 5 presents the land use divided by five major groups. Barren land, the smallest group representing mainly beaches and dunes, covers more than 5 % of the total flooded area only up to +0.2 m MSL. This proportion decreases to 0.7 % by +5 m MSL. Grasslands, which are mostly swampy meadows and pastures, constitute more than 85 % of the total area at water levels below +0.2 m MSL. It remains the biggest land use group at risk until +1.25 m MSL, when croplands come forward. The latter constitute around 43 % of all hazards zones above +1.25 m MSL. Natural vegetation represented by woodlands and bushes contribute to about 11–16 % of land at risk, except for areas below the +0.4 m MSL. That is less than half of their share in the region (32 %).

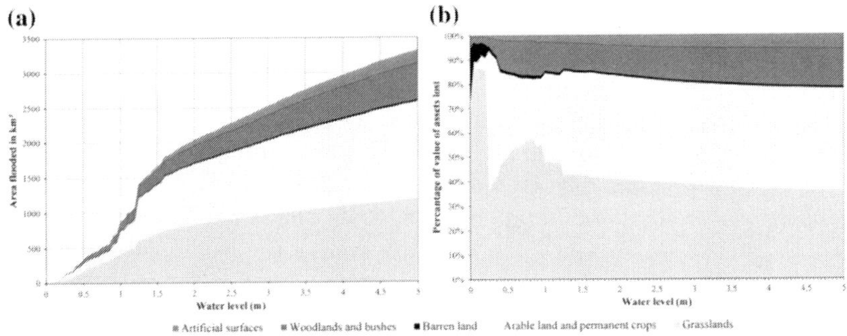

Figure. 5 Exposed area by major groups of land use classes: total area (*left*) and percentage breakdown (*right*)

Finally, artificial surface category is very diversified, as it comprises areas covered by buildings, infrastructure, landfills and similar human-made structures. Its share slowly increases from around 1 to 6 % by +5 m MSL. In general, industrial areas are more exposed than services, especially industrial installations and utilities. Agricultural build-up areas are slightly more at risk than single-family residential zones located further away from croplands. Multi-family residential areas are among the least exposed, with only 1.2 % potentially at risk at +1.5 m MSL (single-family—5.8 %).

It can be added that national parks, landscape parks, natural reserves and Natura 2000 areas also cover a substantial part of the region and are disproportionately exposed: 45 % of their land area is below +5 m MSL. It is caused by the specifics of protected land, which consists predominantly of low-lying wetlands, grasslands and forests. In total, 52 % of land in the +1 m MSL zone is under some form of protection, including 77 % of grasslands and 90 % of forests.

3.1.3 Buildings and infrastructure

The topographic objects database used for this analysis distinguishes 22 categories of buildings. Figure 6 compares the exposure of selected groups of buildings. Below +0.7 m MSL, farmhouses are the main type of buildings in the hazard zone, followed by single-family houses. The latter are the biggest group at risk by floor area for virtually every other scenario. Multi-family houses grow in number quickly above +1.2 m MSL, but it is not until +2.5 m that their floor area exceeds that of farmhouses. This illustrates how areas outside urban centres, where single-family houses and farmhouses dominate, are vastly more at risk than densely populated urban areas. The proportion of inhabitants living in exposed single-family residences is almost 90 % at +1 m MSL and decreases at more extreme scenarios, before converging with houses containing two and more flats at +4.1 m MSL. Healthcare buildings are at the beginning among the least endangered group, with barely 2 % located within the +2 m MSL hazard zone. This percentage, however, increases vastly thereafter, and buildings used for health services end up as one of the most exposed groups. Other assets disproportionately at risk include warehouses and other storage facilities as well as industrial buildings. On the other end of the scale, schools and trade/services buildings are among the least exposed in both number and floor area.

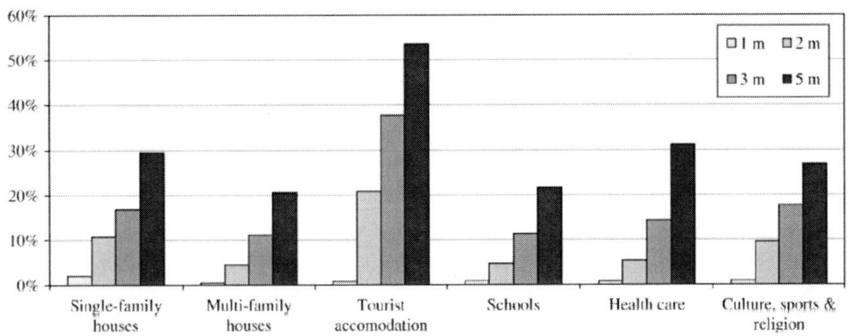

Figure. 6 Exposed buildings as a percentage of all buildings in the category in the study area

Tourism is unlikely to be much affected directly. Virtually no tourist traffic (measured both by the number of nights spend or number of visitors) is endangered below +0.9 m MSL, near the upper range of SLR projections up to 2100. Tourist accommodation other than hotels is safe at low water levels, but ultimately becomes the most exposed in all scenarios above +1.45 m MSL. In case of hotels, the percentage is lower, 48 % compared to

68 % for other accommodation establishments, though it still comes at second place in the 5-m hazard zone. In total, 62 % of tourist nights in the region during 2011 were spent in buildings lying below +5 m MSL.

Many buildings in the hazard zones have substantial intangible value. Buildings used for religious practices (mainly catholic churches) are among the least endangered of all 22 categories of buildings analysed. Almost 4400 buildings in the region (1 % of the total) are included in the national heritage sites' list. Below +2.4 m MSL, the fraction under threat is narrowly smaller than buildings in overall, but for higher levels, the opposite is true. Museums and libraries are slightly less likely to be in the hazard zone below +1.7 m MSL. However, at +5 m MSL, 47 % of them are at risk; only tourist accommodation buildings have a larger fraction in danger.

Transport infrastructure is generally more exposed than buildings. Paved roads are the most exposed type up to +2.3 m MSL; above that value tracks come first, while airstrips are of least concern in all scenarios. For roads, the magnitude of exposure decreases with the increase in its importance. Motorways can only be flooded by water levels rising above +2.6 m MSL. Meanwhile, 5 % of expressways could be affected in the +2.5 m MSL hazard zone, followed by national and regional roads (both 12 %), county roads (15 %), municipal roads (16 %), local and private roads (19 %) and company-owned roads (39 %). Finally, 18 % of railways are exposed at +2.5 m MSL, albeit the figure is somewhat exaggerated due to the numerous sidings located in ports.

3.2 Economic impact of sea level rise

Sea level rise causes total loss of land and immovable assets in the affected area. At the lower end of IPCC's SLR projections (+0.3 m MSL), potential losses are estimated to amount to 1.1 bln PLN (0.07 % of 2011 GDP), while on the higher end (+1 m MSL), they soar to 9.6 bln PLN (0.63 %), as indicated in Fig. 7a. Total value of assets in the +5 m MSL zone is 160 bln PLN (10.5 % of 2011 GDP). Notwithstanding, the structure of assets at risk does not change significantly with rising water levels. Figure 7c presents the breakdown by economic activity for SLR. Assets used for agricultural and forestal production dominate at water levels below +0.9 m MSL, as low-lying grasslands, arable land and forests near estuaries and coastal lakes would be flooded first. At higher elevations, housing dominates as the main exposed asset, with the proportion reaching 40 % at +0.9 m MSL and gradually rising to a maximum of 53 %. Industrial assets fluctuate around 10 %, while the percentage associated with the services sector soars to slightly above 20 % at +1.8 m MSL and maintains that level for the remaining scenarios. In agriculture, land is the main source of losses. For other economic activities, losses are predominantly to buildings, with those used for storage worth most up to +1.3 m MSL; for manufacturing and utilities between +1.3 m and +2.4 m MSL; and tourist accommodation above that value. In total, land and buildings are almost equal in value up to +0.85 m MSL, when the losses associated with the latter increase sharply. Buildings constitute more than 60 % of losses at water levels above +1.45 m MSL and more than 70 % above +2.4 m MSL, before peaking at 74 %. The percentage of potential losses connected with infrastructure slowly increases with the water level, but generally stays around 10 %.

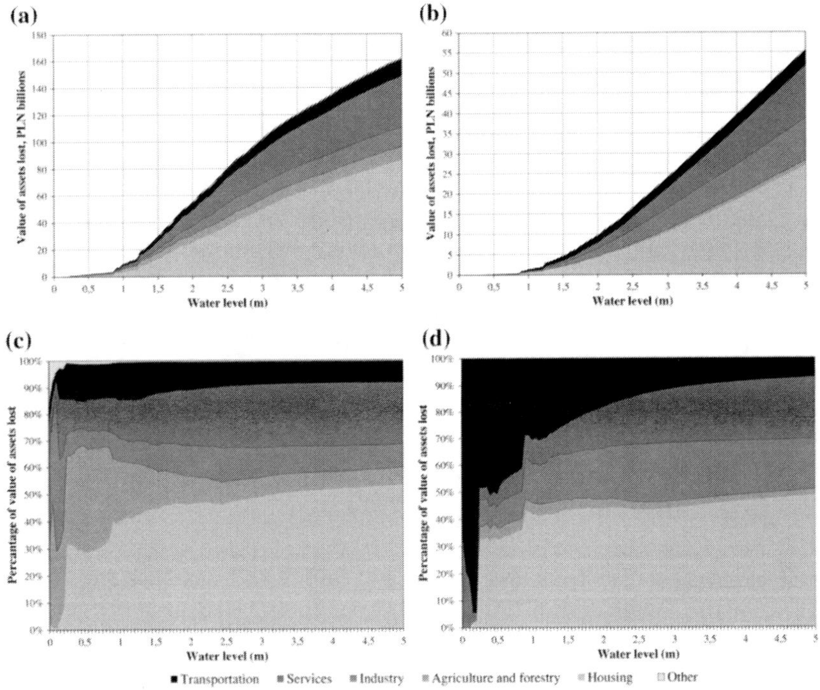

Figure. 7 Potential losses due to sea level rise: cumulative total values (**A**) and structural breakdown by activity (**C**). Potential losses due to coastal floods: cumulative total values (**B**) and structural breakdown by activity (**D**)

3.3 Economic impact of coastal floods

Coastal floods have a different impact on assets than sea level rise. As noticed in Table 3, losses of fixed assets depend on damage curves, while no land is lost permanently. Therefore, economic losses due to coastal floods exhibit a rather exponential behaviour (Fig. 7b, d). By contrast, damages caused by SLR resemble more a logarithmic curve. The total losses are estimated at 1.4 bln PLN (0.09 % of 2011 GDP) for a 1-m surge and 16.6 bln PLN (1.1 %) for a 2.5-m surge, provided that it occurs in the entire coastal zone (Fig. 7b). The breakdown of those losses by economic sectors is even more stable at various water levels than in case of SLR (Fig. 7d). Apart from water levels below the uncertainty range, where merely a few buildings or roads could be affected, housing dominates at around 40 % of all losses. Industry comprises around 20 % of total, while the services sector hovers up a growing fraction from less than 10 % (below +1 m MSL) to 23 % (+5 m MSL). Assets in the agricultural sector are worth little and therefore are responsible for only a small percentage of losses (only 4 % at 1 m, 2 % at 3 m). Infrastructure contributes significantly at low water levels (more than half of losses below +0.5 m MSL), but less so for other scenarios; the percentage value drops below 30 % at +1.25 m and below 20 % at +1.9 m, ultimately reaching 7 % at +5 m MSL.

Additional feature of coastal floods that is different than SLR is its recurring character. In Fig. 8, we present number of affected inhabitants and assets, where surge heights have been recalculated as return periods. Those return periods were combined with different SLR scenarios. It can be clearly seen that even small values of sea level rise decrease significantly the return periods of coastal floods. A total of 100,000 persons are in the hazard zone of a 500-year coastal flood, but with a 30-cm SLR (at lower end of IPCC projections), it becomes a 90-year event. The return period further decreases to 30 years with a 50-cm SLR and as little as 2.5 years with a 1-m SLR; at 2 m, SLR itself would affect that number of people. Even a 10,000-year event becomes only a 50-year event at the upper range of IPCC projections. Potential economic losses also increase significantly—a 100-year event could cause 6 bln PLN (0.4 % of 2011 GDP) of damages under current conditions, but 19 bln PLN (1.2 %) with 1 m SLR.

3.4 Temporal dimension of losses

Previous sections looked at the total potential losses due to SLR and coastal floods at various scenarios, based on the distribution of land, population and assets in 2011. In this section, we analyse how those losses would be distributed over time according to IPCC sea level rise projections, and what would be the effect of future demographic evolution.

Sea level rise scenarios from the IPCC range from +0.28 (low) through +0.53 (medium) to +0.98 m (high) increase in global MSL by 2100. In Fig. 9a, the average yearly loss of assets (land, buildings, infrastructure) per decade is presented. It is assumed that value of assets will be constant relative to GDP. For comparison, annual expenditure on coastal protection (1) and flood defence (2) is provided (2004–2014 average; CSO 2015). The former corresponds to a mere 38 million PLN, but the lowest of the IPCC projections does not cause damage higher than that. In the medium scenario, damages exceed those costs so only in the 2050 s. In the most severe projection, damage from SLR increases sharply in the end of the century, to about 0.03 % of GDP per year, which equals about half of the yearly central government expenditure on disaster relief in Poland (2004–2014 average; CSO 2015).

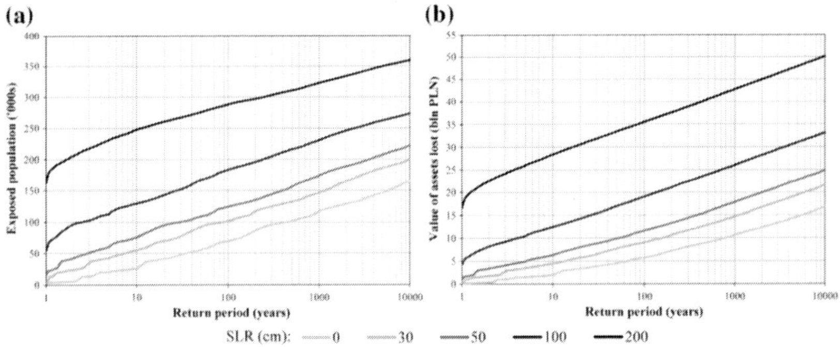

Figure.8 Population (*left*) and assets (*right*) exposed to coastal flood by return period and sea level rise (SLR) scenario

Adding demographic evolution to the picture (Fig. 9b) shows that those values could change. The graph presents demographic evolution from 1960 to 2100 by hazard zone, using data by basic administrative unit for 1960–1988 from Eurostat (2015) and 1999–2014 from CSO (2015), together with future projections. Until recently, the population in the coastal zone increased faster than the national total. However, CSO (2015) projects that the population in the hazard zone corresponding to the low and medium scenarios will decrease substantially by the end of the century. Meanwhile, the population within the reach of the high scenario of SLR will increase in the medium term, decreasing slightly towards year 2100. Additionally, the population currently endangered by +1 m MSL is 49 % urban, but this value will drop to 39 % by 2050. This is due to urban sprawl, i.e. the migration of wealthier people from cities to the surrounding rural municipalities, particularly around the Tricity. It should be noted that we assume here that the population in an administrative unit will change by the same factor both in "endangered" and "safe" zones. It is difficult to conclude whether the population is likely to relocate towards more flood-prone areas. However, 3 % of buildings located below +1 m

MSL were under construction at the time of the last update of the topographic objects database, slightly more than the percentage for the entire study area.

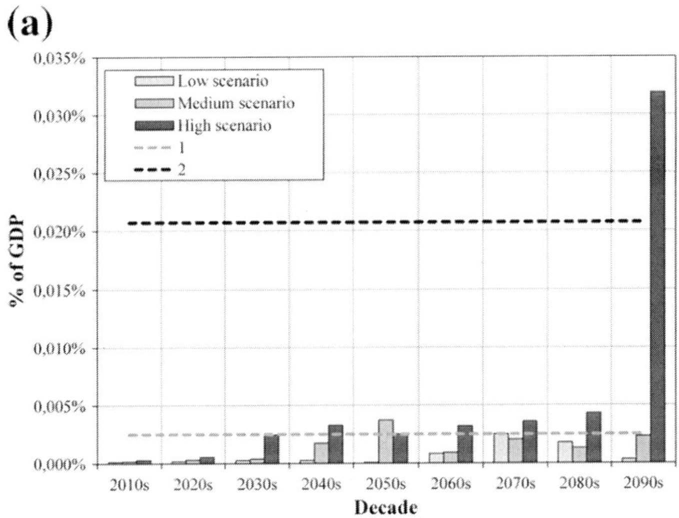

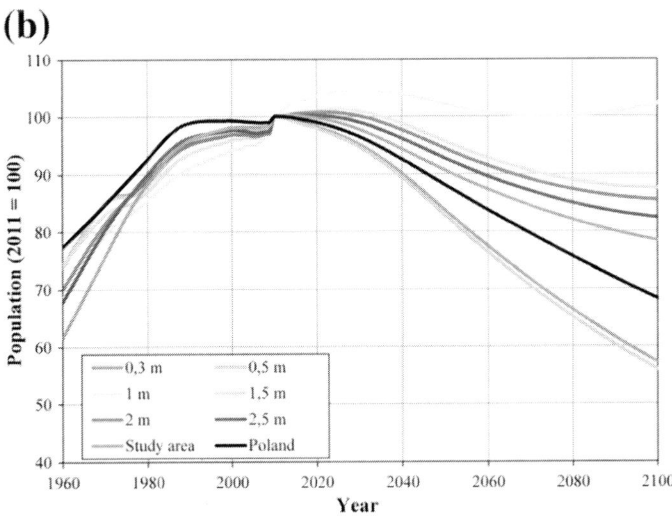

Figure. 9 *Left* potential losses due to inundation from sea level rise (annual averages by decade) with a comparison to average expenditure on (1) coastal protection (2) flood defences (2004–2014; CSO 2015). *Right* estimated population change by hazard zone (water level in m) relative to 2011 census data

3.5 Adaptation

Adaptation to inundation caused by SLR could be done by raising flood defences. Full protection can be cost-effective, as indicated in Fig. 10a. At +0.5 m MSL, investing 83 million PLN saves almost 1.5 bln PLN of damages (almost 0.1 % of GDP), though loss of land worth 267 million PLN still occurs. By +1 m MSL, investments rise to 1.9 bln PLN, but 7.3 bln PLN (0.5 % of GDP) is spared. Annualized, the costs of flood defences and loss of land are small. In Fig. 10b, they are compared with annual coastal protection expenditure (grey line), and until the 2080s, the adaptation costs are smaller. Also, even the costs in the 2090s in the high scenario of SLR are a quarter of current yearly investments on dikes (relative to GDP), and only 7 % of annual disaster relief spending in Poland. Full protection at +1 m MSL would require 654 km of new/upgraded dikes (Fig. 10c), of which about a third in the area of responsibility of the maritime administration and the rest around inland waters maintained by water management authorities. This proportion mostly holds for other water levels, while the length of defences for coastal flood protection increases to 1000 km at +1.2 m and 2000 km at +1.8 m MSL. Still, the defences could be more streamlined, also by enclosing some smaller channels and building pumping stations. At least a tenth of the dikes can be replaced with around 20 stations, saving investment costs, though increasing maintenance costs.

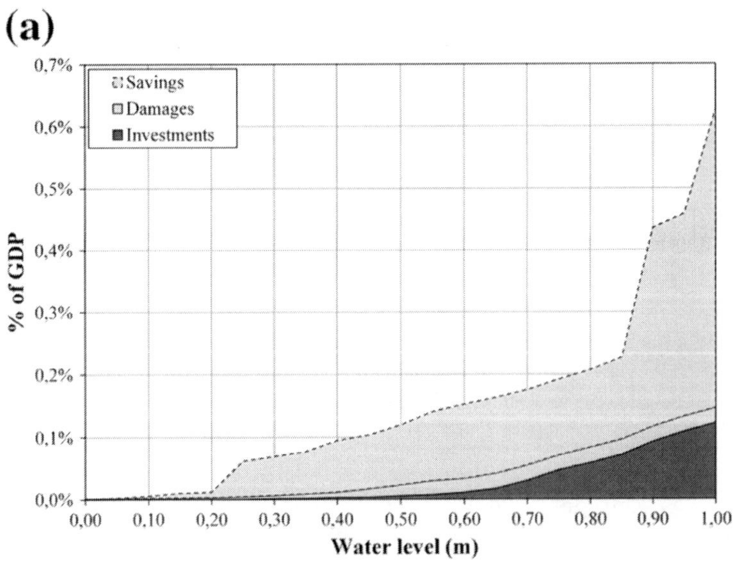

(b)

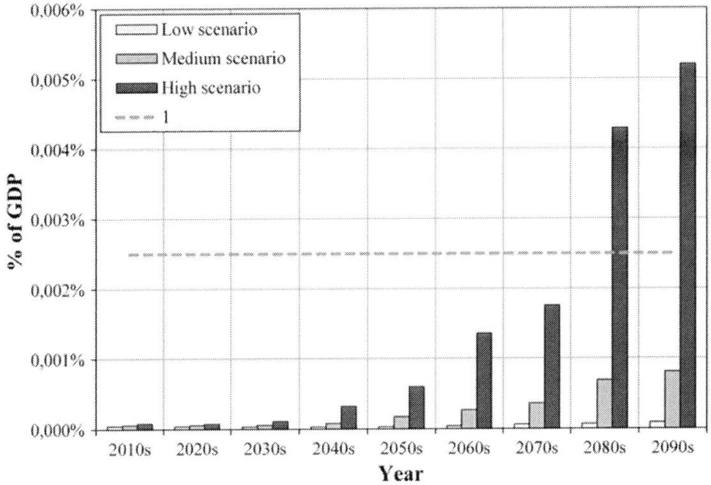
0,006%
0,005%
0,004%
0,003%
0,002%
0,001%
0,000%
% of GDP
Low scenario
Medium scenario
High scenario
1
2010s
2020s
2030s
2040s
2050s
2060s
2070s
2080s
2090s
Year

(c)

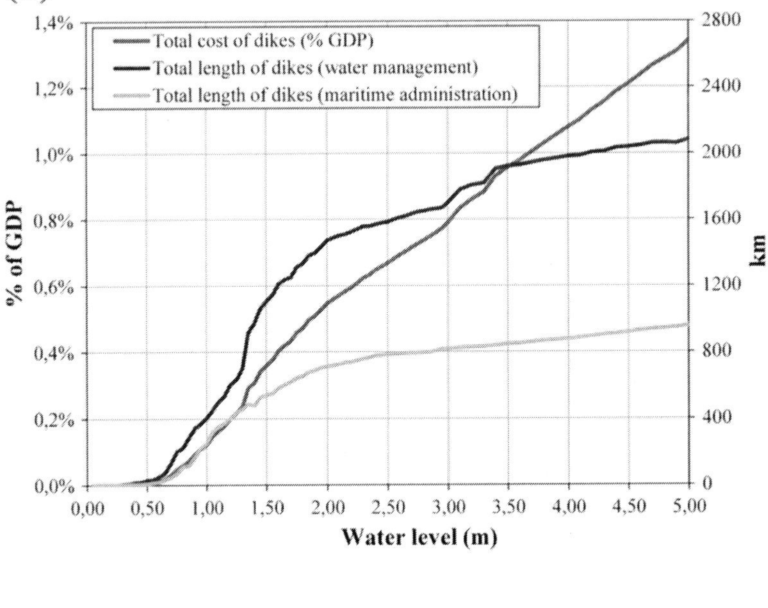
1,4%
1,2%
1,0%
0,8%
0,6%
0,4%
0,2%
0,0%
% of GDP
Total cost of dikes (% GDP)
Total length of dikes (water management)
Total length of dikes (maritime administration)
2800
2400
2000
1600
1200
800
400
0
km
0,00
0,50
1,00
1,50
2,00
2,50
3,00
3,50
4,00
4,50
5,00
Water level (m)

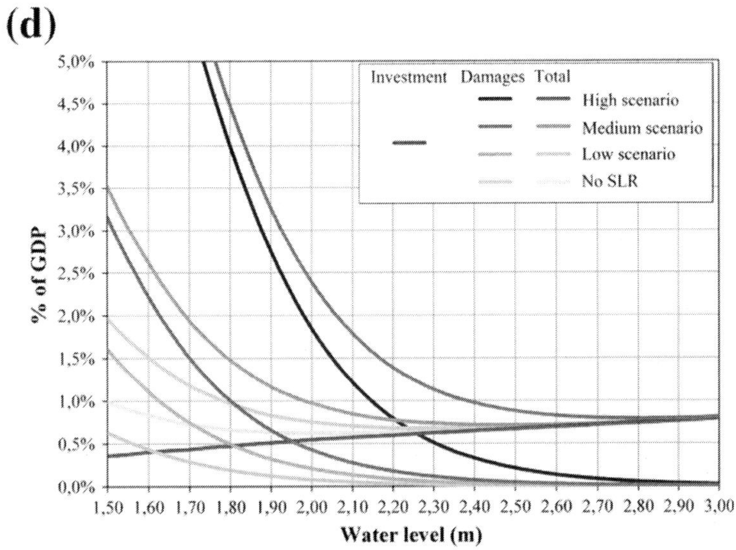

Figure. 10 Investments in defences against sea level rise, remaining damages (losses) and savings (**A**), annualized investments plus remaining losses for SLR scenarios; 1—annual coastal protection expenditure (**B**), length of flood defences and its costs needed for full protection (**C**), expected damages (2016–2100 total) and investments in flood defences (**D**)

In Fig. 10d, the investments and expected losses during 2016–2100 are presented. Under current conditions, the economically optimal flood defence protection level for the entire coast is +2.06 m MSL (excluding the safety margin). When SLR is taken into account, the protection level increases to +2.35 m MSL in case of low SLR and +2.81 m MSL in the highest scenario. The cost estimate is also a higher: 0.79 % of GDP in the highest scenario instead of 0.62 % (both figures including about 0.05 % of GDP in damages over 85 years). Those costs could be optimized further, if instead of one protection standard for the entire coast, we make this calculation at the level of basic administrative units. Under present climate, the adaptation costs and losses amount to 0.39 % of GDP, and with SLR, they increase to 0.49–0.67 %. However, this calculation does not include potential loss of life to coastal floods, as it is not possible to assign an economic value to this factor. Nevertheless, the described protection levels reduce the expected number of affected inhabitants to 80–120 per year.

4. DISCUSSION

4.1 Uncertainties

As with all studies on SLR and coastal floods, also the results presented in the previous section have their limitations; there are also several sources of uncertainty. Many of those are described in Sect. 2. The most crucial is the accuracy of the representation of the terrain and flood defence structures. The effects of the DEM's nominal accuracy (20 cm in most of the area) are presented in Fig. 11. The uncertainty is large around +1 m MSL, but still improved compared to some other, similar studies.

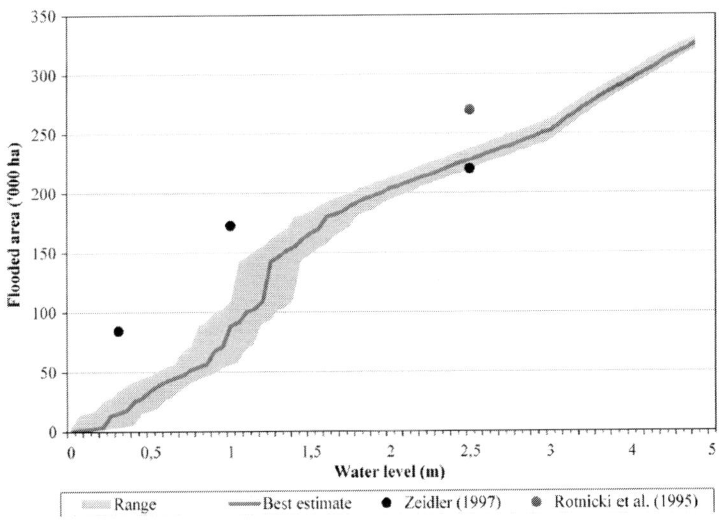

Figure. 11 Uncertainty in delimitation of the area affected by sea level rise, with comparison to two previous studies

The data used to derive the exposure of assets also carry some uncertainty. Actual damage depends largely on buildings' construction characteristics and type of preparations made before the flood (Apel et al. 2009). Additionally, the equations used here were developed in Germany, since there is no methodology based on Polish empirical data. For the sake of illustration, in Fig. 12 we compare the results of coastal flood damage estimation using Polish government methodology (Law 2013/104), generic damage functions for European countries from Huizinga (2007) and a Dutch methodology made by Rijkswaterstaat (Kok et al. 2004). Some large discrepancies could be observed, even though it is partially a result of a different classification of assets in those methods.

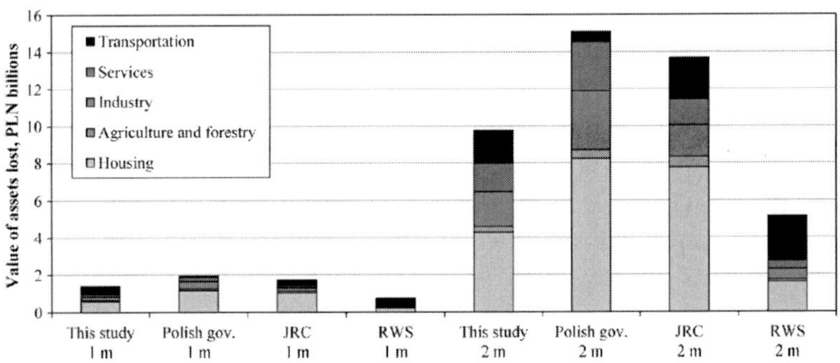

Figure. 12 Comparison of potential damage estimates from a 1- and 2-m storm surge using three alternate methodologies from the Polish government (Law 2013/104), *JRC* Joint Research Center (Huizinga 2007) and *RWS* Rijkswaterstaat (Kok et al. 2004)

We also use a simplification commonly applied in both to local and global studies that SLR is the only influence of climate change on extreme water levels (Xu and Huang 2013; Hinkel et al. 2014; Muis et al. 2015). Yet, recent modelling studies indicate that changes in wind regime will mostly cause a further increase in storm surge maxima, also in the Baltic (Gräwe and Burchard 2012; Vousdoukas et al. 2016). Furthermore, some dynamic effects of the sea level rise were omitted. Erosion and sedimentation could substantially change the level of hazard. In the context of flood hazard, the important information is whether erosion will destroy the barriers separating the inland from seawater. We did not analyse those aspects for a number of reasons. For one thing, in previous studies (e.g. Zeidler 1997; Bosello et al. 2012), the issue was investigated as a separate problem than flooding, i.e. changes in topography were not included in the delimitation of hazard zones. Thus, a "static" approach gives us comparability with the results of those previous studies, especially since the focus of the paper is on inundation caused by sea level rise and coastal floods. Additionally, there is substantial variation along the Polish coast of erosion and sedimentation rates, and their impact on local morphology is still being investigated (Deng et al. 2014; Furmańczyk et al. 2014). Furthermore, the aforementioned studies considered coastal erosion as a much smaller source of losses than inundation of land by sea level rise and, foremost, coastal floods.

Similarly, the increase in groundwater levels caused by SLR is not analysed here, though this may result in inundation of low-lying areas otherwise fully protected from seawater. This aspect has not yet been properly investigated (Rotzoll and Fletcher 2013), especially in relation to the Polish coast. Salt intrusion is yet another issue, though the previous studies (Pluijm et al. 1992; Zeidler 1997; Bosello et al. 2012) considered potential damages to be minimal, and this aspect does not influence the level of flood hazard.

Another concern is the use of only eight tide gauges for the calculation of return periods the analysis of flood protection. In the open coast is not a large concern, as the difference between gauges in the 100-year return period is only 12 cm, but inaccuracies occur in the Vistula and Odra rivers mouths as well as the coastal lagoons. Furthermore, the "bathtub fill" method exaggerates storm surge hazard. An actual overtopping of flood defences would mostly cause inundation of a much smaller area than indicated here, since the peak of the surge lasts only a short period of time. Breilh et al. (2013) estimates that the static method performs well for areas located no more than 3 km from the coast or estuary. In our case, around a third of the flood zone is within that limit for most scenarios (35 % for +0.5 m, 31 % for +1.5 m, 33 % for +2.5 m MSL). However, the proportion of exposed buildings in the 3-km zone increases from 8 % for +0.5 m to 39 % for +1.5 m and 53 % for +2.5 m MSL.

4.2 Comparison with previous studies

Our results can be contrasted with two studies that analysed impacts of sea level rise specifically on the Polish coast. Table 4 provides such a comparison, together with the estimated uncertainty of the results of this study (i.e. ±20 cm accuracy of the DEM and ±20 % accuracy of assets value and population). Even taking the maximum values from our results, there is a substantial difference in estimates of exposure of inhabitants and assets. The primary reason is methodological: flood zones in Rotnicki and Borówka (1991) were delimitated without taking into account numerous barriers between the sea and low-lying areas, which was not possible at the resolution of their source material. Rotnicki et al. (1995) provided results for only one hazard zone of +2.5 m MSL apart from identifying assets lying below sea level. According to the authors, the 2.5-m contour represents the water level of a 100-year storm surge (1.5 m) superimposed on top of 1 m SLR. As presented in Table 4, the flooded area calculated by Rotnicki et al. is about 18 % larger than our results indicate. The difference increases when analysing the number of exposed inhabitants. Rotnicki et al. put it at almost 300,000 persons, while in this study, it is nearly 50 % lower. It should be noted that study refers to demographic situation in the late 1980s, but the Polish population numbers barely changed since then. Such a big difference can be explained by the fact that densely populated areas are mostly well protected by flood defences, which were not taken into account in that study.

Table 4 Impacts of sea level rise in Poland: comparison of results from this analysis with VA'92 study as reported by Zeidler (1997) and a study by Rotnicki et al. (1995)

Category	Water level	This study		Zeidler (1997)		Rotnicki et al. (1995)
Reference period	(m)	2011		1995	2025	ca. 1990
Flooded area (km^2)	0.30	155.2	(16.4–341.8)	845.1	845.1	—
	1.00	885.0	(541.3–1,093.4)	1727.7	1727.7	—
	2.50	2278.4	(2178.4–2376.5)	2203.3	2203.3	2696.7
Persons affected	0.30	1777	(0–3362)	40,860	64,600	—
	1.00	19,966	(4245–33,432)	146,040	230,600	—
	2.50	167,503	(112,497–230,672)	234,840	371,600	297,135
Value of assets at risk (bln PLN at 2011 prices)	0.30	1.1	(0.1–2.2)	47.8	78.1	—
	1.00	9.6	(2.5–15.9)	223.2	366.7	—
	2.50	80.9	(55.3–107.6)	366.7	574.0	—
Value of assets at risk (% GDP)	0.30	0.1 %	(0.0–0.1 %)	7.5 %	3.1 %	—
	1.00	0.6 %	(0.2–1.0 %)	35.0 %	14.6 %	—
	2.50	5.2 %	(3.6–6.9 %)	57.5 %	22.9 %	—
Railway lines (without sidings) affected (km)	0.30	2.4	(1.4–3.1)	35	35	—
	1.00	25.5	(7.7–28.6)	180	180	—
	2.50	160	(138–185)	219	219	347
Cost of dikes for full protection (bln PLN at 2011 prices)	0.30	0.02	(0.003 to −0.1)	15.9	27.9	—
	1.00	1.9	(0.7–3.7)	28.7	62.1	—
	2.50	10.2	(7.6–13.1)	—	—	—
Cost of dikes for full protection (% GDP)	0.30	0.001 %	(0–0.007 %)	2.5 %	1.1 %	—
	1.00	0.1 %	(0.05–0.2 %)	4.4 %	2.5 %	—
	2.50	0.7 %	(0.5–0.9 %)	—	—	—

Cost of dikes in Zeidler's study was calculated using its earlier iteration (Pluijm et al. 1992)

A complementary study called VA'92 (Ziedler 1997) provides data for three flood zones, namely +0.3, 1 and 2.5 m MSL and for two time periods: 1995 and 2025. As given in Table 4, the inundated area is six times larger at +0.3 m MSL in VA'92 compared to this study, but the difference decreases with higher water levels, converging at +2.5 m MSL. Number of inhabitants at risk is substantially lower than in VA'92, while the value of assets is even more so. At the time of that study, Polish population was forecasted to increase in the future, hence the big difference for the 2025 scenario; currently, the population is expected to decrease. Particularly striking is the difference in relative economic losses, estimated at 0.6 % of

GDP, while VA'92 study gives an extremely high value of 35 % for 1995, forecasted to decrease to 15 % by 2025. That study also looked at adaptation and estimated that very large investments will be needed: Table 4 compares the cost of providing full protection by building and upgrading dikes. Zeidler's study does not give all information, but notes that 350 km of defences would be necessary to protect the Odra estuary against 30 cm SLR, and 604 km for 1 m SLR and at least the same amount for the Żuławy area; in the latter scenario, Pluijm et al. (1992) gave a value of 699 km. In our analysis, the figures are 10 and 654 km, respectively, for the entire coast. This is because we include in the delimitation of hazard zones the majority of existing defences, in contrast to the other study which omitted most of them.

More recently, an analysis of impacts of SLR in Europe using DIVA model was carried out by Bosello et al. (2012). For a comparable scenario (+0.95 m MSL), it gave an estimate of inundated area that is four time higher than in this study. Yet, it has put the value of inundated land at only 0.03 % GDP in contrast to our estimate of 0.15 %, though the former figure refers to the projected economic situation in 2085 instead of the current situation. That study also expects that rising waters will cause the GDP to drop by 0.046 %, which is in line with our estimate of a 0.041 % loss in commercial fixed assets. Meanwhile, Hinkel et al. (2010) estimated the expected number of inhabitants affected by coastal floods (without adaptation, but including demographic developments) as 30,000–60,000 per year in 2100 assuming a modest sea level rise of 35–45 cm. For comparable water levels, we estimate the expectation at about 29,000–35,000 persons.

In order to further assess the impacts of quality of data on our results, we have juxtaposed them with official flood hazard maps (KZGW 2015). Those studies were prepared for selected parts of the coast, utilizing two-dimensional hydrodynamic modelling and locally collected data, such as detailed surveys on flood protection structures and administrative registers and databases. They mostly cover a single 500-year flood scenario, which translates into water levels of 1.7–3.0 m above MSL depending on location. Most of those maps present very similar flood zones, with the main exception being the Żuławy Wiślane area, which is protected by a more elaborate flood defence system of pumps, sluices and weirs. Table 5 presents estimates of inhabitants exposed to coastal floods in the main coastal cities in Poland. Scenario A denotes our approximation of the KZGW's storm surge scenarios. In the KZGW maps, water level is spatially variable, because a hydrodynamic model was applied to calculate the effects of surges. The results show that the difference between our results and the official maps is mostly small, though in every case, our estimates of the number of inhabitants exposed are higher. However, part of the difference could be the effect of different sources of demographic data used in both studies. KZGW used detailed data from the population register instead of disaggregated estimates that are used here. In Table 5, we also include the number of exposed inhabitants using our calculations of return periods (B). They are mostly lower than KZGW scenario, which is probably caused by inclusion of a larger number of water level observations and different methodology of return period calculations in that study. In effect, our estimate of the number of affected inhabitants for a comparable return period is lower.

Table 5 Impacts of coastal floods in selected major coastal cities

City	Population in 2011 ('000s)	Water level (m)		Exposed population ('000s)		
		A	*B*	This study (*A*)	This study (*B*)	KZGW maps
Szczecin	410.1	1.85	1.63	2.5	1.7	1.8
		1.60	0.99	1.6	1.1	1.3
Police	34.0	1.75	1.63	0.2	0.2	0.1
Świnoujście	41.5	2.60	1.93	15.4	7.7	14.1
		2.30	1.21	11.2	4.2	9.6
Kołobrzeg	47.1	2.00	1.98	0.7	0.6	0.3
Ustka	16.4	2.60	1.85	1.2	0.5	0.8
Gdynia	249.1	2.35	1.96	0.2	Nil	0.0
Sopot	38.7	2.50	1.96	1.3	Nil	0.2
Gdańsk	460.3	2.50	1.96	38.4	14.7	34.7
		2.10	1.20	22.3	7.0	20.0
Elbląg	124.7	1.40	1.96	2.0	3.0	0.7

Water level in the table is (*A*) an approximation of KZGW's 100- or 500-years flood scenarios and (*B*) our calculation of 100- or 500-year coastal flood

5. CONCLUSIONS

In this paper, we have undertaken a broad analysis of potential impacts of inundation caused by sea level rise and coastal floods in the Polish coastal zone. Its aim was to increase the accuracy and level of detail of country-wide risk estimates, covering both population and economics, compared to currently available studies. This was made possible by recent developments in the national stock of spatial data. The most important findings of the study are:

- Impacts of SLR, as well as adaptation costs to SLR, will be much smaller than indicated in previous studies. 1 m SLR would affect directly 20,000 inhabitants and 9.6 bln PLN of assets (0.6 % of GDP), while the cost of increasing flood defences for full protection of the population would be 1.9 bln PLN.

- Annualized, adaptation costs and remaining losses would be equivalent of a fraction of average coastal protection budget (about 0.0025 % of GDP); only in the highest IPCC projection, the budget would have to be more than doubled by the 2080s.
- Coastal floods will increase substantially in frequency due to sea level rise. Half-meter SLR will double the number of inhabitants and assets in the 100-year flood zone, while 1 m SLR will more than triple it. Economically optimal flood defences would cost 0.39 % of GDP at current climate conditions, and 0.49–0.67 % of GDP within SLR scenarios indicated by IPCC.
- Rural population would be proportionately much more exposed than inhabitants in urban areas. Assets used in agricultural production are the most exposed category within the range of IPPC projections of SLR. For higher SLR values and coastal floods, housing comprises the biggest percentage of assets at risk, at about 40 %.
- SLR will have very limited direct impact on the regionally important tourist industry. However, the beaches and natural reserves which draw the visitors are among the most vulnerable: more than half of the land in the +1 m MSL hazard zone is under some kind of environmental protection.

But again, one should bear in mind that several simplifications have been used in this study and more research is required, especially on coastal flood scenarios, damage curves, future climate, population and economic projections. More advanced modelling techniques that take into account dynamics of storm surges and coastal processes (erosion and sedimentation) are yet to be used in the Polish coastal zone.

ACKNOWLEDGMENTS

The authors would like to thank the Central Surveying and Cartographic Documentation Center (CODGiK) in Warsaw, Poland, for kindly providing the topographic objects database and digital elevation models used in this study. Special thanks should be extended to S. N. Jonkman for his comments on the draft of this paper, as well as to the anonymous referees for their detailed suggestions which helped to improve this paper significantly.

REFERENCES

1. Apel H, Aronica GT, Kreibich H, Thieken AH (2009) Flood risk analyses—how detailed do we need to be? Nat Hazards 49:79–98. doi:10.1007/s11069-008-9277-8
2. Bates P, De Roo AP (2000) A simple raster-based model for flood inundation simulation. J Hydrol 236:54–77. doi:10.1016/S0022-1694(00)00278-X
3. Bednarczyk S, Jarzębińska T, Mackiewicz S, Wołoszyn E (2006) Vademecum ochrony przeciwpowodziowej. Krajowy Zarząd Gospodarki Wodnej, Gdańsk
4. Bosello F, Nicholls RJ, Richards J, Roson R, Tol RSJ (2012) Economic impacts of climate change in Europe: sea-level rise. Clim Change 112:63–81. doi:10.1007/s10584-011-0340-1
5. Breilh JF, Chaumillon E, Bertin X, Gravelle M (2013) Assessment of static flood modeling techniques: application to contrasting marshes flooded during Xynthia (western France). Nat Hazards Earth Syst Sci 13:1595–1612. doi:10.5194/nhess-13-1595-2013
6. Cardona OD, van Aalst MK, Birkmann J, Fordham M, McGregor G, Perez R, Pulwarty RS, Schipper ELF, Sinh BT (2012) Determinants of risk: exposure and vulnerability. In: Field CB, Barros V, Stocker TF, Qin D, Dokken DJ, Ebi KL, Mastrandrea MD, Mach KJ, Plattner GK, Allen SK, Tignor M, Midgley PM (eds) Managing the risks of extreme events and disasters to advance climate change adaptation. Cambridge University Press, Cambridge and New York, pp 65–108
7. Church JA, White NJ (2011) Sea-level rise from the late 19th to the early 21st century. Surv Geophys 32:585–602. doi:10.1007/s10712-011-9119-1
8. Church JA, Clark PU, Cazenave A, Gregory JM, Jevrejeva S, Levermann A, Merrifield MA, Milne GA, Nerem RS, Nunn PD, Payne AJ, Pfeffer WT, Stammer D, Unnikrishnan AS (2013) Sea level change. In: Stocker TF, Qin D, Plattne GK, Tignor M, Allen SK, Boschung J, Nauels A, Xia Y, Bex V, Midgley PM (eds) Climate change 2013: the physical science basis contribution of working group I to the fifth assessment report of the intergovernmental panel on climate change. Cambridge University Press, Cambridge and New York, pp 1137–121
9. CODGiK (2014) Numeryczne Dane Wysokościowe. http://www.codgik.gov.pl/index.php/zasob/numeryczne-dane-wysokosciowe.html. Accessed 11 Dec 2014
10. CODGiK (2016) Baza Danych Obiektów Topograficznych BDOT10k http://www.codgik.gov.pl/index.php/zasob/baza-danych-obiektow-topograficznych.htm l. Accessed 27 Jul 2016
11. CSO (2015). http://stat.gov.pl/en/. Accessed 12 Sept 2015
12. Deng J, Zhang WY, Schneider R, Harff J, Dudzińska-Nowak J, Terefenko P, Giza A, Furmańczyk K (2014) A numerical approach for approximating the historical morphology of wave-dominated coasts—a case study of the Pomeranian Bight, southern Baltic Sea. Geomorphology 204:425–443. doi:10.1016/j.geomorph.2013.08.023
13. Eijgenraam CJJ (2006) Optimal safety standards for dike-ring areas. CPB Netherlands Bureau for Economic Policy Analysis Discussion Paper 62
14. Ekman M (2009) The changing level of the Baltic Sea during 300 years: a clue to understanding the Earth. Summer Institute for Historical Geophysics, Åland Islands
15. Emschergenossenschaft/Hydrotec (2004) Hochwasser-Aktionsplan Emscher—Methodik der Schadensermittlung. http://www.eglv.de/fileadmin/EmscherGenossenschaft/2.3_Flussgebietsmanagement/Karten_Anlage/Methodik_der_Schadensermittlung/methodik_schadensermittlung.pdf. Accessed 12 Sept 2015
16. EU (2007) Directive 2007/60/EC of the European Parliament and of the Council of 23 October 2007 on the assessment and management of flood risks. OJ L 288:27–34
17. Eurostat (2015). http://epp.eurostat.ec.europa.eu/portal/page/portal/statistics/themes. Accessed 27 Aug 2015
18. Furmańczyk K, Andrzejewski P, Benedyczak R, Bugajny N, Cieszyński Ł, Dudzińska-Nowak J, Giza A, Paprotny D, Terefenko P, Zawiślak T (2014) Recording of selected effects

and hazards caused by current and expected storm events in the Baltic Sea coastal zone. J Coast Res Special Issue 70:338–342. doi:10.2112/SI70-057.1
19. Gräwe U, Burchard H (2012) Storm surges in the Western Baltic Sea: the present and a possible future. Clim Dyn 39:165–183. doi:10.1007/s00382-011-1185-z
20. Hallegatte S (2012) A framework to investigate the economic growth impact of sea level rise. Environ Res Lett 7:015604. doi:10.1088/1748-9326/7/1/015604
21. Hinkel J, Nicholls RJ, Vafeidis AT, Tol RS, Avagianou T (2010) Assessing risk of and adaptation to sea-level rise in the European Union: an application of DIVA. Mitig Adapt Strateg Glob Change 15:703–719. doi:10.1007/s11027-010-9237-y
22. Hinkel J, Linckea D, Vafeidis AT, Perrettec M, Nicholls RJ, Tol RSJ, Marzeion B, Fettweis X, Ionescu C, Levermann A (2014) Coastal flood damage and adaptation costs under 21st century sea-level rise. Proc Natl Acad Sci USA 111:3292–3297. doi:10.1073/pnas.1222469111
23. Huizinga H (2007) Flood damage functions for EU member states. JRC-Institute for Environment and Sustainability/HKV Consultants: Lelystad
24. Hunter J (2010) Estimating sea-level extremes under conditions of uncertain sea-level rise. Clim Change 99:331–350. doi:10.1007/s10584-009-9671-6
25. Jonkman SN, Hillen MM, Nicholls RJ, Kanning W, van Ledden M (2013) Costs of adapting coastal defences to sea-level rise—new estimates and their implications. J Coast Res 29:1212–1226
26. Kok M, Huizinga H, Vrouwenvelder A, Barendregt A (2004) Standaardmethode2004—Schade en Slachtoffers als gevolg van overstromingen. DWW-2005-005, RWS Dienst Weg- en Waterbouwkunde
27. Kron F (2005) Flood risk = hazard * values * vulnerability. Water Int 30(1):58–68. doi:10.1080/02508060508691837
28. Kummu M, de Moel H, Salvucci G, Viviroli D, Ward PJ, Varis O (2016) Over the hills and further away from coast: global geospatial patterns of human and environment over the 20th–21st centuries. Environ Res Lett 11(3):034010. doi:10.1088/1748-9326/11/3/034010
29. KZGW (2011) Raport z wykonania wstępnej oceny ryzyka powodziowego. KZGW, Warszawa. http://www.kzgw.gov.pl/pl/Wstepna-ocena-ryzyka-powodziowego.html. Accessed 10 Feb 2015
30. KZGW (2015) Informatyczny System Osłony Kraju. http://mapy.isok.gov.pl/imap/. Accessed 10 Feb 2015
31. Łonyszyn P, Terefenko O (2014) Creation of an alternative season based on sustainable tourism as an opportunity for Baltic Sea Region. J Coast Res Special Issue 70:454–459. doi:10.2112/SI70-077.1
32. Merz B, Kreibich H, Thieken A, Schmidtke R (2004) Estimation uncertainty of direct monetary flood damage to buildings. Nat Hazards Earth Syst Sci 4:153–163. doi:10.5194/nhess-4-153-2004
33. Ministry of Treasury (2012) Sprawozdanie o stanie mienia Skarbu Państwa. Warsaw. http://bip.msp.gov.pl/download/1/5554/Sprawozdanie2011.pdf. Accessed 10 Feb 2015
34. Mokrech M, Kebede AS, Nicholls RJ, Wimmer F, Feyen L (2015) An integrated approach for assessing flood impacts due to future climate and socio-economic conditions and the scope of adaptation in Europe. Clim Change 128:245–260. doi:10.1007/s10584-014-1298-62015
35. Muis S, Guneralp B, Jongman B, Aerts JCJH, Ward PJ (2015) Flood risk and adaptation strategies under climate change and urban expansion: a probabilistic analysis using global data. Sci Total Environ 538:445–457. doi:10.1016/j.scitotenv.2015.08.068
36. Nicholls RJ, Klein RJT (2005) Climate change and coastal management on Europe's coast. In: Vermaat JE, Bouwer L, Turner K, Salomons W (eds) Managing European coasts: past, present and future, environmental science monograph series. Springer, Berlin, pp 199–226
37. Nicholls RJ, Mimura N (1998) Regional issues raised by sea-level rise and their policy implications. Clim Res 11:5–18

38. NOAA (2016) Sea level rise maps. http://www.star.nesdis.noaa.gov/sod/ lsa/SeaLevelRise/LSA_SLR_maps.php. Accessed 27 Jul 2016
39. Paprotny D (2014) Trends in storm surge probability of occurrence along the Polish Baltic Sea coast. arXiv: 1410.2547
40. Peltier WR, Argus DF, Drummond R (2015) Space geodesy constrains ice-age terminal deglaciation: the global ICE-6G_C (VM5a) model. J Geophys Res Solid Earth 120:450–487. doi:10.1002/2014JB011176
41. Permanent Service for Mean Sea Level (2015). http://www.psmsl.org/data/obtaining/map.html. Accessed 8 Jan 2015
42. Pfeffer WT, Harper JT, O'Neel S (2008) Kinematic constraints on glacier contributions to 21st century sea level rise. Science 321:1340–1343. doi:10.1126/science.1159099
43. Piketty T, Zucman G (2014) Capital is back: wealth-income ratios in rich countries, 1700–2010. Q J Econ 129:1155–1210. doi:10.1093/qje/qju018
44. Pluijm M, Toms G, Zeidler RB, van Urk A, Misdorp R (1992) Vulnerability assessment to accelerated sea level rise: case study Poland. Ministry of Transport, Public Works and Water Management, The Hague
45. Poulter B, Halpin PN (2008) Raster modelling of coastal flooding from sea-level rise. Int J Geogr Inf Sci 22:167–182. doi:10.1080/13658810701371858
46. Pruszak Z, Zawadzka E (2005) Vulnerability of Poland's coast to sea-level rise. Coast Eng J 47:131–155. doi:10.1142/S0578563405001197
47. Pruszak Z, Zawadzka E (2008) Potential implications of sea-level rise for Poland. J Coast Res 24:410–422. doi:10.2112/07A-0014.1
48. Rahmstorf S (2007) Sea-level rise a semi-empirical approach to projecting future. Science 315:368–370. doi:10.1126/science.1135456
49. Richter A, Groh A, Dietrich R (2012) Geodetic observation of sea-level change and crustal deformation in the Baltic Sea region. Phys Chem Earth 53–54:43–53. doi:10.1016/j.pce.2011.04.011
50. Rotnicki K, Borowka RK (1991) Impact of a future sea level rise in the Polish Baltic coastal zone. EPA International Sea Level Rise Studies Project Report. Division of Coastal and Environmental Studies. Rutgers University Institute of Coastal and Marine Sciences, New Brunswick
51. Rotnicki K, Borówka RK, Devine N (1995) Accelerated sea level rise as a threat to the Polish Coastal Zone—quantification of risk. J Coast Res Special Issue 22:111–135
52. Rotzoll K, Fletcher CH (2013) Assessment of groundwater inundation as a consequence of sea-level rise. Nat Clim Change 3:477–481. doi:10.1038/nclimate1725
53. Statistisches Bundesamt (2015) Bau- und Immobilienpreise. https://www.destatis.de/DE/ZahlenFakten/GesamtwirtschaftUmwelt/Preise/BauImmobilienpreise/BauImmobilienpreise.html. Accessed 21 Jan 2015
54. Tomczak A (1995) Geological structure and Holocene evolution of the Polish coastal zone. J Coast Res Special Issue 22:16–31Urbański J (2012) GIS w badaniach przyrodniczych. University of Gdańsk, Gdańsk
55. Van Dantzig D (1956) Economic decision problems for flood prevention. Econometrica 24:276–287
56. Vousdoukas MI, Voukouvalas E, Annunziato A, Giardino A, Feyen L (2016) Projections of extreme storm surge levels along Europe. Clim Dyn. doi:10.1007/s00382-016-3019-5
57. Webster TL (2010) Flood risk mapping using LiDAR for Annapolis Royal, Nova Scotia, Canada. Remote Sens 2:2060–2082. doi:10.3390/rs2092060
58. Wiśniewski B, Kowalewska-Kalkowska H (2007) Water level fluctuations in the Odra River mouth area in relation to passages of deep low-pressure systems. Oceanol Hydrobiol Stud 36:69–82. doi:10.2478/v10009-007-0009-2
59. Wiśniewski B, Wolski T (2009) Katalogi wezbrań i obniżeń sztormowych poziomów morza oraz ekstremalne poziomy wód na polskim wybrzeżu. Maritime University of Szczecin, Szczecin
60. Wiśniewski B, Wolski T, Musielak S (2011) A long-term trend and temporal fluctuations of the sea level at the Polish Baltic coast. Oceanol Hydrobiol Stud 40:96–107. doi:10.2478/s13545-011-0020-9

61. Wolski T, Wiśniewski B, Giza A, Kowalewska-Kalkowska H, Boman H, Grabbi-Kaiv S, Hammarklint T, Holfort J, Lydeikaitė Ž (2014) Extreme sea levels at selected stations on the Baltic Sea coast. Oceanologia 56:259–290. doi:10.5697/oc.56-2.259
62. Wróblewski A (1994) Analysis and forecast of long-term sea level changes along the Polish Baltic Sea coast. Part II. Annual mean sea levels—forecast to the year 2100. Ocealonogia 36:107–120
63. Xu S, Huang W (2013) Effects of sea level rise on frequency analysis of 1 % annual maximum water levels in the coast of Florida. Ocean Eng 71:96–102. doi:10.1016/j.oceaneng.2013.01.013
64. Zeidler RB (1997) Climate change vulnerability and response strategies for the coastal zone of Poland. Clim Change 36:151–173. doi:10.1023/A:1005394909182

CHAPTER 11

Coastal infrastructure vulnerability: an integrated assessment model

Komali Kantamaneni[1,2]

[1] *Faculty of Architecture, Computing and Engineering, University of Wales, Trinity Saint David, Mount Pleasant, Swansea SA1 6ED, UK*

[2] *School of Maritime and Engineering, Southampton Solent University, E Park Terrace, Southampton, Hampshire SO14 0YN, UK*

ABSTRACT

The significance of coastal regions to the infrastructure and the need to protect such assets are crucial to the economy of countries. Therefore, there is a real need to enhance the understanding of coastal infrastructure susceptibility as well as to develop methodologies to estimate vulnerability. A review of the literature regarding coastal vulnerability reveals that the focus has been on geomorphological and physical parameters but not infrastructure and the associated fiscal factors. In order to address this knowledge gap, an innovative model is developed, i.e., the Coastal Infrastructure Vulnerability Index (CIVI). Then the model is applied to the case of the Aberystwyth coast demonstrating how the model estimates the vulnerability of the coastal infrastructure (comprising population, commercial and residential properties). Subsequently, the CIVI scores were used to rank coastal sections into five classes, ranging from extremely low to extremely high, based on the relative magnitude of the vulnerability. The rankings for each parameter were combined, and then an index value was calculated. Results revealed that Aberystwyth contains more than £40 billion of coastal infrastructure vulnerability and more than 10,000 inhabitants are at the high coastal risk posed by flooding, erosion, storm surges, and strong winds.

Keywords: Integrated model, Coastal infrastructure, vulnerability index, (CIVI) Aberystwyth Costs

1 INTRODUCTION

Several coastal regions across the world are endangered to both natural and anthropogenic hazards, which are expected to increase in the near future (Nicholls et al. 2007; Kron 2008; Weisse et al. 2012). The construction of a variety of infrastructures such as properties (commercial and residential), roads, ports, and breakwaters completely dominates the natural habitats and leads to further rapid coastal damage. Humans have changed coastal regions by introducing artificial constructions in 2580 BCE on the Red Sea shores in the Egypt (Tallet and Marouard 2014), and these structures affect geomorphology and coastal systems (Bulleri and Chapman 2010) in a negative way; however, this impact severity depends on the particular geographical area. Climate change induced elements such as sea-level rise, coastal flooding, erosion, and storm surge are the main reasons for coastal infrastructure damage as well as vulnerability (Dolan and Walker 2006; Phillips and Jones 2006; Bosello and De Cian 2014). Increased weather events also affect the socio-economic circumstances of coastal regions significantly (Hinkel et al. 2010). Therefore, coastal infrastructure vulnerability needs assessment to a greater degree to ameliorate existing problems and to prevent further decline.

1.1 Coastal vulnerability appraisal methods

Since three decades several works have been made to establish strategies and procedures for evaluating coastal vulnerability to climate change and other related aspects, accompanied with economics (Cutter et al. 2003; Lewsey et al. 2004; Vincent 2004; Rygel et al. 2006; Phillips and Jones 2006; Hinkel et al. 2009; Torresan et al. 2012; Addo 2013; Tang et al. 2013; Wolters and Kuenzer 2015; Denner et al. 2015; Wu et al. 2016). A summary of several methodologies established and applied globally is provided here. The four primary methods (Ramieri et al. 2011) to evaluate coastal vulnerability to climate change are as follows:

1) Index-based methodology

2) Indicator-based methodology

3) GIS (geographical information systems) based decision support systems

4) Dynamic computer models

1.1.1 Index-based methodology

Index-based methods evaluate coastal vulnerability by a single magnitude and are normally unit less. This method measures by the quantitative or semi-quantitative assessment as well as an amalgamation of diverse variables. These methods are not directly transparent since the final index does not allow for the understanding of the expectations and combinations that led to its measurement. The CVI outcomes can be shown on vulnerability maps at various scales to identify regions where the elements that add to coastal changes make greatest contributions to coastline retreat (Harvey and Woodroffe 2008; Pendleton et al. 2010). First coastal vulnerability index was developed by Gornitz (1990) followed by several researchers developing diverse CVI indices across the globe (McLaughlin and Cooper 2010; Palmer et al. 2011; Yin et al. 2012; Denner et al. 2015).

1.1.2 Indicator-based methodology

An indicator-based index is a popular tool for measuring the intensity of exposure of communities to hazards and coastal vulnerability. The index comprises several indicators, which are interlinked with the specific formula. In recent decades, some researchers have established several vulnerability indicators within the socio-economic and ecological system context (King and MacGregor 2000; Brooks et al. 2005; Barnett et al. 2008; Torresan et al. 2008; Abson et al. 2012; Balica et al. 2012).

1.1.3 GIS-based decision support systems (GIS-DSS)

GIS-based methodologies are useful for evaluating the physical condition of a particular system and the development of risk maps. It is useful to acquire further information about the physical geography of particular region through computerisation to transmute vast databases into thematic maps. The GIS-DSS is two types; development of an information technology tool for the management of Southern European lagoons under the influence of river-basin runoff (DITTY-DSS) (Agnetis et al. 2006; Mocenni et al. 2009; Casini et al. 2015) and decision support system for coastal climate change impact assessment (DESYCO-DSS) (Santoro et al. 2013; Zanuttigh et al. 2014).

1.1.4 Dynamic computer models

Dynamic computer simulations are useful for analysing and mapping susceptibility and risks of coastal systems (Cowell et al. 1995; Brown et al.

2006). Available methods for this procedure can be divided into two parts; sector models and integrated assessment models. Sector models focus on the examination of coastal vulnerability linked to a specific coastal system, and integrated assessment models appraise the coastal vulnerability systems to multiple climate change impacts (Mcleod et al. 2010).

1.2 Common framework for evaluation of coastal vulnerability

In 1990, IPCC published a standard methodology for assessing the vulnerability of coastal areas to sea-level rise; it contains seven systematic stages (seven indicators) (Table 1) that permit for the identification of population, natural and physical resources at risk and costs and possibility of potential responses to adverse impacts (Nicholls 1995).

Though several other methods have been developed for coastal vulnerability assessment, most of the researchers focused their interest on the index- and indicator-based methods. However, there is no standard evaluation method on town/city scale to estimate the current infrastructure vulnerability of the Aberystwyth at current scenarios. Therefore, this study developed an integrated CIVI and subsequently analysed the infrastructure vulnerability of the population, commercial and residential properties of Aberystwyth, UK.

2 DESCRIPTION OF STUDY AREA

Aberystwyth (52°25′N 4°05′W) is a small sea-side (Irish Sea) town in the county of Ceredigion in the Wales, UK (Aberystwyth Guide 2014) (Fig. 1), located towards the centre of the falcate of Cardigan Bay and also positioned between three hills. Aberystwyth is the main touristic spot and administrative region of the west coast of Wales; though it has a small coastline (>2 km) (Aberystwyth Guide 2014), it has a high socio-economic value because the town's economy is based mainly on tourism, education, and retail sectors.

2.1 Coastal infrastructure damage

Frequent storm strikes in Aberystwyth are not unusual phenomena. Starting several decades ago, repeated storms ravaged this region and damaged several £million to £billion worth of infrastructure (Fig. 2a–d as evidenced in 1927, 2008, 2013 and 2014 (major events). In particular, the 2014 massive tides (>20 feet) damaged >2 km of railway track between

Aberystwyth and Machynlleth severely, and nearly 2 weeks were needed for repairs. In addition, the widespread destruction of the seawall and walking path and the flooding of more than ten houses occurred (Welsh Government 2014). Future climatic conditions and levels of damage due to various coastal hazards (Table 2) will worsen the situation if strict adaptation and coastal defence procedures are not implemented in the near future (Slingo et al. 2014).

Table 1 Indicators of the Coastal vulnerability—IPCC common approach (Gilbert and Vellinga 1990)

Indicator	Explanation
Affected Population	Number of people living in the hazard region affected by sea-level rise
Population at risk	The average annual number of people inundated by storm surge
Investment value at loss	Current market value of infrastructure which could be lost due to sea-level rise
Land at loss	Size of land that would be lost due to rise of sea-level rise
Wetland at loss	Size of area of wetland that would be lost due to sea-level rise
Adaptation costs	Adaptation costs to sea-level rise, with an overwhelming importance on defence
Population at risk	The average number of people flooded by storm surge per year, supposing the cost of adaptation to be in residence

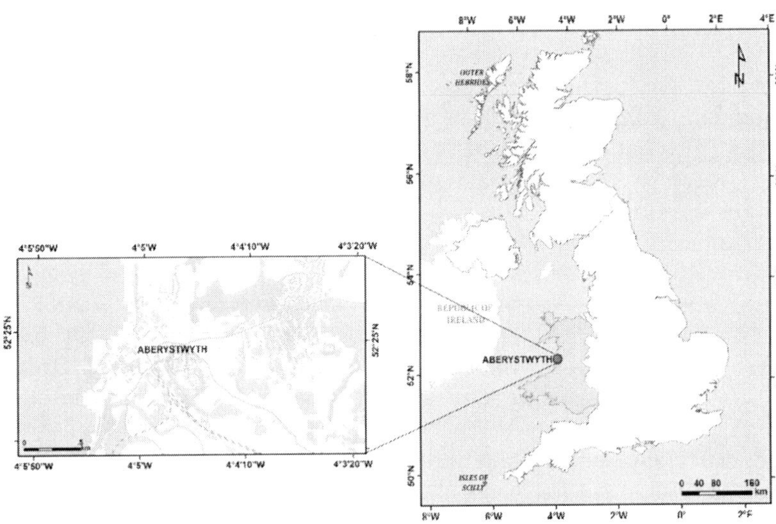

Figure. 1 Map of the study area

Figure. 2 Decadal Infrastructure Damage in Aberystwyth. a Residential property damage in 1927 (British Geological Survey 2014), b residential property damage in 1938 (British Geological Survey 2014), c commercial property damage in 2012 (Morris 2012), d commercial and residential property damage in 2014 (The Times 2014)

3 DATA

Population, commercial and residential properties data were obtained from ONS (Office for National Statistics), local and sub-local Councils of Wales and Aberystwyth; fiscal data of commercial and residential properties obtained from HM Revenue and Customs (HMRC) offices of Wales and the Agricultural Mortgage Corporation (AMC). Along with the information as mentioned above, this study also utilised the data obtained by multiple observations of the coastal site of the Aberystwyth over 3 years (2012–2015) period. Parameter's statistics of each coastal cell (at 0.5 km resolution) was determined by using orthophotographs of Ordnance Survey, Welsh Assembly Government—Aerial Photographs and Google Earth maps. However, current study only used the data of rateable properties and did not take into consideration some heritage properties such as church and museums or massive structures like bridges and other constructions. While SPSS (statistical package for social sciences) (21st version) was used for analysis and exploration of CIVI values and furtherer construction of CIVI. ArcGIS (10.3 version) and Welsh Assembly Government Arial and Google (Open Street) maps were used in the development of coastal vulnerability maps in various scales.

Table 2 Coastal hazards of the Aberystwyth (Adopted from Ramieri et al. 2011)

Principal Marine regions	Coastal hazards	Vulnerability/impacts
Celtic Sea	Coastal flooding	Infrastructure damage
	Various storm events	Community loss
	Coastal erosion	Land degradation
	Storm surge	Damage to the coastline
	High waves, tides and winds	Damage to infrastructure
Irish Sea	Coastal flooding	Infrastructure damage
	Various storm events	Community loss
	Coastal erosion	Land degradation
	Storm surge	Damage to infrastructure
	High waves	Damage to coastal wall and infrastructure

4 METHODOLOGY

4.1 Development of an integrated model

There is a real need to evaluate and compare the intensity of vulnerability of different sites, zones, and nations across the globe. The familiarity of coastal vulnerability will allow the researchers, policy, as well as decision makers to predict and perform on the adverse scenarios of existing and upcoming vicissitudes ensuing from global and regional sea-level rise and other impacts of climate change. A generalised and simple framework is required to illuminate explicit communication regarding coastal vulnerability and expressive comparison among susceptibility appraisals. Several kinds of technical, natural and social methods have already been employed (mentioned in the introduction section), but the process of applying the framework regionally and globally (from a fiscal perspective) is still in the embryonic stage. A definite procedure is then required to categorise the vulnerability of the coastal infrastructure; accordingly, a novel integrated model has been developed for the evaluation of coastal infrastructure vulnerability of the Aberystwyth coast, i.e., CIVI.

Two coastal vulnerability index (CVI) approaches were adopted for this study, based on an adaptation of the work of Balica et al. (2012) and Palmer et al. (2011). Accordingly, an integrated model (Fig. 3) was established to evaluate the vulnerability of the Aberystwyth coast by amalgamating indicator and index-based methods. The fiscal parameters were selected using the indicator-based method of Balica et al. (2012), and the concept of development of CIVI was taken from the index-based approach of Palmer et al. (2011). The fiscal values threshold for parameters was inspired by Aberystwyth fiscal consequences.

4.2 Fiscal parameters selection

Current study scrutinised various events in relation to a coastal vulnerability in the UK, such as population, commercial and residential properties, storm conditions, rainfall trends, coastal erosion, etc. Based on the analysis of various conditions, which are presented in Table 3, twelve parameters were selected, taking into account the UK coastal regions and their susceptibility and exposure to the coastal vulnerability events.

4.2.1 Reduction of parameters

Reduction of parameters for an evaluation of fiscal coastal vulnerability at city or town scale is necessary. A large number of parameters (12) does not offer factual results in this particular scenario, so to simplify the methodological process, they are reduced and restricted to 3, based on the potentiality of the parameters (Table 4). Parameter reduction is not a new procedure in coastal vulnerability assessment studies, and several researchers have already implemented this technique successfully. Balica et al. (2012) initially considered 71 indicators and then reduced their number to 12, and the Canadian Council of Ministries of Environment (2003) selected nearly 100 indicators, which were reduced to 12 as well.

4.2.2 Parameters description

The population is widely accepted imperative parameter in both physical and socio-economic sections of coastal research, and it also considers as one of the vital infrastructures (Simone 2004). Current study measures the population in monetary terms and sets a cost to the human life based on US—2011—Environment Protection Agency estimations, i.e., £6.9 million (adjusted for 2015 inflation £rates) (Appelbaum 2011). However, Aberystwyth population has diverse age groups and communities with different economic status; therefore, this study offers on average £4 m to the life of the UK (Aberystwyth) people at current scenarios.

Residential and commercial properties are also important parameters in the coastal vulnerability studies. Using these structures as parameters in coastal vulnerability studies is not new; several researchers used in their studies to evaluate vulnerability in both physical and socio-economic studies throughout the world (Klein et al. 2003; Jacob et al. 2007; Kubal et al. 2009; Thatcher et al. 2013; Arkema et al. 2013; Wu et al. 2016;

Mazumdar and Paul 2016). The Economic threshold was offered by identifying a number of properties in 0.5 km cells and then estimated the commercial value of those properties and then provided the range of values from extremely low to extremely high.

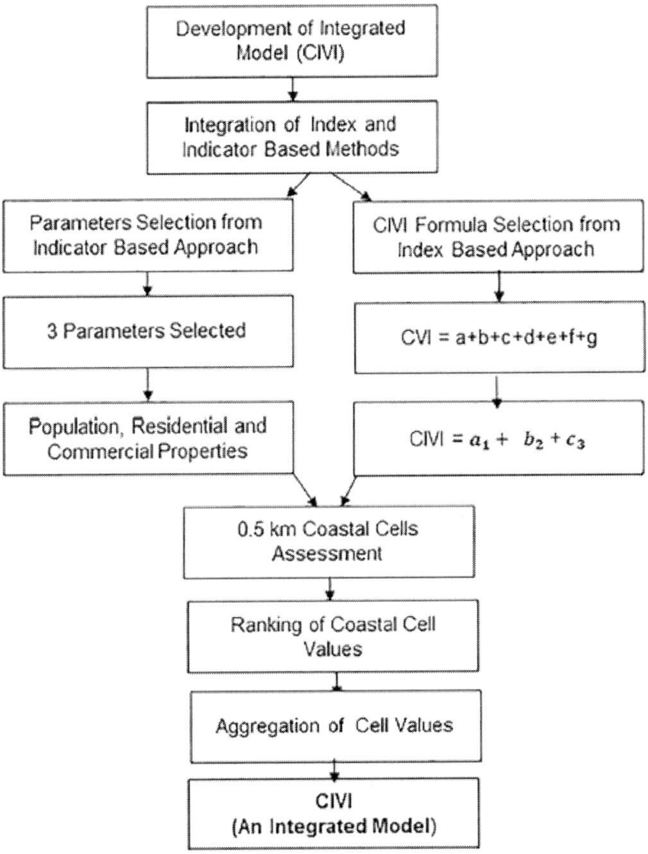

Figure. 3 Flow chart of development of integrated model

4.3 Technical description and calculation of CIVI

A certain length of transect line was drawn on the Aberystwyth coastline, and then a 0.5 km square measurement was placed on the transect line from the coast point to outside of coast, i.e., towards the civilisation/communities (Fig. 4) to appraise coastal infrastructure vulnerability within the economic perspective, while, as shown in Fig. 4, second and third cells are overlapped. Therefore, these overlapped properties did not take into consideration for an evaluation. However, uncovered properties which are located in-between the cells of first and second as well as third and fourth are taken into account for an assessment. This consideration helps to attain factual fiscal figures.

Table 3 Parameters selection procedure

Number	Parameter	Selection process and reason
1	Population in coastal vulnerability zones	More than 50 % of population is living near the coastline (Small and Nicholls 2003)
2	Infrastructure (properties, roads, etc.)	Nearly 6 million properties (one in six) are at risk of coastal flooding and erosion (Ramsbottom et al. 2012)
3	Land use	Around 60 % of the best agricultural land is 5 m or less above sea level (Zsamboky et al. 2011)
4	Rainfall	Heavy rainfall trends across the country in recent decades (Osborn and Hulme 2002; Maraun et al. 2008)
5	Flood/storm impact	Increased severity of flood/storm impact in recent periods (Reynard et al. 2001; Woodworth et al. 2007; Stevens et al. 2014)
6	Fiscal value of the place	The economic value of the area plays a vital role in fiscal studies, as well as disaster management studies (Swarbrooke 1999; Haigh and Amaratunga 2010)
7	Coastal erosion	High coastal erosion at some regions of the UK (Phillips and Jones 2006; Kantamaneni and Phillips 2016)
8	High growth of population alongside coasts	High growth of population expected alongside coasts (Small and Nicholls 2003).
9	Drainage system	A poor drainage system causes severe problems particularly when flooding strikes (Coulthard and Frostick 2010; Butler and Davies 2004)
10	Warning system	The robustness of the UK's warning system for natural disasters, e.g. no warning system during tornado strikes (Kantamaneni et al. 2015)
11	Marine Industry Growth	Maritime industries and the service sector annually contributes >£17 billion to the UK economy and it will be £25 billion by 2020 (Marine Industries Leadership Council 2011)
12	Politics and Policies	Changing the political situation also plays a vital role in the assessment of coastal vulnerability (White and Howe 2002; Patt et al. 2009)

Table 4 Fiscal (coastal infrastructure vulnerability) parameters

No.	Parameter	Designated symbol for CIVI	Measurable units (£)
1	Population	$a1$	Million–billion (economic)
2	Residential property	$b2$	Million–billion (economic)
3	Commercial property	$c3$	Million–billion (economic)

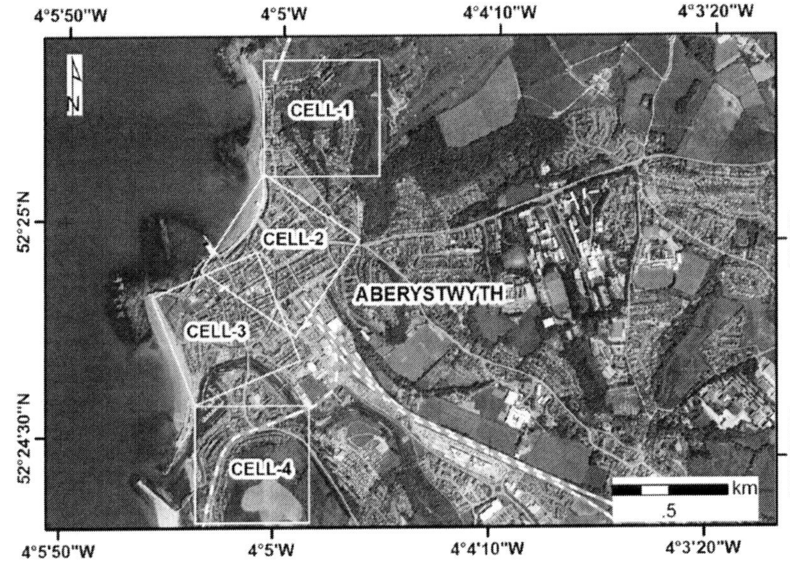

Figure. 4 Coastal cell (0.5 km) on transect line

Besides that, fiscal parameter ranking was (Table 5) used to measure the coastal vulnerability and subsequently each cell was assigned a CIVI score and then all the cells of the parameters were calculated.

This study categorised the CIVI scores into five categories: extremely low (1), low (2), moderate (3), high (4) and extremely high (5). The scores of all cells of three parameters aggregated to rank the coastal infrastructure vulnerability. With rankings applied these values were then put into a simple equation (Eq. 1) to analyse CIVI score for each coastal section. Simple summation of individual rankings provided a total relative vulnerability score. The minimum possible score was 3, and the maximum was 15.

$$\text{Relative CIVI} = a_1 + b_2 + c_3 \tag{1}$$

a_1—Population	5 (max)
b_2—Residential property	5 (max)
c_3—Commercial Property	5 (max)

Maximum CIVI score	15
Minimum CIVI score	3

Table 5 CIVI Parameter Ranking

Fiscal Parameter	1 Extremely Low £	2 Low £	3 Moderate £	4 High £	5 Extremely High £
Population	<1 bn	1 bn–6 bn	>6 bn–12 bn	>12 bn–20 bn	>20 bn–28 bn
Residential property	<30 m	30 m–80 m	>80n–30 m	>130 m–180 m	>180 m–1 bn
Commercial property	<2 m	2 m–10 m	>10 m–30 m	>30 m–70 m	>70–300 m

(*m* million, *bn* billion)

Coastal sections scoring within the mid-range (7–9) were ranked as moderate vulnerability, and coastal sections scoring in between 13 and 15 were categorised as an extremely high vulnerability. However, the ranking system as follows (Table 6);

5 RESULTS AND DISCUSSION

More than two km transect line was drawn on Aberystwyth coast and accordingly, four 0.5 km cells placed and then measured. Currently, this town has thousands of coastal population and very expensive properties (commercial and residential) (Table 7).

Moreover, Aberystwyth is critically vulnerable to wave attacks and high tides (Fig. 5). For several decades, it has been affected severely by a series of storms with high waves, tides, and storm surges, particularly in 2008, 2010, 2013, and 2014. Specifically, 2014 storm ravaged this region and caused an astonishing >£1.5 m worth of damage to the infrastructure (Ceredigion County Council 2014). The tidal range in the area is always greater than on other sites, with the largest increase of waves coming from the south-west, which is also the direction of the most frequent storms.

Though Aberystwyth has a shorter coastline, it has a high coastal infrastructure vulnerability and comparatively uneven across the coast. Some coastal cells are extremely high, and some are high to moderate vulnerability. Extremely high CIVI scores were recorded at the second and third cells, high at the fourth cell, and moderate at the first cell (Fig. 6). Currently, Aberystwyth has >40 bn worth of CIV (Fig. 7) with 4613 residential properties, 530 commercial properties, and >10,017 people are at high coastal risk.

Table 6 Vulnerability level ratings classified by total relative vulnerability score

Relative vulnerability score	Rating of vulnerability
Extremely high vulnerability	13–15
High vulnerability	10–12
Moderate vulnerability	7–9
Low vulnerability	4–6
Extremely low vulnerability	1–3

Table 7 Infrastructure details of Aberystwyth

Infrastructure name	Number
Commercial properties	758
Residential properties	6591
Population	15,139

Figure. 5 a Construction of new coastal wall (during 2014 flood strike, coastal wall was severally damaged), **b** commercial and residential properties near the coastline in Aberystwyth (photographs were taken by author—2015)

6 LIMITATIONS

Due to the lack of recent literature (most of the research information is more than 10 years old) on Aberystwyth coastal vulnerability in both physical and fiscal aspects, there is not much scope to compare with other similar existing studies, especially at regional and city/town scales. With these concerns in mind, the subsequent vital subjects in any effort of simplification of the findings would need to be certified carefully, such as GDP, local economy, and redevelopment procedures.

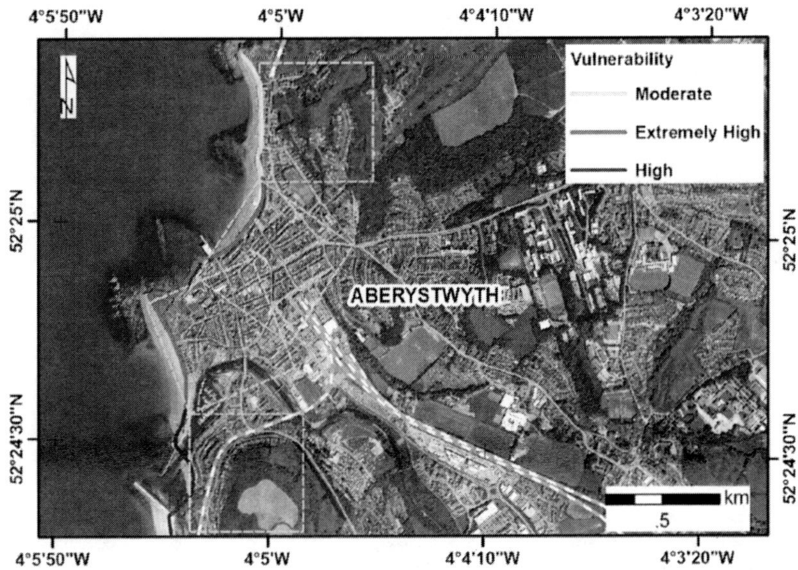

Figure. 6 Coastline vulnerability rates superimposed upon the aerial photograph

7 CONCLUSION

This study revealed the coastal infrastructure vulnerability of the Aberystwyth by the establishment of a novel integrated model, i.e., CIVI. Fiscal parameters that considered the existing economic conditions of the population, commercial and residential properties were used to appraise the relative economic coastal infrastructure vulnerability. Results showed that Aberystwyth consists of >£40 billion worth of CIV. Efficient and factual results for CIVI computation are intensely reliant on the quality and varied type of data used, which influence the vulnerability of a particular coastal stretch. This is a statistical and objective approach to illustrate the intensity of coastal infrastructure vulnerability at Aberystwyth coast. This integrated method enables the production of statistics and quantification of different levels of vulnerability to fulfil standards substitute to regional, local and sub-local authorities in the nationalised policy for control of climate change related coastal hazards in North Wales, UK. This technique of assessing infrastructure vulnerability can purpose as a primary susceptibility appraisal from which a map of probable intensities of vulnerability can be generated to allow cost–benefit scrutinise. The use of an appraisal of coastal infrastructure vulnerability can also be employed to define fiscal viability of coastal defence and the distribution of compelled funding. This model will be very useful to coastal economists, engineers and planning managers for better planning to reduce the coastal vulnerability.

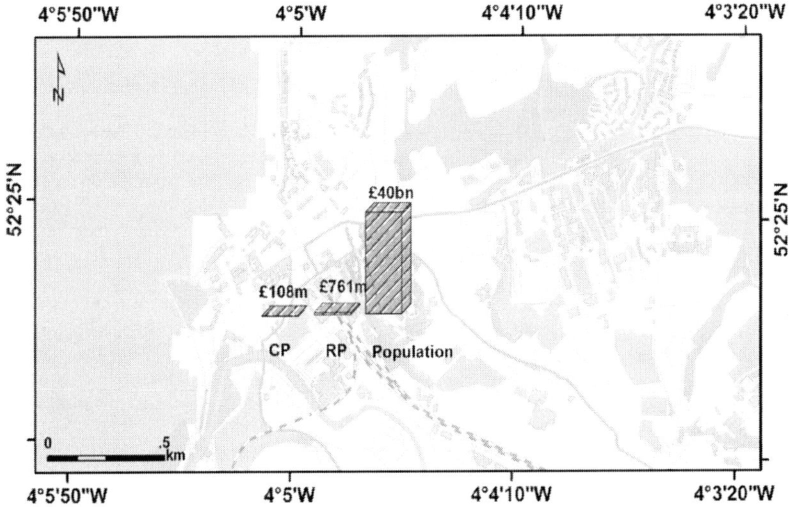

Figure. 7 Fiscal coastal vulnerability map of Aberystwyth (CP-commercial properties; RP-residential properties)

ACKNOWLEDGMENTS

I would like to thank Professor Mike Phillips, Dr. Rhian Jenkins (University of Wales Trinity Saint David, UK) and Professor Nagesh Kumar (Indian Institute of Science, Bengaluru, India) for much valuable discussion, who supported my work with vital insights from their vast experience. A particular acknowledgement is made to my work associates Dr. Talib Butt and Dr. Tony Thomas for their valuable comments on the conceptual framework. I wish to give special thanks for the comments of anonymous reviewers on an earlier version, which contributed significantly to the improvement of the manuscript. I am also very grateful to the staff of Aberystwyth Council for providing updated statistics of population, commercial and residential properties and to Welsh Assembly Government, Aerial Photographs Unit, Cardiff, Wales, for the aerial photographs used in the study.

REFERENCES

1. Aberystwyth Guide (2014) Introduction. http://www.aberystwythguide. org.uk/aberpix.shtml. Accessed 13 Jan 2016
2. Abson DJ, Dougill AJ, Stringer LC (2012) Using principal component analysis for information-rich socio-ecological vulnerability mapping in Southern Africa. Appl Geogr 35:515–524
3. Addo KA (2013) Assessing coastal vulnerability index to climate change: the case of Accra-Ghana. J Coastal Res 2:1892
4. Agnetis A, et al (2006) Development of a decision support system for the management of Southern European lagoons Center for Complex Systems Studies University of Siena, Siena, Italy Disponível em. http://www.dittyproject.org/Article/WP8_DSS_ structure_final.pdf
5. Appelbaum B (2011) As US agencies put more value on a life, businesses fret. The New York Times, 16. http://www.nytimes.com/2011/02/17/business/economy/ 17 regulation.html?_r=0. Accessed 10 Mar 2014
6. Arkema KK et al (2013) Coastal habitats shield people and property from sea-level rise and storms. Nat Climate Change 3:913–918
7. Balica S, Wright NG, van der Meulen F (2012) A flood vulnerability index for coastal cities and its use in assessing climate change impacts. Nat Hazards 64:73–105
8. Barnett J, Lambert S, Fry I (2008) The hazards of indicators: insights from the environmental vulnerability index. Ann As Am Geogr 98:102–119
9. Bosello F, De Cian E (2014) Climate change, sea level rise, and coastal disasters. A review of modeling practices. Energy Econ 46:593–605. doi:10.1016/j.eneco.2013.09.002
10. British Geological Survey (2014) BGS Geo-heritage—images from the collections http://britgeoheritage.blogspot.co.uk/2013/06/storm-damage-in-aberystwyth-october.html. Accessed 25 July 2015
11. Brooks N, Adger WN, Kelly PM (2005) The determinants of vulnerability and adaptive capacity at the national level and the implications for adaptation. Global Environ Change 15:151–163
12. Brown I, Jude S, Koukoulas S, Nicholls R, Dickson M, Walkden M (2006) Dynamic simulation and visualisation of coastal erosion. Comput Environ Urban Syst 30:840–860
13. Bulleri F, Chapman MG (2010) The introduction of coastal infrastructure as a driver of change in marine environments. J Appl Ecol 47:26–35
14. Butler D, Davies J (2004) Urban drainage. CRC Press, Boca Raton
15. Canadian Council of Ministries of the Environment (2003) Climate, nature, people: indicators of Canada's climate change. www.ccme.ca. Accessed 28 June 2013
16. Casini M, Mocenni C, Paoletti S, Pranzo M (2015) Decision support system development for integrated management of European coastal lagoons. Environ Model Softw 64:47–57

17. Ceredigion County Council (2014) 2014 storm damage and repair costs. http://www.ceredigion.gov.uk/English/Your-Council/Pages/default.aspx. Accessed 22 June 2015
18. Coulthard T, Frostick L (2010) The Hull floods of 2007: implications for the governance and management of urban drainage systems. J Flood Risk Manag 3:223–231
19. Cowell P, Roy P, Jones R (1995) Simulation of large-scale coastal change using a morphological behaviour model. Mar Geol 126:45–61
20. Cutter SL, Boruff BJ, Shirley WL (2003) Social vulnerability to environmental hazards*. Soc Sci Q 84:242–261
21. Denner K, Phillips M, Jenkins R, Thomas T (2015) A coastal vulnerability and environmental risk assessment of Loughor Estuary, South Wales. Ocean Coast Manag 116:478–490
22. Dolan AH, Walker I (2006) Understanding vulnerability of coastal communities to climate change related risks. J Coastal Res 3:1316–1323
23. Gilbert J, Vellinga P (1990) Strategies for adaption to sea level rise. Report of the Coastal Zone Management Subgroup, Intergovernmental Panel on Climate Change World Meteorological Organization and UN Environmental Programme: the Netherlands
24. Gornitz V (1990) Vulnerability of the East Coast, USA to future sea level rise. J Coastal Res 9:201–237
25. Haigh R, Amaratunga D (2010) An integrative review of the built environment discipline's role in the development of society's resilience to disasters. Int J Disaster Resilience Built Environ 1:11–24
26. Harvey N, Woodroffe CD (2008) Australian approaches to coastal vulnerability assessment. Sustain Sci 3:67–87
27. Hinkel J, Klein RJ (2009) Integrating knowledge to assess coastal vulnerability to sea-level rise: the development of the DIVA tool. Global Environ Change 19:384–395
28. Hinkel J, Nicholls RJ, Vafeidis AT, Tol RS, Avagianou T (2010) Assessing risk of and adaptation to sea-level rise in the European Union: an application of DIVA. Mitig Adapt Strat Global Change 15:703–719
29. Jacob KH, Gornitz V, Rosenzweig C (2007) Vulnerability of the New York City metropolitan area to coastal hazards, including sea-level rise: inferences for urban coastal risk management and adaptation policies. Manag Coastal Vulnerability: Global Reg Local Elsevier Publ 141–158
30. Kantamaneni K, Phillips M (2016) Transformation of climate: will floods and coastal erosion crumble the UK economy? Int J Climate Ch Impacts Responses 8:45–59
31. Kantamaneni K, Phillips M, Jenkins R, Oakley J, Ibeabuchi KO (2015) Could the UK economy be impacted by an increase in Tornado occurrence?: a consequence of climate change in the 21st century. Int J Climate Ch Impacts Responses 7:29–39
32. King D, MacGregor C (2000) Using social indicators to measure community vulnerability to natural hazards. Aust J Emerg Manag The 15:52
33. Klein RJ, Nicholls RJ, Thomalla F (2003) Resilience to natural hazards: how useful is this concept? Global environmental change. Part B Environ Hazards 5:35–45
34. Kron W (2008) Coasts—The riskiest places on Earth. In: Proceedings of the 31st international conference on coastal engineering (ICCE), World Scientific. Hamburg, Germany, pp 3–21
35. Kubal C, Haase D, Meyer V, Scheuer S (2009) Integrated urban flood risk assessment–adapting a multicriteria approach to a city. Nat Hazards Earth Syst Sci 9:1881–1895
36. Lewsey C, Cid G, Kruse E (2004) Assessing climate change impacts on coastal infrastructure in the Eastern Caribbean. Mar Policy 28:393–409
37. Maraun D, Osborn T, Gillett N (2008) United Kingdom daily precipitation intensity: improved early data, error estimates and an update from 2000 to 2006. Int Climatol 28:833–842
38. Marine Industries Leadership Council (2011) A strategy for growth for the UK marine industries. https://www.gov.uk/government/uploads/system/uploads/ attachment_data/file/31813/11-1310-strategy-for-growth-uk-marine-industries.pdf. Accessed 12 Aug 2014

39. Mazumdar J, Paul SK (2016) Socioeconomic and infrastructural vulnerability indices for cyclones in the eastern coastal states of India. Nat Hazards 1–23
40. McLaughlin S, Cooper JAG (2010) A multi-scale coastal vulnerability index: a tool for coastal managers. Environ Hazards 9:233–248
41. Mcleod E, Poulter B, Hinkel J, Reyes E, Salm R (2010) Sea-level rise impact models and environmental conservation: a review of models and their applications. Ocean Coast Manag 53:507–517
42. Mocenni C, Casini M, Paoletti S, Giordani G, Viaroli P, Comenges J-MZ (2009) A decision support system for the management of the Sacca di Goro (Italy). In: Marcomini A, Suter GW, Critto A (eds) Decision support systems for risk-based management of contaminated sites. Springer, pp 1–24
43. Morris K (2012) Summer 2012 Part 1: The Great Flood of June 9th 2012: off the scale. http://www.geologywales.co.uk/storms/summer12a.htm. Accessed 15 Mar 2014
44. Nicholls RJ, Leatherman SP, Dennis K, Volonte C (1995) Impacts and responses to sea-level rise: qualitative and quantitative assessments. J Coast Res 1:26–43
45. Nicholls RJ, Wong PP, Burkett VR, Codignotto JO, Hay JE, McLean RF, Ragoonaden S, Woodroffe CD (2007) Coastal systems and low-lying areas. In: Parry ML, Canziani OF, Palutikof JP, van der Linden PJ, Hanson CE (eds) Climate change 2007: impacts, adaptation and vulnerability. Contribution of working group II to the fourth assessment report of the intergovernmental panel on climate change. Cambridge University Press, Cambridge, pp 315–356
46. Osborn TJ, Hulme M (2002) Evidence for trends in heavy rainfall events over the UK. Philos Trans R Soc Lond A Math Phys Eng Sci 360:1313–1325
47. Palmer B, et al (2011) Preliminary coastal vulnerability assessment for KwaZulu-Natal, South Africa. J Coastal Res 1390
48. Patt AG, Schroter D, Vega-Leinert ACDL et al (2009) Vulnerability research and assessment to support adaptation and mitigation: common themes from the diversity of approaches. In: Patt AG, Schroter D, Klein RJT, Vega-Leinert ACDL (eds) Assessing vulnerability to global environmental change: making research useful for adaptation, decision making and policy. Earthscan, London
49. Pendleton EA, Thieler ER, Williams SJ (2010) Importance of coastal change variables in determining vulnerability to sea-and lake-level change. J Coastal Res 176–183
50. Phillips MR, Jones AL (2006) Erosion and tourism infrastructure in the coastal zone: problems, consequences and management. Tour Manag 27:517–524
51. Ramieri E et al (2011) Methods for assessing coastal vulnerability to climate change European environment agency, European topic centre on climate change impacts, vulnerability and adaptation. pp 1–91. https://www.researchgate.net/profile/ Andrea_Barbanti/publication/301296277_Methods_for_assessing_coastal_vulnerability_to_climate_change/links/5710bd7008ae68dc790a2421.pdf. Accessed 15 June 2014
52. Ramsbottom D, Sayers P, Panzeri M (2012) Climate change risk assessment for the floods and coastal erosion sector Defra project code GA0204 report to Defra, London, UK
53. Reynard NS, Prudhomme C, Crooks SM (2001) The flood characteristics of large UK rivers: potential effects of changing climate and land use. Climat Change 48:343–359
54. Rygel L, O'sullivan D, Yarnal B (2006) A method for constructing a social vulnerability index: an application to hurricane storm surges in a developed country. Mitig Adapt Strat Global Change 11:741–764
55. Santoro F, Tonino M, Torresan S, Critto A, Marcomini A (2013) Involve to improve: a participatory approach for a decision support system for coastal climate change impacts assessment. The North Adriatic case. Ocean Coast Manag 78:101–111
56. Simone AM (2004) People as infrastructure: intersecting fragments in Johannesburg Public culture 16:407–429
57. Slingo J et al (2014) The recent storms and floods in the UK. UK Met. Office and Centre for Ecology and Hydrology. http://www.metoffice.gov.uk/research/news/2014/uk-storms-and-floods. Accessed 10 Mar 2015
58. Small C, Nicholls RJ (2003) A global analysis of human settlement in coastal zones. J Coast Res 584–599

59. Stevens AJ, Clarke D, Nicholls RJ (2014) Trends in reported flooding in the UK. Hydrol Sci J 61:1884–2013
60. Swarbrooke J (1999) Sustainable tourism management. CABI 1–61
61. Tallet P, Marouard G (2014) The harbor of Khufu on the Red Sea Coast at Wadi al-Jarf. Egypt Near East Archaeol (NEA) 77:4–14
62. Tang HS et al (2013) Vulnerability of population and transportation infrastructure at the east bank of Delaware Bay due to coastal flooding in sea-level rise conditions. Nat Hazards 69:141–163
63. Thatcher CA, Brock JC, Pendleton EA (2013) Economic vulnerability to sea-level rise along the northern US Gulf Coast. J Coast Res 63:234–243
64. The Times (2014) Flood warnings as more rain arrives. http://www.thetimes.co.uk/tto/weather/article3965949.ece. Accessed 21 Mar 2015
65. Torresan S, Critto A, Dalla Valle M, Harvey N, Marcomini A (2008) Assessing coastal vulnerability to climate change: comparing segmentation at global and regional scales. Sustain Sci 3:45–65
66. Torresan S, Gallina V, Giannini V, Rizzi J, Zabeo A, Critto A, Marcomini A (2012) DESYCO: a decision support system to provide climate services for coastal stakeholders dealing with climate change impacts Sea 8:10
67. Vincent K (2004) Creating an index of social vulnerability to climate change for Africa Tyndall Center for Climate Change Research Working Paper 56:41
68. Weisse R, von Storch H, Niemeyer HD, Knaack H (2012) Changing North Sea storm surge climate: an increasing hazard. Ocean Coast Manag 68:58–68
69. Welsh Government (2014) Written statement—coastal flooding—January 2014. http://gov.wales/about/cabinet/cabinetstatements/2014/coastalflooding/?lang=en. Accessed 12 Apr 2015
70. White I, Howe J (2002) Flooding and the role of planning in England and Wales: a critical review. J Environ Plan Manag 45:735–745
71. Wolters ML, Kuenzer C (2015) Vulnerability assessments of coastal river deltas-categorization and review. J Coast Conservat 19:345–368
72. Woodworth P, Flather R, Williams J, Wakelin S, Jevrejeva S (2007) The dependence of UK extreme sea levels and storm surges on the North Atlantic Oscillation. Cont Shelf Res 27:935–946
73. Wu T, Hou X, Chen Q (2016) Coastal economic vulnerability to sea level rise of Bohai Rim in China. Nat Hazards 80:1231–1241
74. Yin J, Yin Z, Wang J, Xu S (2012) National assessment of coastal vulnerability to sea-level rise for the Chinese coast. J Coast Conserv 16:123–133
75. Zanuttigh B et al (2014) The Seus decision support system for coastal risk management. Coast Eng 87:218–239
76. Zsamboky M, Fernández-Bilbao A, Smith D, Knight J, Allan J (2011) Impacts of climate change on disadvantaged UK coastal communities York. Joseph Rowntree Found, UK

CHAPTER 12

Ultra-thin clay layers facilitate seismic slip in carbonate faults

Luca Smeraglia[1], Andrea Billi[2], Eugenio Carminati[1,2], Andrea Cavallo[3], Giulio Di Toro[4,5,6], Elena Spagnuolo [5] & Federico Zorzi[6]

ABSTRACT

Many earthquakes propagate up to the Earth's surface producing surface ruptures. Seismic slip propagation is facilitated by along-fault low dynamic frictional resistance, which is controlled by a number of physico-chemical lubrication mechanisms. In particular, rotary shear experiments conducted at seismic slip rates (1 ms^{-1}) show that phyllosilicates can facilitate co-seismic slip along faults during earthquakes. This evidence is crucial for hazard assessment along oceanic subduction zones, where pelagic clays participate in seismic slip propagation. Conversely, the reason why, in continental domains, co-seismic slip along faults can propagate up to the Earth's surface is still poorly understood. We document the occurrence of micrometer-thick phyllosilicate-bearing layers along a carbonate-hosted seismogenic extensional fault in the central Apennines, Italy. Using friction experiments, we demonstrate that, at seismic slip rates (1 ms^{-1}), similar calcite gouges with pre-existing phyllosilicate-bearing (clay content ≤3 wt.%) micro-layers weaken faster than calcite gouges or mixed calcite-phyllosilicate gouges. We thus propose that, within calcite gouge, ultra-low clay content (≤3 wt.%) localized along micrometer-thick layers can facilitate seismic slip propagation during earthquakes in continental domains, possibly enhancing surface displacement.

1. INTRODUCTION

Many earthquakes propagate up to the Earth's surface producing surface ruptures, possibly associated with tsunami generation, infrastructure damage, and fatalities[1,2,3,4]. Seismic slip propagation is facilitated by low dynamic frictional resistance along faults[5,6,7]. Therefore, understanding the mechanisms that lower rock resistance and facilitate seismic slip propagation along faults[8] is relevant for hazard assessment within seismically active regions. To understand such a mechanism, rotary shear experiments simulating natural seismic deformation conditions have been conducted on different natural fault rocks at seismic slip rates (≥1 ms^{-1}) in the past 20 years[5, 9]. Results show that a number of processes triggered by high mechanical work rates can induce fault weakening[10, 11], especially within thermally-unstable rocks such as carbonates (i.e., thermochemical pressurization and grain-size dependent processes[11,12,13,14,15,16,17,18]). In particular, experiments have revealed the weakening behavior of phyllosilicates, especially in water-saturated conditions, due to their low steady-state dynamic friction and to the absence of peak friction[19,20,21,22,23]. For this reason, previous studies[24,25,26,27] suggested that pelagic clay commonly present along offshore subduction zones participate to co-seismic fault lubrication, possibly facilitating and enhancing upward co-seismic slip propagation, sometimes associated with tsunamigenic seafloor displacement. Less understood is the origin of surface displacements associated with moderate ($M_w \leq 7.0$) earthquakes in continental seismically-active regions[4, 28, 29] (see Supplementary Table S1), as in such settings clay occurrence can be very modest. Here we propose that less than 3 wt.% of phyllosilicates localized within pre-existing ultra-thin layers along carbonate-hosted faults can facilitate co-seismic slip propagation, possibly enhancing surface displacement. In particular, we first document the occurrence in nature of hitherto unknown phyllosilicate-bearing micrometer-thick layers along the carbonate-hosted seismically-active Tre Monti Fault[30,31,32] and then we use a rotary shear apparatus to simulate the behavior of this natural setting during co-seismic slip propagation.

The Tre Monti Fault is a shallow (exhumed from depths <3 km[31]) seismogenic extensional fault belonging to the carbonate domain of the central Apennines, Italy[30,31,32,33] (Fig. 1a). The central Apennines is a fold-thrust belt (Oligocene-Quaternary) characterized by imbricate thrust sheets consisting of ~4–5 km thick pre-orogenic massive platform carbonates overlain by ~2 km thick phyllosilicate-rich syn-orogenic deposits (i.e., hemipelagic marls and foredeep clayey-sandstones)[34,35,36]. Massive carbonate rocks are the main lithology of the central Apennines down to at least 7–8 km depth[37, 38]. During post-orogenic extension (Pliocene to present time), the fold-thrust belt has been dissected by a system of extensional faults, which bound several intramountain basins filled by Plio-Quaternary continental deposits[39,40].

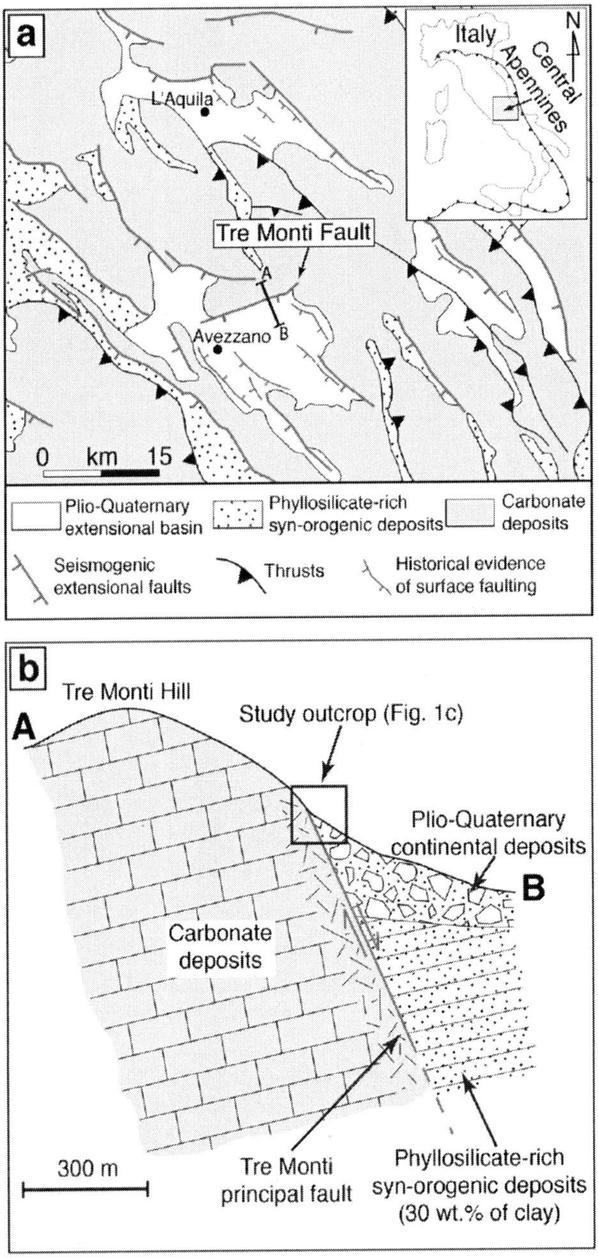

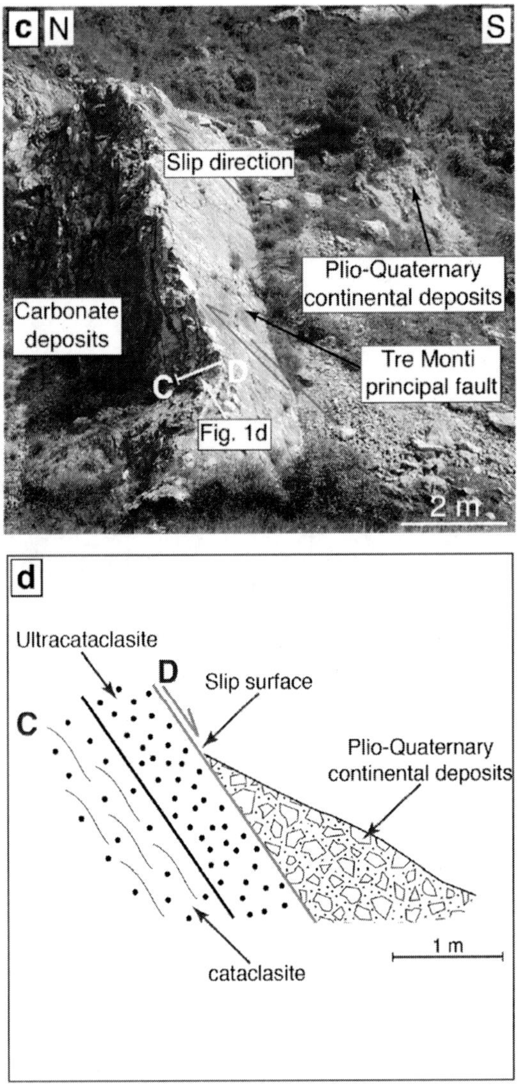

Figure 1 Geological setting. (**a**) Geological setting of the Tre Monti Fault and evidence of surface faulting from historical earthquakes in the central Apennines, Italy. Geological map and base map has been drawn using Adobe Illustrator CS5. (**b**) Simplified geological cross-section through the Tre Monti Fault. Cross-section trace in (**a**). Cross-section has been drawn using Adobe Illustrator CS5. **(c)** The study outcrop of the Tre Monti principal fault (Latitude 42°04′34″ N, Longitude 13°29′58″ E). **(d)** Simplified structural cross-section through the Tre Monti principal fault showing the ultracataclasite above the foliated cataclasite. Cross-section has been drawn using Adobe Illustrator CS5.

These faults have generated historical and instrumental seismicity up to M_w 7.0[41] (Fig. 1a; e.g., Avezzano, 1915, M_w 7.0 earthquake; L'Aquila, 2009, M_w 6.1 earthquake; Amatrice-Norcia, 2016, M_w 6.0 and 6.5 earthquakes). In this region, earthquakes mainly nucleate at depths ≤8 km and co-seismic slip propagates upward along carbonate-hosted faults, often generating surface displacement[4, 30, 42, 43] (Fig. 1a), as also observed in other seismically-active areas worldwide (see Supplementary Table S1). Therefore, geological and mechanical studies of faults exhumed from shallow depths, such as the Tre Monti Fault, are significant as the small-scale compositional heterogeneity of shallow fault zones can promote or inhibit seismic slip propagation up to the Earth's surface.

2. RESULTS

2.1 Natural fault zone structures

The Tre Monti Fault juxtaposes pre-orogenic carbonate deposits (footwall) with both syn-orogenic phyllosilicate-rich deposits (Fig. 1b; average phyllosilicate content ~30%, see Supplementary Table S2) and overlying continental (fluvial, lacustrine, and slope) deposits (hangingwall). The fault displays a dip of ~60° and a maximum displacement of ~1,500–2,000 m[31, 44]. The principal fault surface is excellently exposed along a ~8 km long carbonate-hosted fault scarp[30,31,32] (Fig. 1c) exhumed by multiple surface faulting events[26]. The principal fault is associated with a ~1 m-thick cataclastic fault core, which consists of an ultracataclasite lying above a cataclasite[27] (Figs 1d and 2a–c). The ultracataclasite shows more intense grain comminution than the cataclasite (Fig. 2c) and forms discontinuous patches and plagues above the cataclasite owing to fault scarp erosion. X-ray Powder diffraction (XRPD) and energy dispersive spectroscopy (EDS) analyses revealed that the cataclasite consists solely of calcite, whereas the ultracataclasite consists of calcite, phyllosilicates (~1.5 wt.%; mainly smectite, illite, and kaolinite; see Supplementary Fig. S1 and Supplementary Table S2), detrital micas, quartz, K-feldspar, and plagioclase (<1%; see Supplementary Fig. S2 and Supplementary Table S3), which are also present within the syn-orogenic deposits (see Supplementary Table S2)[45, 46]

Phyllosilicates are concentrated within slip-parallel micrometer-thick (from ~30 μm up to ~1 mm in thickness) layers at the boundary between the ultracataclasite and the cataclasite (Fig. 2b and c). The average phyllosilicate content within these layers is ~5 wt.% (see Supplementary Fig. S2 and Table S2). Phyllosilicate-bearing layers have sharp boundaries, which truncate calcite clasts (Fig. 2b and c). Within these layers, calcite clasts are scattered within a phyllosilicate matrix and are usually wrapped by phyllosilicates creating a continuous film (Fig. 2d). At the nanoscale, phyllosilicates consist of granular zones and elongated fibers comprising clumped and chained ~50-nm-thick nanospherules and nanotubes (Fig. 2e and f).

Due to lack of evident dissolution seams and teeth-shaped margins (i.e., typical of stylolites[47, 48]) close to the phyllosilicate-bearing layers, we exclude that these layers formed by residual processes connected with pressure-solution of carbonates. Moreover, XRPD and EDS analyses show that the non-carbonate minerals of the ultracataclasite (Supplementary Tables S2 and S3) are identical to those of syn-orogenic deposits in the fault hangingwall (Supplementary Table S2). Hence, we infer that the phyllosilicates derived from the phyllosilicate-rich syn-orogenic deposits at the fault hangingwall, and that they were smeared and mixed into the ultracataclasite during fault activity[49]. Although we cannot totally exclude phyllosilicates derived by pressure-solution of carbonates, we suggest that the contribution of phyllosilicates derived by such processes (if any) can be neglectable[49]. However, independently from the mode of phyllosilicate concentration along thin layers (i.e., through clay smearing, pressure-solution of carbonates, and so on) in the TMF or in another carbonate-hosted faults, in the following section we emphasize the mechanical role of pre-existing phyllosilicate-bearing layers during co-seismic slip propagation.

2.2 Laboratory experiments and comparison with natural fault products

Based on the previously-reported field, mineralogical, and microscopic observations, we conducted low to high velocity rotary shear friction experiments (see Methods) to investigate the frictional behavior of micrometer-thick phyllosilicate-bearing layers within synthetic calcite gouges reproducing those observed along the Tre Monti Fault (Fig. 2b and c). We performed experiments on 3 mm-thick gouge layers obtained by crushing and sieving (<125 μm) the natural foliated cataclasite (100% calcite) and the phyllosilicate-rich syn-orogenic deposits (~30% of phyllosilicates), both under dry and under water-saturated (wet) conditions.

Experimental slip rates were set at 0.001 ms^{-1} and 1 ms^{-1} to simulate subseismic and co-seismic slip rates, respectively. We held the normal stress constant at 8.5 MPa and 5 MPa in dry and wet experiments, respectively, consistently with lithostatic stress at shallow depth (<0.5 km). The SHIVA apparatus does not allow the measurement/control of pore pressure during the wet experiments. Accordingly the measured friction should be considered as apparent (or effective), as suggested by Chen *et al.*

(2017), for the unknown pore pressure within the gouge. We set normal stress to 5 MPa under wet conditions to avoid gouge extrusion from the gouge holder. In all experiments, gouges were initially pre-sheared at 0.01 ms^{-1} for 10 cm of total slip under 0.5 MPa of normal stress. This procedure was done to properly simulate the behaviour of our natural fault rocks (Fig. 2b), which were (pre-) sheared by multiple slip events and to limit the pressurization effects due to gouge compaction, as also done in previous experiments in rotary shear apparatuses[50]. We sheared the gouges imposing a total slip of 0.5 meter, both during low- (0.001 ms^{-1}) and high-velocity (1 ms^{-1}) experiments, to simulate, in the latter case, the amount of co-seismic slip occurring at depth <0.5 km during a M_w 6.0–6.5 earthquake[51]. We carried out experiments using three different gouge configurations (Fig. 3a) aimed at understanding the mechanical effect of a small percentage of clay concentrated along ultra-thin layers within the calcite gouge. The configurations are: (1) 10% of phyllosilicate-rich deposits randomly mixed with 90% of calcite ("mixed gouge experiments"; total phyllosilicate content of ~3 wt.%); (2) 2 mm-thick layer of 100% calcite overlain by a 1 mm-thick layer consisting of a random mixture of 10% of phyllosilicate-rich deposits and 90% of calcite ("layered gouge experiments"; total phyllosilicate content in the 1-mm thick layer of ~3 wt.%); this configuration, in particular, simulates the phyllosilicate-bearing layers observed within natural samples from the Tre Monti Fault (Fig. 2b and c). For this configuration, we also performed a non-shear compression experiment to study the pre-shear microstructures (compare Supplementary Fig. S3 with Fig. 3a "layered gouge" configuration); (3) 100% calcite gouge ("calcite gouge experiments"). Complete experimental procedures and results are listed in the method section and Supplementary Table S4.

At subseismic slip rates (0.001 ms^{-1}), all gouges have a similar friction/apparent friction coefficient evolution from the first peak value to a steady-state value (0.7–0.8; Fig. 3b). At seismic slip rates (1 ms^{-1}), all experiments are characterized by a dynamic weakening behavior. In particular, during the decay of friction/apparent friction (i.e., the weakening phase) towards a steady-state value (~0.4) occurs after a strengthening phase defined as the distance (i.e., slip) to the onset of dynamic weakening (D_{ow} [52]; Fig. 3c and d). We calculated the D_{ow} values considering the maximum distance reached by the strengthening phase until the start of the weakening phase (Fig. 3c and d). Remarkably, we recognize a general trend showing that layered gouges weaken faster and have lower fracture energy than mixed gouges, which, in turn, weaken faster and have lower fracture energy than carbonate gouges both in dry and in wet conditions (Fig. 3c and d, and Supplementary Table S4). In particular, layered gouges have a D_{ow} shorter than those measured for mixed and calcite gouges, both in dry and in wet conditions (Fig. 3c and d).

Overall, wet gouges show a lower steady-state dynamic friction values than dry gouges, as previously observed within water-saturated gouges[20,21,22] (Fig. 3d).

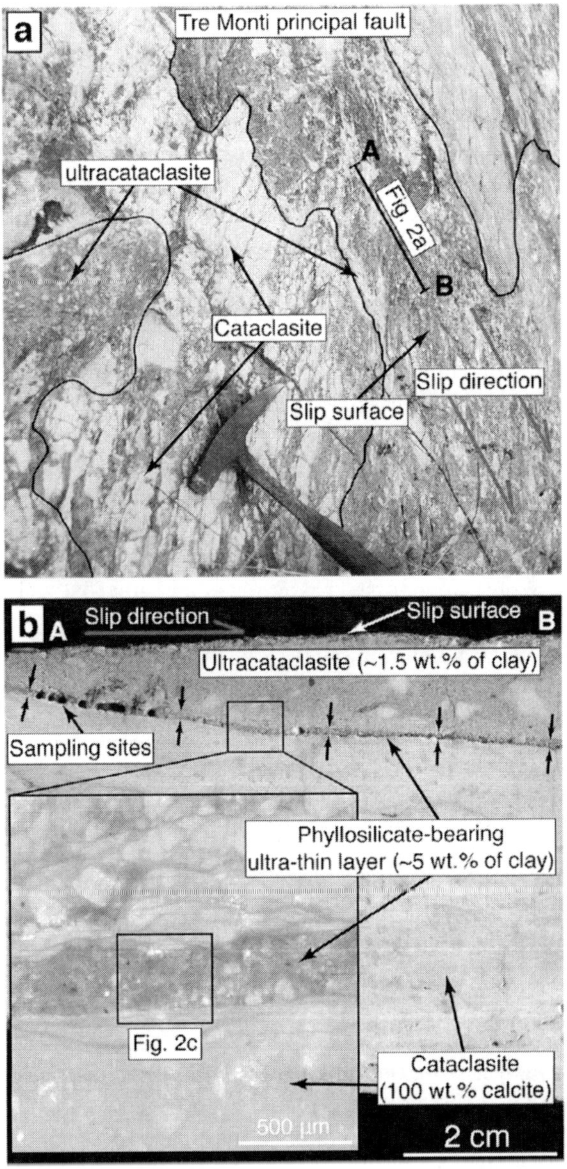

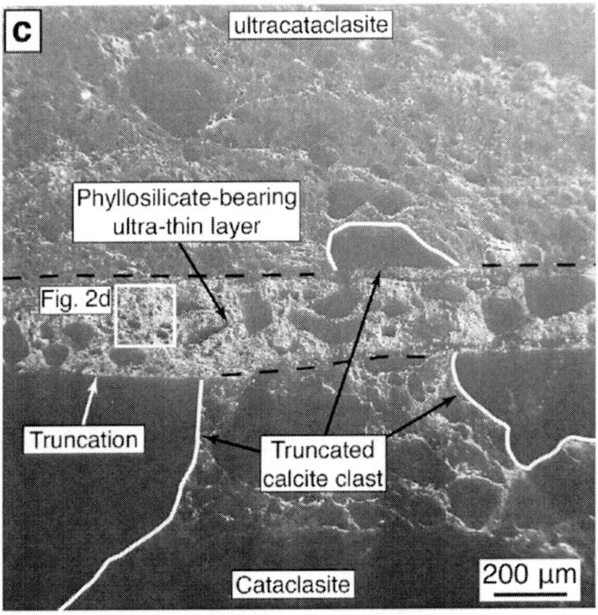
c
ultracataclasite
Phyllosilicate-bearing ultra-thin layer
Fig. 2d
Truncation
Truncated calcite clast
Cataclasite
200 µm

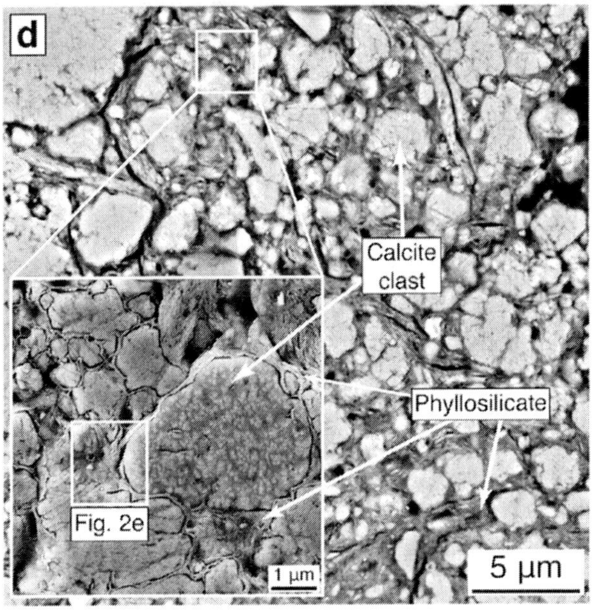
d
Calcite clast
Phyllosilicate
Fig. 2e
1 µm
5 µm

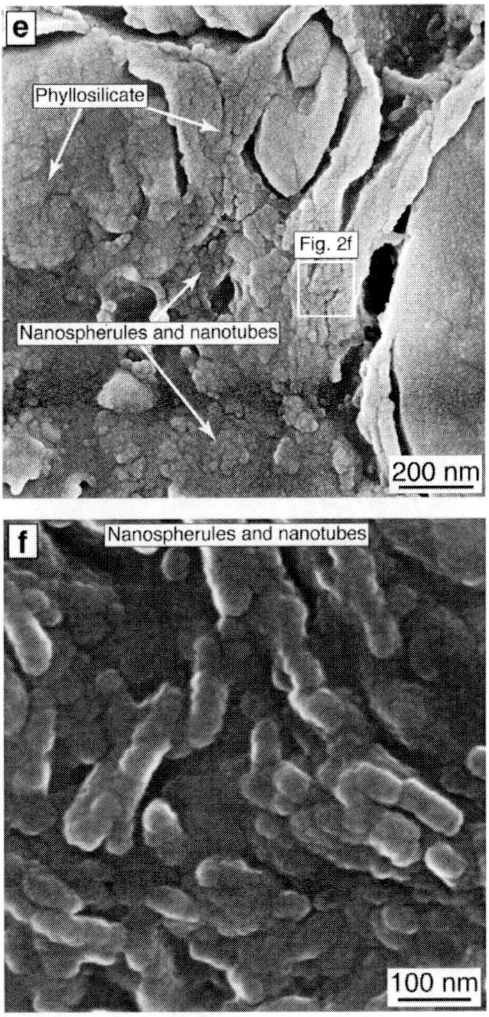

Figure 2 Phyllosilicate occurrence within the carbonate-hosted Tre Monti Fault. **(a)** Detail of the Tre Monti principal fault showing patches of ultracataclasite above the foliated cataclasite. **(b)** Phyllosilicate-bearing layer at the boundary between the ultracataclasite and the foliated cataclasite. Inset shows a microphotograph under optical microscope showing a detail of the phyllosilicate-bearing layer. **(c)** FE-SEM image of micrometer-thick phyllosilicate-bearing layer, which truncates calcite clasts. **(d)** FE-SEM image of carbonate clasts within a phyllosilicate matrix. Inset shows calcite clasts wrapped by a continuous phyllosilicates film. **(e–f)** FE-SEM images of nanostructures of natural phyllosilicates from the phyllosilicate-bearing layers characterized by clumped and chained nanospherules and nanotubes.

To verify the reproducibility of the mechanical data, we replicated a set of six additional experiments both in wet (three experiments: mixed, layered, and calcite gouge experiments; Fig. 3e) and in dry conditions (three experiments: mixed, layered, and calcite gouge experiments; Fig. 3e). Despite the high heterogeneity of the tested materials (i.e., natural rocks such as syn-orogenic deposits), these experiments reproduce well the previously observed mechanical trend: layered gouges weaken faster than mixed gouges, which, in turn, weaken faster than calcite gouges both in wet and dry conditions (Fig. 3e).

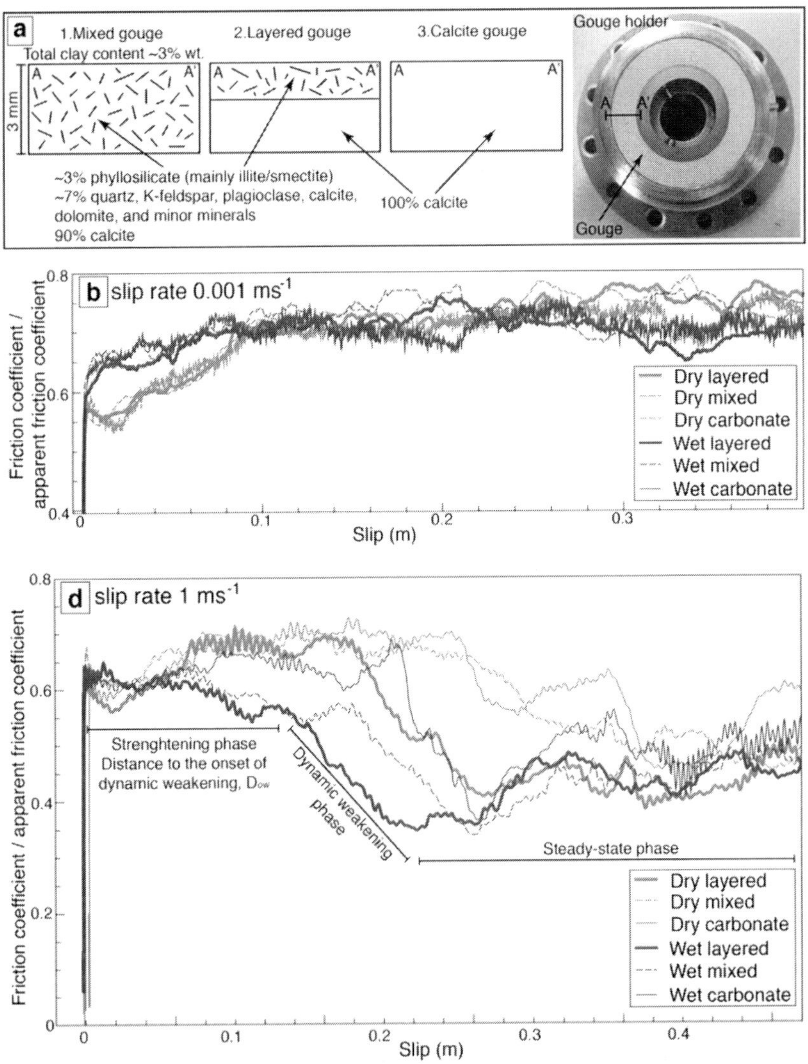

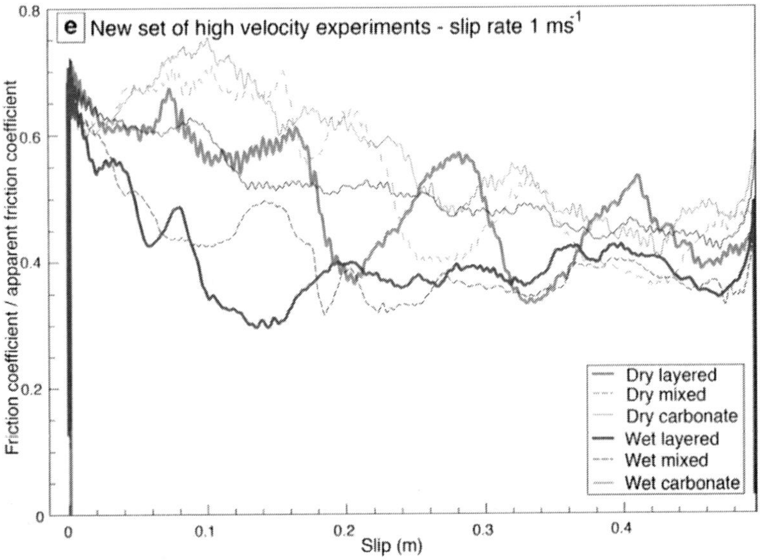

Figure 3 Results from friction experiments. **(a)** experimental setup: (left) the three different configurations of tested gouges and (right) the gouge holder. (**b**) Friction coefficient versus slip during experiments at subseismic (0.001 ms^{-1}) slip rate. **(c)** Distance to the onset of dynamic weakening, D_{ow}, versus clay content. Layered gouges are characterized by a D_{ow} lower than those of mixed and calcite gouges, both in dry and in wet conditions. **(d,e)** Friction coefficient versus slip during experiments at seismic (1 ms^{-1}) slip rate.

Using FE-SEM, we investigated the micro- and nano-structures obtained from the experimental gouges. Layered and mixed gouges sheared at subseismic velocity have a fabric consisting of randomly distributed particles of phyllosilicates and calcite (Fig. 4a). On the contrary, layered and mixed gouges sheared at high velocity (i.e., seismic slip rate, 1 ms^{-1}), both in wet and in dry conditions, have similar microstructures characterized by (1) reduced grain size within a ~500 μm-thick deformation zone along the slip surface and by (2) phyllosilicate segregation and concentration along micrometer-thick layers (Fig. 4b,c and d). These layers were absent within the pre-sheared gouges (see Supplementary Fig. S3). Phyllosilicate-bearing layers form discontinuous patches rather than continuous layers and include carbonate clasts floating within a fine-grained matrix consisting of phyllosilicate lamellae wrapping around clasts (Fig. 4b and c). These microstructures are identical to those of natural micrometer-thick phyllosilicate-bearing layers observed within the Tre Monti Fault ultracataclasite (Fig. 2d).

Mixed and layered gouges in dry conditions show extreme grain size reduction and shear localization along a ~50 μm-thick principal slip zone (Fig. 4d). This principal slip zone is a mixture of phyllosilicates and calcite consisting of slip-parallel ~50 nm thick fibers composed by an array of nanospherules and nanotubes (Fig. 4d). In places, individual nanospherules clump together to form granular bulbous zones. These structures are very similar to phyllosilicate nanostructures observed within the Tre Monti Fault phyllosilicate-bearing layers (Fig. 2e and f). On the contrary, the principal slip zones of experiments conducted at subseismic slip rates are not characterized by nanospherules or nanotubes (Fig. 4a). Given the very slow creep rates and large cumulative slips possibly accommodated by natural slipping zones, we cannot rule out that clay nanospherules and nanotubes within the Tre Monti Fault (Fig. 2e and f) may have formed aseismically, as suggested by Verberne *et al.*53 for nanospherules within calcite gouges sheared at subseismic velocity (e.g., slip velocity 1 μm/s). However, in the case of the Tre Monti Fault, the coexistence of (1) clay nanospherules and nanotubes within phyllosilicate-bearing layers (Fig. 2e and f) similar to those observed within other phyllosilicates sheared at co-seismic slip-rates (compare with Fig. 9c in Ujiie *et al.*, 2011), (2) truncated carbonate clasts (observed only within calcite gouges sheared at co-seismic slip rates54) along phyllosilicate-bearing layers (Fig. 2c), and (3) co-seismic displacements documented by *in-situ*^{36}Cl fault scarp exposure dating30, possibly suggests that phyllosilicate-bearing layers formed during fault slip at co-seismic slip rates. Moreover, we emphasize that no nanospherules and nanotubes were generated during low-velocity experiments both in dry and in wet conditions (Fig. 4a). Accordingly, we hypothesize that the Tre Monti Fault hosted co-seismic slip propagation along phyllosilicate-bearing layers up to shallow levels.

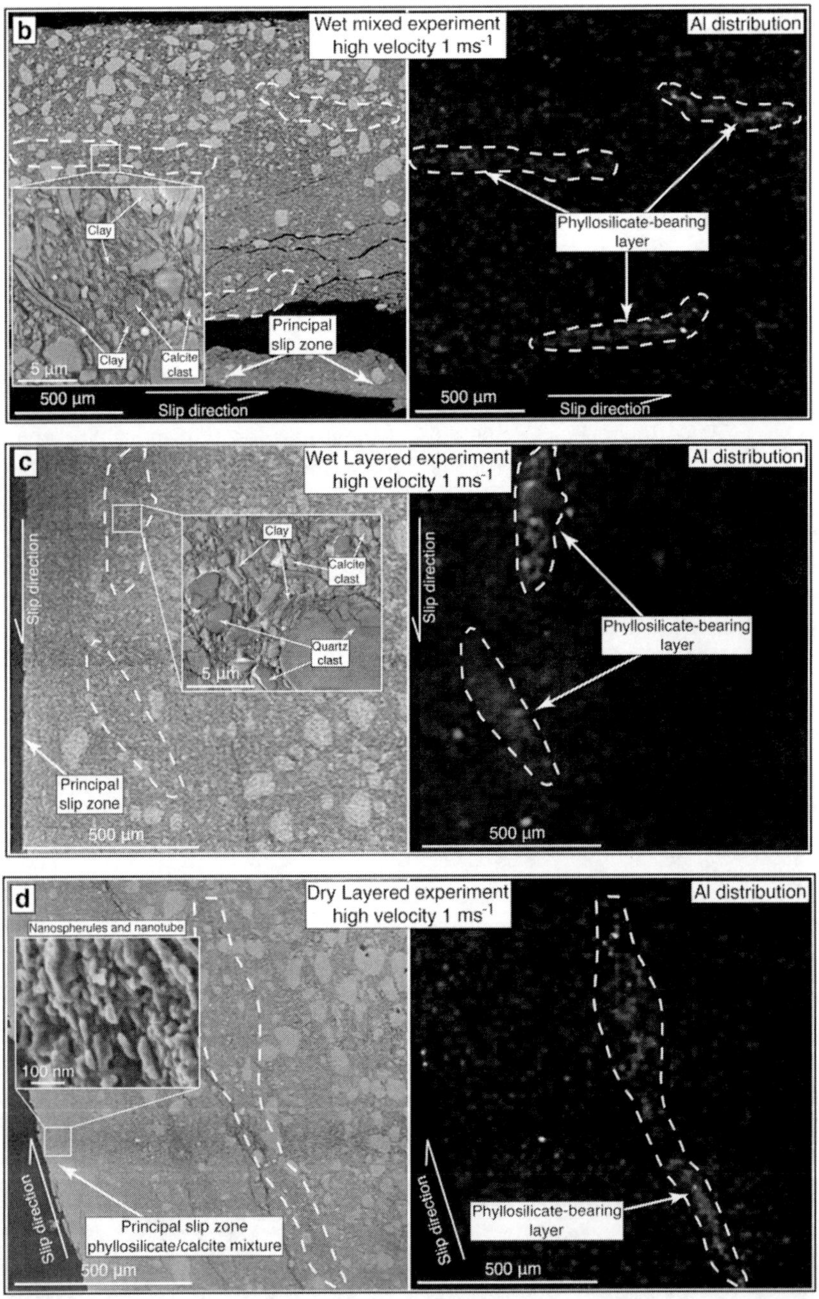

Figure 4 Experimental microstructures. **(a)** Left: FE-SEM image of dry layered gouge experimentally-sheared at subseismic velocity (0.001 ms^{-1}). Grain size reduction occurs toward the principal slip zone. Inset shows a

detail of phyllosilicate lamellae showing no nanospherules or nanotubes that occur, in contrast, in samples from experiments at seismic velocity (see **d**). Right: EDS map showing no concentration and segregation of phyllosilicates along distinct layers that occur, in contrast, in samples from experiments at seismic velocity (see **b–d**). FE-SEM image of **(b** left**)** wet mixed gouge experiment and **(c** left**)** wet layered gouge experiment sheared at seismic-slip velocity (1 ms^{-1}) showing grain size reduction toward the principal slip zone. EDS map shows segregation and concentration of phyllosilicates along micrometer-thick layers both for mixed (**b** right) and for layered (**c** right) gouges under wet conditions. Insets show carbonate clasts within phyllosilicate-bearing layer wrapped by phyllosilicates. **(d)** Left: FE-SEM image of dry layered gouge experiment sheared at seismic-slip velocity (1 ms^{-1}) showing, toward the principal slip surface, grain size reduction and compaction stronger than wet gouges. Inset shows the nanostructures from the principal slip zone. These structures consist of clumped and chained nanospherules and nanotubes identical to those observed within phyllosilicates along the Tre Monti principal fault (see Fig. 2e and f). Right: EDS map showing concentration and segregation of phyllosilicates along a micrometer-thick layer.

3. DISCUSSION AND CONCLUSIONS

We conclude that calcite gouges with an illite/smectite content as low as 3% localized (i.e., due to clay smearing and/or pressure-solution/neoformation processes) along pre-existing micrometer layers (Fig. 3a) weaken faster than calcite and mixed calcite-phyllosilicate gouges (Fig. 3c,d, and e) at seismic slip rates (1 ms^{-1}) both in dry and in wet conditions. Therefore, the occurrence of such phyllosilicate-bearing layers can facilitate fault weakening within carbonate-hosted faults, enhancing co-seismic slip propagation up to the Earth's surface in continental domains. Clay-bearing layers can also be produced during co-seismic slip by clay segregation and localization from an initial mixture of clay and calcite, as shown within gouges sheared at seismic slip rates (1 ms^{-1}; Fig. 4b–d). The segregation of clay into numerous phyllosilicate-bearing layers can act as a self-enhancing process, promoting further fault weakening during earthquakes along newly formed phyllosilicate-bearing layers. Regardless such clay concentration (Fig. 4b–d), mixed and layered gouges sheared at seismic slip rates, both in wet and dry conditions, have similar microstructures characterized by reduced grain size within a ~500 µm-thick deformation zone along the slip surface (Fig. 4b–d). This evidence suggests that the starting layered vs. mixed configurations have no influence on the final gouge microstructure. On the contrary, mechanical data show that layered gouges weaken faster than mixed gouges (Fig. 3c). This evidence suggests that the initial gouge setting (layered vs. mixed) exerts a strong mechanical control on the weakening behaviour of the gouges regardless of the final microstructures.

The steady-state friction coefficient measured in our experiments (~0.4) is somewhat higher than that (~0.2) resulting from previous experiments performed on carbonate gouges at co-seismic slip rates[13,54]. This difference is likely due to the lower applied normal stress (i.e., 8.5 MPa and 5 MPa in dry and wet conditions, respectively) set in our experiments to simulate co-seismic slip at depth <500 m. Moreover, we used a metal gouge holder hardened with WC (i.e., tungsten carbide), which owns a thermal diffusivity of 3.5 10^{-5} m^2s^{-1} about ten times larger than that of calcite gouges and natural rocks[55]. The metal holder allowed easy diffusion of heat, which buffered the temperature rise during shearing of the gouge[56]. Since most co-seismic weakening mechanisms are temperature-dependent[9], the thermal properties of the sample holder influence strain localization, heat production, temperature rise and the measured friction coefficient (i.e., higher the temperature in the slipping zone, lower the friction coefficient)[56]. Similarly, regarding the extrapolation of our experimental measurements of the friction coefficient to natural conditions, the lower thermal diffusivity of rocks and the higher normal stress at depth can further reduce (i.e., below 0.4) the friction coefficient during earthquakes[9, 10, 57].

In summary, we propose that the 3 wt.% of clay concentrated within pre-existing ultra-thin layers could be the lower boundary for facilitating dynamic weakening during earthquake slip at shallow crustal levels in continental domains. The dynamic weakening effect can be even stronger if the phyllosilicate content exceeds 3 wt.%[22].

4. METHODS

4.1 Description of starting materials

We used natural materials collected in the field from the foliated carbonate cataclasite and the phyllosilicate-rich deposits to perform high velocity friction experiments (Fig. 2a). The foliated cataclasite consists of 100% calcite whereas the total phyllosilicate content of the phyllosilicate-rich deposits is ~30% (i.e., the phyllosilicate rich powder used for the experiments contain ~30% of phyllosilicates; Table S2). We used 4.5 g of gouge for each experiments resulting in ring-shaped gouge layers (35/55 mm int./ext. diameters), with initial thicknesses of ~3 mm (Fig. 3a).

4.2 Description of XRD analysis performed on natural rocks

We performed powder X-ray diffraction (XRPD) analyses and energy dispersive spectroscopy (EDS) to determine the mineralogy of natural fault rocks and experiment materials. In particular, we performed XRPD analyses on the insoluble residue from the ultracataclasite and from the natural phyllosilicate-bearing layers. We isolated the insoluble residues using HCl acid to dissolve calcite. XRPD data were obtained using a PANalytical θ-θ diffractometer equipped with a long fine-focus Cu X-ray tube (operating at 40 kV and 40 mA) and a real-time multiple strip (RTMS) detector (X'Celerator). The scan was performed over the 2θ range of 3–80°, with a virtual 2θ step size of 0.017°, and a counting time of 100 s/step. The program High Score Plus (PANalytical) was used for phase identification and quantitative phase analysis with Rietveld refinement[58].

4.3 Descriptions of experimental rotary shear apparatus and gouge holder

We performed friction experiments on the Slow to High Velocity rotary shear Apparatus (SHIVA) at INGV, Roma[59]. The apparatus has a horizontal setup, with the rotary column on one end, the sample chamber in the middle, the axial column and loading system side at the other end. Rotary motion is supplied to the sample pair by two brushless engines of different size. The larger engine operates in a velocity range of 1–3,000 RPM, and owns a peak torque of 932 Nm and a nominal maximum power output of 280 kW. We performed experiments using a dedicated metal sample holder (Fig. 3a) appropriately designed for shearing gouges at high velocities[13]. Ring-shaped layers of gouge (55 mm external diameter, 35 mm internal diameter) are confined by outer and inner rings of tungsten carbide that rotate and slide over a base disc of the same material. The base disc and the rotary base plate are etched with a cross-hatch pattern to provide roughness at the boundaries of the gouge layer. The normal load, exerted by an air actuator, was applied on the gouge layer by the stationary column of SHIVA. Normal load applied to the outer and inner rings is independent of that applied to the gouge layer, and is modulated by five outer springs and one inner spring: normal load is proportional to the spring stiffness and to the amount of compression during each experiment. Given that we used the same layer thickness for each experiment, we consider the normal load on the outer and inner rings to be the same for each experiment. Calibration of the gouge holder is discussed in detail in Smith *et al.* (2013).

4.4 Description of experimental conditions

We conducted dry experiments at room humidity conditions, whereas, in water-saturated experiments, we water damped gouges with 0.5 ml of distilled water. Gouge materials for tests were oven-dried at 40 °C for days before the experiments. Experimental slip rates were set at 0.001 ms^{-1} and 1 ms^{-1}, for a total sliding displacement of 0.5 meter under water-saturated (wet) and dry conditions. We held normal stress constant at 8.5 MPa and 5 MPa for dry and wet experiments, respectively. The SHIVA apparatus does not allow the measurement/control of pore pressure during the wet experiments. Accordingly the measured friction should be considered as apparent (or effective), as suggested by Chen *et al.* (2017), for the unknown pore pressure within the gouge. The experimental sequence progressed as follows: gouges were initially pre-sheared at 1 cms^{-1} for 10 cm of total slip under 0.5 MPa of normal load, both in low velocity and in high velocity experiments and both in dry and in wet conditions. In all experiments, gouges were initially pre-sheared at 0.01 ms^{-1} for 10 cm of total slip under 0.5 MPa of normal load. This procedure was done to properly simulate the behaviour of our natural fault rocks (Fig. 2b), which were (pre-) sheared by multiple slip events and to limit the pressurization effects due to gouge compaction, as also done in previous experiments in rotary shear apparatuses[50]. After, gouges were loaded to the target normal stress and were sheared up to the target displacement. For high velocity experiments, acceleration and deceleration rates were of 24 ms^{-2}. After the experiments, gouges were cohesive and flinty. We impregnated gouge fragments under vacuum using a LR-white resin. We then prepared petrographic sections from the impregnated gouges cut perpendicular to the gouge layers and approximately parallel to the slip direction. We studied micro- and nano-structures (both natural and experimental) through scanning electron microscope on polished thin sections ~30 μm thick. All samples were uniformly coated with 1–2 nm thick chrome particles. We used three different facilities. (1) A thermal FE-SEM Zeiss Merlin Gemini 2 with 0.7 nm of resolution and four detectors (two in lens detector for high resolutions) at CERTEMA (Multidisciplinary technology laboratory, Grosseto, Italy). This device is equipped with a high sensitivity EDS detector (active area of 50 mmq) and with a 5 crystals WDS spectrometer. (2) A thermal FE-SEM JEOL (jsm6500f) at INGV (National Institute of Geophysics and Volcanology), Rome, with 2.5 nm of resolution and two detectors. EDS has an active area of 25 mmq. (3) A FE-SEM Zeiss Auriga at CNIS (Center for Nanotechnologies Applied to Engineering, Sapienza University of Rome, Italy) with 1 nm of resolution and three detectors (one in lens detector for high resolutions) All thin-sections were examined under optical microscope at various magnifications prior than SEM analyses.

ACKNOWLEDGEMENTS

This work has been funded by an European Research Council Consolidator Grant (614705) to G. Di Toro, by PRIN2010/11 (Project 20107ESMX9) and Progetti di Ateneo Sapienza 2014 to E. Carminati, by PRIN 2015 (Project 2015EC9PJ5_001) to C. Doglioni, and by Progetti di Ateneo Sapienza 2015 to C. Collettini. We thank D. Mannetta for thin section preparation, P. Scarlato and E. Nazzari for laboratory support at INGV, S. Sarto and F. Mura for help and access to CNIS (Center for Nanotechnologies Applied to Engineering, Sapienza University) laboratory, CERTEMA (Multidisciplinary technology laboratory, Grosseto, Italy) staff for electron microscopy analyses, L.-W. Kuo for help during experiments, L. Aldega, S. Aretusini, C. Collettini, M. Fondriest, and T. Tesei for constructive discussions, and C. Doglioni for initiating and encouraging this research. We thank M. Ikari and K. Oohashi for constructive comments on an early version of this manuscript. We thank the Editor and two anonymous reviewers for their editorial and scientific work on our manuscript.

REFERENCES

1. Wells, D. L. & Coppersmith, K. J. New empirical relationships among magnitude, rupture length, rupture width, rupture area, and surface displacement. *Bull. Seismol. Soc. Am.* **84**, 974–1002 (1994).
2. Ambraseys, N. N. & Jackson, J. A. Faulting associated with historical and recent earthquakes in the Eastern Mediterranean region. *Geophys. J. Int.* **133**, 390–406 (1998).
3. Martino, S., Prestininzi, A. & Romeo, R. W. Earthquake-induced ground failures in Italy from a reviewed database. *Nat. Hazards Earth Syst. Sci.* **14**, 799–814 (2014).
4. Mildon, Z.K. *et al.* Active normal faulting during the 1997 seismic sequence in Colfiorito, Umbria: did slip propagate to the surface? *J. Struct. Geol.* (2016).
5. Di Toro, G., Goldsby, D. L. & Tullis, T. E. Friction falls towards zero in quartz rock as slip velocity approaches seismic rates. *Nature* **427**, 436–439 (2004).
6. Kanamori, H. & Brodsky, E. E. The physics of earthquakes. *Reports on Progress in Physics* **67**, 1429 (2004).
7. Rice, J. R. Heating and weakening of faults during earthquake slip. *J. Geophys. Res.* **111**, B05311 (2006).
8. Chang, J. C., Lockner, D. A. & Reches, Z. Rapid acceleration leads to rapid weakening in earthquake-like laboratory experiments. *Science* **338**, 101–105 (2012).
9. Di Toro, G. *et al.* Fault lubrication during earthquakes. *Nature* **471**, 494–498 (2011).
10. Di Toro, G., Hirose, T., Nielsen, S., Pennacchioni, G. & Shimamoto, T. Natural and experimental evidence of melt lubrication of faults during earthquakes. *Science* **311**, 647–649 (2006).
11. Han, R., Shimamoto, T., Hirose, T., Ree, J.-H. & Ando, J. Ultralow friction of carbonate faults caused by thermal decomposition. *Science* **316**, 878–881 (2007).
12. De Paola, N. *et al.* Fault lubrication and earthquake propagation in thermally unstable rocks. *Geology* **39**, 35–38 (2011).
13. Smith, S. A. F. *et al.* Co-seismic recrystallization during shallow earthquake slip. *Geology* **41**, 63–66 (2013).

14. Siman-Tov, S., Aharonov, E., Sagy, A. & Emmanuel, S. Nanograins form carbonate fault mirrors. *Geology* **41**, 703–706 (2013).
15. De Paola, N., Holdsworth, R., Viti, C., Collettini, C. & Bullok, R. Can grain size sensitive flow lubricate faults during the initial stages of earthquake propagation? *Earth Planet Sc. Lett* **431**, 48–58 (2015).
16. Green, H. W. II, Shi, F., Bozhilov, K., Xia, G. & Reches, Z. Phase transformation and nanometric flow cause extreme weakening during fault slip. *Nat. Geos* **8**, 484–489 (2015).
17. Sulem, J. & Famin, V. Thermal decomposition of carbonates in fault zones: slip-weakening and temperature-limiting effects. *J. Geophys. Res.* **114**, B03309 (2009).
18. Mitchell, M. T. *et al.* Catastrophic emplacement of giant landslides aided by thermal decomposition: Heart Mountain, Wyoming. *Earth and Planetary Science Letters* **411**, 1–9 (2015).
19. Brantut, N., Schubnel, A., Rouzaud, J. N., Brunet, F. & Shimamoto, T. High-velocity frictional properties of a clay-bearing fault gouge and implications for earthquake mechanics. *J. Geophys. Res.* **113**, B10401 (2008).
20. Ujiie, K., Tsutsumi, A. & Kameda, J. Reproduction of thermal pressurization and fluidization of clay-rich fault gouges by highvelocity friction experiments and implications for seismic slip in natural faults. *Geol. Soc. London, Spec. Publ* **359**, 267–285 (2011).
21. Ujiie, K. *et al.* Low coseismic shear stress on the Tohoku-Oki megathrust determined from laboratory experiments. *Science* **342**, 1211–1214 (2013).
22. Bullock, R. J., De Paola, N. & Holdsworth, R. E. An experimental investigation into the role of phyllosilicate content on earthquake propagation during seismic slip in carbonate faults. *J. Geophys. Res.* **120**, 3187–3207 (2015).
23. Oohashi, K., Hirose, T., Takahashi, M. & Tanikawa, W. Dynamic weakening of smectite-bearing faults at intermediate velocities: implications for subduction zone earthquakes. *J. Geophys. Res.* **120**, 1572–1586 (2015).
24. Faulkner, D. R., Mitchell, T. M., Behnsen, J., Hirose, T. & Shimamoto, T. Stuck in the mud? Earthquake nucleation and propagation through accretionary forearcs. *Geophys. Res. Lett.* **38**, L18303 (2011).
25. Ujiie, K. & Kimura, G. Earthquake faulting in subduction zones: Insights from fault rocks in accretionary prisms. *Progress in Earth and Planetary Science* **1** (2014).
26. Kameda, J. *et al.* Pelagic smectite as an important factor in tsunamigenic slip along the Japan Trench. *Geology* **43**, 155–158 (2015).
27. Kirkpatrick, J. D. *et al.* Structure and lithology of the Japan Trench subduction plate boundary fault. *Tectonics* **34**, 53–69 (2015).
28. Gudmundsson, A. Great challenges in structural geology and tectonics. *Frontiers in Earth Science* **1**, 1–10 (2013).
29. Balsamo, F., Aldega, L., De Paola, N., Faoro, I. & Storti, F. The signature and mechanics of earthquake ruptures along shallow creeping faults in poorly lithified sediments. *Geology* **42**, 435–438 (2014).
30. Benedetti, L. *et al.* Earthquake synchrony and clustering on Fucino faults (central Italy) as revealed from *in situ* 36Cl exposure dating. *J. Geophys. Res.* **118**, 4948–4974 (2013).
31. Smeraglia, L. *et al.* Origin and role of fluids involved in the seismic cycle of extensional faults in carbonate rocks. *Earth and Plan. Sci. Lett.* **450**, 292–305 (2016a).
32. Smith, S. A. F., Billi, A., Di Toro, G. & Spiess, R. Principal Slip Zones in Limestone: Microstructural Characterization and Implications for the Seismic Cycle (Tre Monti Fault, Central Apennines, Italy). *Pure Appl. Geophys.* **168**, 2365–2393 (2011).
33. Roberts, G. P. Multi-seismic cycle velocity and strain fields for an active normal fault system, central Italy. *Earth and Plan. Sci. Lett.* **251**, 44–51 (2007).
34. Doglioni, C. A proposal of kinematic modelling for W-dipping subductions - Possible applications to the Tyrrhenian – Apennines system. *Terra Nova* **3/4**, 423–434 (1991).
35. Cosentino, D., Cipollari, P., Marsili, P. & Scrocca, D. Geology of the central Apennines; a regional review. *J. Virtual Explor.* **36**, 12 (2010).
36. Carminati, E., Lustrino, M. & Doglioni, C. Geodynamic evolution of the central and western Mediterranean: tectonics vs. igneous petrology constraints. *Tectonophysics* **579**, 173–192 (2012).

37. Billi, A. *et al.* First results from the CROP-11 deep seismic profile, central Apennines, Italy: evidence of mid-crustal folding. *J. Geol. Soc. London* **163**, 583–586 (2006).
38. Patacca, E., Scandone, P., Di Luzio, E., Cavinato, G. P. & Parotto, M. Structural architecture of the central Apennines: Interpretation of the CROP 11 seismic profile from the Adriatic coast to the orographic divide. *Tectonics* **27**, 1–36 (2008).
39. Doglioni, C., D'Agostino, N. & Mariotti, G. Normal faulting vs regional subsidence and sedimentation rate. *Mar. Pet. Geol* **15**, 737–750 (1998).
40. Cavinato, G. P. & De Celles, P. G. Extensional basins in tectonically bimodal central Apennines fold-thrust belt, Italy: Response to corner flow above a subducting slab in retrograde motion. *Geology* **27**, 955–958 (1999).
41. Galli, P., Galadini, F. & Pantosti, D. Twenty years of paleoseismology in Italy. *Earth-Science Rev* **88**, 89–117 (2008).
42. Valoroso, L. *et al.* Radiography of a normal fault system by 64,000 high-precision earthquake locations: The 2009 L'Aquila (central Italy) case study. *J. Geophys. Res.* **118**, 1156–1176 (2013).
43. Demurtas, M. *et al.* Structure of a normal seismogenic fault zone in carbonates: The Vado di Corno Fault, Campo Imperatore, Central Apennines (Italy). *J. Struct. Geol.* **90**, 185–206 (2016).
44. Cavinato, G. P., Carusi, C., Dall'asta, M., Miccadei, E. & Piacentini, T. Sedimentary and tectonic evolution of Plio-Pleistocene alluvial and lacustrine deposits of Fucino Basin (central Italy). *Sediment. Geol.* **148**, 29–59 (2002).
45. Critelli, S. *et al.* Interpreting siliciclastic-carbonate detrital modes in foreland basin system: an example from Upper Miocene arenites of the central Apennines, Italy. *Geol. Soc. Am. Special Paper* **420**, 107–133 (2007).
46. Smeraglia, L., Trippetta, F., Carminati, E. & Mollo, S. Tectonic control on the petrophysical properties of foredeep sandstones in the central Apennines, Italy. *J. Geophys. Res.* **119**, 9077–9094 (2014).
47. Tesei, T., Collettini, C., Viti, C. & Barchi, M. R. 2013. Fault architecture and deformation mechanisms in exhumed analogues of seismogenic carbonate-bearing thrusts. *Journal of Structural Geology* **55**, 167–181 (2013).
48. Viti, C., Collettini, C. & Tesei, T. Pressure solution seams in carbonatic fault rocks: mineralogy, micro/nanostructures and deformation mechanism. *Contribution to Mineral. Petrol* **167**, 970 (2014).
49. Smeraglia, L., Aldega, L., Billi, A., Carminati, E. & Doglioni, C. Phyllosilicate injection along carbonate-hosted faults and implications for co-seismic slip propagation: Case studies from the central Apennines. *Journal of Structural Geology* **93**, 29–50 (2016b).
50. Chen, J., Niemeijer, A., Yao, L. & Ma, S. Water vaporization promotes coseismic fluid pressurization and buffers temperature rise. *Geophysical Research Letters*, doi: 10.1002/2016GL071932 (2017).
51. Anzidei, M. & Pondrelli, S. (Eds) The Amatrice seismic sequence: preliminary data and results. *Annals of Geophysics* **59** (2016).
52. Smith, S. A. F., Nielsen, S. & Di Toro, G. Strain localization and the onset of dynamic weakening in calcite fault gouge. *Earth and Plan. Sci. Lett.* **413**, 25–36 (2015).
53. Verberne, B. A., Pluumper, O., De Vinter, M. D. A. & Spiers, C. J. Superplastic nanofibrous slip zones control seismogenic fault friction. *Science* **346**, 1342–1344 (2014).
54. Fondriest *et al.* Mirror-like faults and power dissipation during earthquakes. *Geology* **41**, 1175–1178 (2013).
55. Clauser C. & Huenges E. Thermal conductivity of rocks and minerals, In Rock *Physics and Phase Relations* T. Ahrens Ed. American Geophysical Union, Washington D.C. USA (1995).
56. Yao, L. *et al.* The crucial role of temperature in high-velocity weakening of faults: Experiments on gouge using host blocks with different thermal conductivities. *Geology* **44**, 63–66 (2016).
57. Boneh, Y., Sagy, A. & Reches, Z. Frictional strength and wear-rate of carbonate faults during high-velocity, steady-state sliding. *Earth and Planetary Science Letters* **381**, 127–137 (2013).

58. Rietveld, H. M. A profile refinement method for nuclear and magnetic structures. *Journal of applied crystallography* **2**, 65–71 (1969).
59. Di Toro, G. *et al.* From field geology to earthquake simulation: a new state-of-the-art tool to investigate rock friction during the seismic cycle (SHIVA). *Rendiconti dei Lincei* **21**, 95–114 (2010).

CHAPTER 13

A new, long-term daily satellite-based rainfall dataset for operational monitoring in Africa

Ross I. Maidment, David Grimes, Emily Black, Elena Tarnavsky, Matthew Young, Helen Greatrex, Richard P. Allan, Thorwald Stein, Edson Nkonde, Samuel Senkunda & Edgar Misael Uribe Alcántara

[1]*Department of Meteorology, University of Reading, Reading RG6 6BB, UK.*

[2]*International Research Institute for Climate and Society (IRI), Columbia University, New York, NY 10964-1000, USA*

[3]*National Centre for Earth Observation (NCEO), Reading RG6 6BB, UK*

[4]*Zambian Meteorological Department, P.O. Box 30200, Lusaka, Zambia*

[5]*Uganda National Meteorological Authority, P.O. Box 7025, Kampala, Uganda.*

[6]*Universidad Autonoma del Estado de Hidalgo, Pachuca 42039, Mexico. Correspondence and requests for materials should be addressed to R.I.M*

CONTRIBUTIONS

R.I.M. did the bulk of the work both creating the TAMSAT datasets, carrying out the evaluations and writing the paper. D.G. had the original insight that skilful daily estimates could be provided by weighting dekadal rainfall estimates by daily CCD. E.B. worked with R.I.M. to design the evaluations and create TAMSAT version 3.0, and drafted some of the discussion sections of the paper. E.T. worked with R.I.M. to create TAMSAT version 2.0, and provided intellectual input to the evaluation of the daily data, as well as operational leadership of the TAMSAT pan-African calibration and validation system, on which the disaggregated daily products are based. M.Y. worked with R.I.M. to create TAMSAT version 3.0. H.G., R.P.A. and T.S. had intellectual input into the evaluations and analysis. E.N., S.S. and E.M.U.A. processed and provided rain gauge records from Zambia, Uganda and Mozambique respectively.

ABSTRACT

Rainfall information is essential for many applications in developing countries, and yet, continually updated information at fine temporal and spatial scales is lacking. In Africa, rainfall monitoring is particularly important given the close relationship between climate and livelihoods. To address this information gap, this paper describes two versions (v2.0 and v3.0) of the TAMSAT daily rainfall dataset based on high-resolution thermal-infrared observations, available from 1983 to the present. The datasets are based on the disaggregation of 10-day (v2.0) and 5-day (v3.0) total TAMSAT rainfall estimates to a daily time-step using daily cold cloud duration. This approach provides temporally consistent historic and near-real time daily rainfall information for all of Africa. The estimates have been evaluated using ground-based observations from five countries with contrasting rainfall climates (Mozambique, Niger, Nigeria, Uganda, and Zambia) and compared to other satellite-based rainfall estimates. The results indicate that both versions of the TAMSAT daily estimates reliably detects rainy days, but have less skill in capturing rainfall amount—results that are comparable to the other datasets.

Design Type(s)	observation design • data integration objective
Measurement Type(s)	hydrological precipitation process
Technology Type(s)	meterological observation
Factor Type(s)	
Sample Characteristic(s)	atmospheric water vapour • Mozambique • Niger • Nigeria • Uganda • Zambia

1. BACKGROUND AND SUMMARY

High spatial and temporal rainfall variability is a major challenge when it comes to managing agricultural activities across Africa, as above or below average rainfall can lead to crop losses and failure[1]. A notable recent example was the occurrence of widespread drought conditions across the Horn of Africa during 2010–2011 which affected over 10 million people[2,3]. To help mitigate these climate-related risks, access to reliable rainfall information, both historic and near-real time, is a necessity. Historic data allows climate risks (e.g., the probability of drought) and long-term changes in the rainfall climate to be assessed, while near-real time data is important to evaluate the present day weather in a historical context. The latter is especially important in monitoring the evolution of hydrological hazards, allowing timely responses from governments and organizations before major crises occurs. Although temporally coarse data (for example, dekadal or monthly) can be useful for evaluating climatic trends and monitoring above or below average rainfall[4], information at fine time scales (e.g., daily) provides information valuable in a range of other applications such as crop modelling, water management and weather index-based insurance[4,5].

Conventionally, rain gauge records provide the most accurate means to obtain information about the rainfall climate. However, the spatially sparse network and often temporally incomplete records at many stations across Africa leaves large parts of the continent unobserved[6]. This problem is exacerbated by the high spatial variability associated with convective rainfall at the daily time-step that makes a rain gauge measurement only representative of rainfall over several square kilometres surrounding the gauge[7]. Except in the vicinity of a continually reporting weather station, gauge observations alone are impractical for the routine assessment of rainfall. Africa-wide, near-real time gauge records are only available via the Global Telecommunications System (GTS) network, usually through automatic weather stations. Although over 700 stations are registered on the GTS network, only a small proportion of these report daily[6,8]. Moreover, access to country-level records that often contain more data than is publicly available, is often only possible via direct contact with African meteorological and hydrological agencies.

The limitations associated with gauge measurements have elevated the importance of satellite-based rainfall estimates in many applications across Africa, especially in agriculture and drought monitoring[8]. Satellite-based algorithms have the advantage of providing full spatial coverage and have been demonstrated to be skilful in many locations over Africa[9-17].

Table 1: Summary of the rain gauge data used in the validation of the daily rainfall estimates.

Country	No. of Stations	Years	Months	No. of Records	Station data coverage			
					< 50%	50–75%	75–90%	>90%
Mozambique	31	1983–2009	Oct-Apr	153,387	2	4	6	19
Niger	107	1992	May-Sep	15,600	0	9	9	89
Nigeria	38	1991–2000	Mar-Oct	86,951	0	1	10	27
Uganda	56	2000–2005	Feb-Jun	34,864	18	8	10	20
Zambia	33	1983–2013	Oct-Apr	192,873	1	4	6	22

While there is an ever growing collection of satellite-based datasets capable of providing near-real time estimates (a selection of which are listed in Table 1 in Maidment *et al.*18), only a handful of publicly available high resolution satellite-based datasets providing historic data (at least 30 years) at the daily time-step and which are continually updated in real time or near-real time, exist for Africa. These are the National Oceanic Atmospheric Administration (NOAA) African Rainfall Climatology version 2.0 (ARC19) and the Climate Hazards Group InfraRed Precipitation with Station data version 2.0 (CHIRPS20) and are described in the Technical Validation section. Given the dearth of Africa-wide long-term (30 years or more) daily rainfall information and large uncertainties in existing observational records over Africa[21-23], the addition of daily satellite-based rainfall datasets with contrasting estimation approaches are extremely valuable for rainfall monitoring and climate research. Moreover, Africa's population is expanding rapidly and it is expected that this trend will continue throughout this century[24]. The pressures such growth is putting on agricultural and water resources, combined with changes in the rainfall climate[21], are encouraging the use of climate-based services such as Enhancing National Climate Services (ENACTS)[25] and Rainwatch[26] in many African countries. These services provide easily accessible historic and near-real time information on the local climate that is useful to a wide range of stakeholders. Such platforms, however, require skilful, long-term and regularly updated rainfall information.

Here, we describe and evaluate two versions (2.0 and 3.0) of the long-term daily TAMSAT (Tropical Applications of Meteorology using SATellite and ground based observations) rainfall dataset (Data Citation 1: University of Reading http://dx.doi.org/10.17864/1947.108 and Data Citation 2: University of Reading http://dx.doi.org/10.17864/1947.112; hereinafter TAMSAT-2 and TAMSAT-3 respectively), based on high resolution Meteosat thermal-infrared (TIR) observations for all of Africa, available from 1983 to the present and updated in near-real time. TAMSAT-2 and TAMSAT-3 are based on the disaggregation of the TAMSAT version 2.0 dekadal18 and TAMSAT version 3.0 pentadal rainfall estimates respectively, to a daily time-step using daily calibrated cold cloud duration (CCD) observations (see Methods section for algorithm details).

In January 2017, the TAMSAT Group released TAMSAT version 3.0—which is produced operationally alongside version 2.0 (ref. 18). Given that the daily rainfall estimates derived from TAMSAT v2.0 have been in the public domain for several years and are used by many users, this paper formally evaluates both TAMSAT-2 and TAMSAT-3. The rainfall estimates have been validated using daily rain gauge measurements from five Africa countries (Mozambique, Niger, Nigeria, Uganda, and Zambia) and compared with estimates from six other satellite-based rainfall datasets, some of which are used widely in rainfall monitoring applications across Africa.

2. METHODS

2.1 TAMSAT algorithm

The daily estimates (both TAMSAT-2 and TAMSAT-3) are derived from the TAMSAT rainfall estimation algorithm. The TAMSAT Group have, since the 1980s, produced estimates at the 10-day (dekad) scale. The algorithm, described in Milford *et al.*[27], Dugdale *et al.*[28], Grimes *et al.*[10]and Maidment *et al.*[18], works on the premise that the use of TIR imagery to monitor the cold cloud tops of rain-bearing convective cumulonimbus systems acts as a useful indicator for rainfall in the Tropics. Despite the simplicity of the TAMSAT operational approach, the dekadal estimates have been shown to perform well where rainfall is predominantly convective in origin[9,12,14,29-32]. The TAMSAT-2 estimates, described in this paper, have also been evaluated over the complex terrain of Ethiopia and demonstrated good skill[33]. Such skill, both at the daily and dekadal time-step, underlines the effectiveness of using TIR imagery in rainfall estimation where and when rainfall is convective in origin. The TAMSAT approach to rainfall estimation, however, does have limitations. Where rainfall from warm rain processes is dominant, such as along the coastal parts of West Africa[34]and over mountainous regions[31], the ability to identify rainy cloud is reduced. In addition, since the TAMSAT estimation approach is geared towards drought monitoring where accurately representing low rainfall totals is important, the algorithm in TAMSAT v2.0 (and all previous versions) was calibrated to better capture the more frequent, low rainfall amounts[18]. In doing so, the total rainfall is underestimated, resulting in an inherent dry bias that is more pronounced when the data (both daily and dekadal estimates) are aggregated (in space and/or time).

The aforementioned dry bias in TAMSAT v2.0 dekadal data, along with unrealistic spatial artefacts that originated from the use of rectangular calibration zones, prompted the TAMSAT Group to modify the calibration design, while ensuring the data is still applicable to drought monitoring. Although the principle features of the TAMSAT rainfall estimation approach have remained the same, the calibration used in version 3.0 differs markedly to version 2.0 and is designed to better capture local variations in the rainfall climate while reducing problems associated with version 2.0. Additionally, the time-step for the primary rainfall estimate is now 5-day (pentad), compared to 10-day in version 2.0. Here, we provide an outline of the common features behind the methodology used to create both TAMSAT-2 and TAMSAT-3. Comprehensive details on the version 2.0 pan-African calibration can be found in Maidment *et al.*18 and Tarnavsky *et al.*8, while details on version 3.0 can be found on the TAMSAT website (http://www.tamsat.org.uk).

The TAMSAT algorithm is based on two primary data inputs: Meteosat TIR imagery provided by The European Organisation for the Exploitation of Meteorological Satellites (EUMETSAT) and rain gauge observations for calibration (see Fig. 1 for the estimation process). The rainfall estimation approach is based on TIR imagery obtained every 15 min from July 2006 and every 30 min prior to this. The TAMSAT algorithm is an example of a cloud-indexing method: the duration of cloud tops exceeding a predetermined temperature threshold, known as cold cloud duration (CCD), acts as a proxy for rainfall.

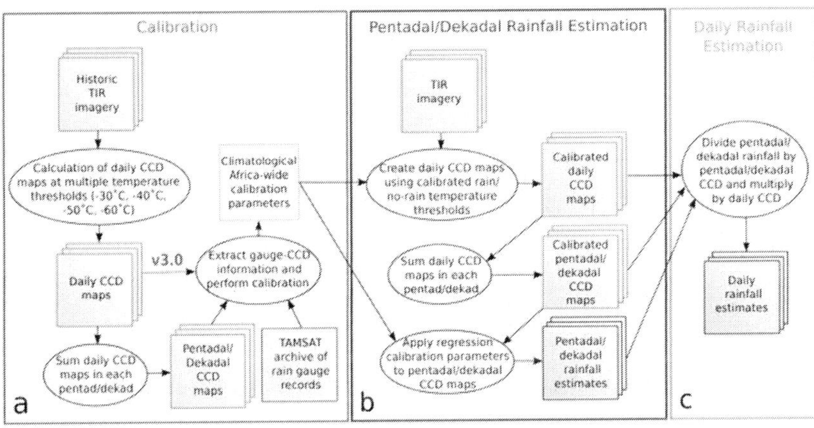

Figure 1: Schematic summarising the TAMSAT daily rainfall estimation process.(a) calibration of the algorithm at the dekadal (version 2.0) and pentadal (version 3.0) time-step, (b) estimation of the pentadal/dekadal rainfall estimates and (c) estimation of the daily rainfall estimates. Squares denote inputs or outputs and ovals denote processes. Note that the process is identical for versions 2.0 and 3.0 except where indicated on the diagram.

The calibration process is divided into two stages. The first stage distinguishes rainy regions from non-rainy regions, while the second stage attempts to assign a rainfall amount for the rainy regions. In the first stage, daily CCD totals are derived at a range of thresholds between −30 °C and −60 °C. These are then summed to the dekadal (in v2.0) or pentadal (in v3.0) time-step and a set of contingency tables are prepared for every threshold, comparing greater than zero CCD at the pixel scale with rainfall occurrence from the collocated rain gauge records. The temperature threshold with the greatest skill for determining rainfall events (greater than 0 mm) is selected based primarily on the rainfall event frequency bias (see Maidment *et al.*[18] for details). In version 2.0, these were determined for large climatologically-similar rectangular zones, whereas in version 3.0, these are derived over 1.0° grid boxes (hence capturing local detail more accurately) where sufficient gauges exist and then interpolated Africa-wide. In the second stage, calibration parameters are obtained by linearly regressing CCD totals for the selected temperature threshold with historical rain gauge accumulations. In version 3.0, a spatially and temporally varying bias adjustment is then made to the calibration parameters. Using the calibration coefficients, rainfall is estimated as a function of CCD, according to equation (1):

$$rain_{timestep} = \begin{cases} a_0 + a_1 CCD_{timestep} & CCD > 0 \\ 0 & CCD = 0 \end{cases} \tag{1}$$

Where *timestep* is either pentad or dekad, depending on the TAMSAT version, and a_0 and a_1 are the linear calibration coefficients. If CCD is equal to zero, rainfall is also assumed to be zero. The TAMSAT method implements a local calibration, hence the linear calibration coefficients vary spatially and monthly to reflect the geographical and temporal variations in the average rainfall climate across Africa8.

The TAMSAT-2 data are derived from TAMSAT dekadal v2.0 estimates that constitute the TAMSAT African Rainfall Climatology And Time-series (TARCAT) dataset[18] which is still routinely updated to the present day. Since the calibrations used in these datasets do not change from year-to-year, the interannual variations in rainfall are dependent only on the satellite observations. The TAMSAT method thus contrasts with other long-term datasets such as CHIRPS and ARC, which merge gauge data in near-real time[19,20]. The inclusion of contemporaneous gauge data arguably makes maximal use of all available data sources, increasing skill where high quality gauge data are available. The African gauge network is, however, not consistent in either time or space, and the inclusion of gauge data may thus introduce artefacts, especially when assessing long term change[21]. The TAMSAT datasets hence can be seen as a complement to the other available products.

2.2 Downscaling to the daily time scale

The currently available TAMSAT dekadal (v2.0) and pentadal (v3.0) rainfall estimates are disaggregated to daily values in proportion to the amount of CCD observed for each day (each daily CCD map is created by considering all TIR images from 06:00 to 06:00 the following day, to coincide with the timing gauge observations are usually taken). This has the advantage that the estimates are constrained to match the dekadal or pentadal rainfall totals which are expected to be reliable. The daily rainfall estimates are thus calculated according to equation (2):

$$rain_{daily} = \frac{rain_{timestep}}{CCD_{timestep}} \times CCD_{daily} \quad (2)$$

where $rain_{daily}$ is the daily rainfall estimate, $rain_{timestep}$ is the dekadal (v2.0) or pentadal (v3.0) rainfall estimate, $CCD_{timestep}$ is the CCD summed over the ten or five days and CCD_{daily} is the daily CCD. The complete process used to create the TAMSAT daily rainfall estimates is illustrated in Fig. 1.

2.3 Data archive

A time-series of daily totals has been generated from 1983 to the present for all of Africa. A day is considered missing if there is a gap of more than six continuous hours in the TIR imagery. For version 2.0, a dekad is considered missing if there are more than two missing days (see Maidment *et al.*[18] for details), whereas for version 3.0, the pentad is considered missing if more than one day is missing. Despite many incomplete or missing TIR images during the 1980s and early 1990s, both version 2.0 and 3.0 have near-complete archives. For version 2.0 for example, based on available data from the EUMETSAT archive, the dataset is approximately 97% complete (as of December 31st 2016). Specifically, of the 12,419 days between January 1st 1983 and December 31st 2016, there are 398 missing days. Of these, in 271 cases, the whole dekad was missing, resulting in no data to disaggregate, and in 127 cases, individual days within the dekad were missing. Of the missing days, 271 were between 1983 and 1989, 114 were between 1990 and 1999, and 13 were after 2000. There have been no missing days since 2007. As expected, the proportion of missing days is similar for TAMSAT-3. The daily estimates are available from January 11th 1983 to the present and are available within two days after the end of each dekad (i.e., 11th, 21st, and 1st of the following month) for version 2.0 and each pentad (i.e., 6th, 11th, 16th, 21st, 26th and 1st of the following month) for version 3.0.

2.4 Data access and format

The daily rainfall estimates (in mm per day) are freely available as netCDF files for each day from the TAMSAT website (http://www.tamsat.org.uk) and the University of Reading Research Data Archive (version 2.0, Data Citation 1: University of Reading http://dx.doi.org/10.17864/1947.108; version 3.0: Data Citation 2: University of Reading http://dx.doi.org/10.17864/1947.112). TAMSAT-2 is also available on the International Research Institute for Climate and Society (IRI) Data Library (https://iridl.ldeo.columbia.edu/SOURCES/.Reading/.Meteorology/.TAMSAT/.TARCAT/.v2p0/.daily/), with TAMSAT-3 expected to be available during 2017.

The spatial resolution is 0.0375° latitude by 0.0375° longitude with estimates provided for all land points in Africa, including Madagascar. In addition, the TAMSAT website contains quicklook images for each day and a time series extraction tool can be used to extract area-average data for countries, administrative districts and user defined rectangular regions or user defined pixels in csv format. The IRI Data Library includes additional subsetting and data analysis tools.

2.5 Technical Validation

2.5.1 Study regions and validation data

The daily satellite rainfall estimates have been evaluated using rain gauge records covering four countries (Mozambique, Nigeria, Uganda and Zambia) and one region over south-west Niger consisting of a dense network (see Fig. 2 and Table 1). These regions of Africa are characterised by contrasting rainfall climates and thus, the validation provides a useful indicator of the expected skill of the TAMSAT daily estimates (and the other satellite estimates used in this study) across Africa. The section below summarises the general climate of each region considered.

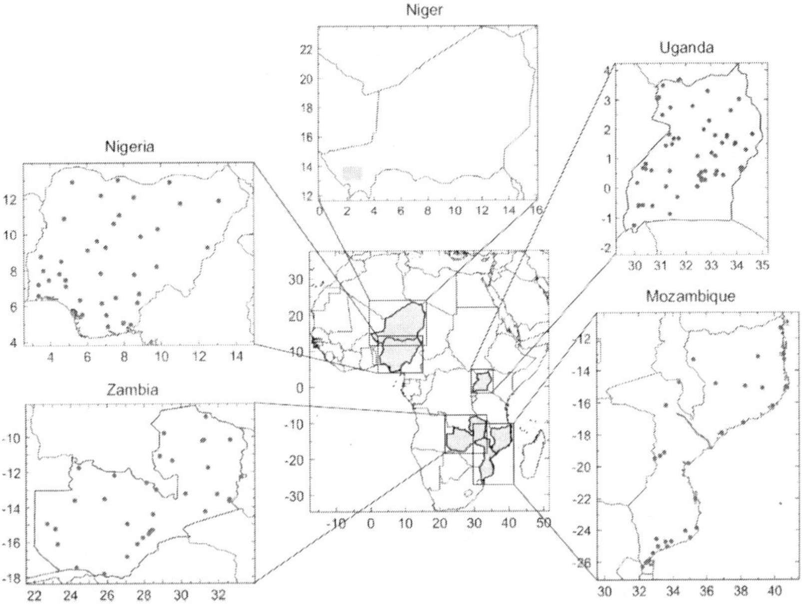

Figure 2: Location of the five rain gauge networks used in this study. The yellow shading over south-west Niger denotes the region covered by the Hapex-Sahel gauge network.

The rainfall over Niger is typical of that experienced over most of the Sahel, characterized by a single rainy season occurring during boreal summer. The main features of the rainy season are the West African Monsoon, which advects moisture-laden air onto the continent and African Easterly Waves that are associated with the passage of westward propagating mesoscale convective systems that are responsible for the majority of rainfall over this part of Africa[35,36]. TIR-based estimation algorithms, including TAMSAT, have demonstrated high skill over the Sahel[9,12,32]. Much of Nigeria's rainfall climate is similar in nature, although rainfall in the coastal regions and areas surrounding the Cameroon Highlands to the east are often modulated by oceanic and orographic effects respectively, complicating the relationship between cloud top temperature and rainfall[31].

Most of Uganda experiences two rainy seasons associated with the seasonal northward and southward migration of the Inter-Tropical Convergence Zone[14]. Whilst rainfall is convective in origin, the presence of mountain chains to the east and southwest of the country and large bodies of water, such as Lake Victoria and Lake Albert, influence the local climate considerably. While this presents a challenge for TIR-based algorithms due to the increased occurrence of rainfall from warm clouds, particularly where local changes to the rainfall processes are pronounced, 10-day total satellite-derived estimates have shown to be skilful over this region[14].

Zambia has one rainy season occurring between October and April. As the country is relatively flat and landlocked and rainfall is primarily a result of convective systems, cold cloud tops of these convective systems and rainfall are usually well correlated, as found across Niger.

Finally, the climate of Mozambique contrasts with the other regions considered in this study. The close proximity to the Indian Ocean and the passage of tropical depressions and cyclones create a varied and complex climate. Such variable weather regimes presents a challenge for TIR-based algorithms, especially when other data (e.g., gauge data) are not incorporated contemporaneously[17].

The daily rain gauge records from Nigeria, Uganda and Zambia were obtained directly from their respective meteorological agencies. Each of these datasets were subject to rigorous quality control measures. These procedures involved checking for erroneous entries, duplicates, and outliers. If outliers were flagged, temporal and spatial checks were then conducted. The high density Niger dataset was created during the Hydrology-Atmosphere Pilot Experiment in the Sahel (Hapex-Sahel) experiment in the early 1990s (refs 37- 40) and has been used in many subsequent studies[10]. Finally, the Mozambique data was sourced from The Mozambique National Institute of Meteorology and quality controlled for The World Bank[41]. Only those records during each region's rainy season were used (for Uganda, records covering the 'long rains' were used). Whilst not all stations used have complete records, each regional dataset had at least 15,000 gauge records available for validation (see Table 1).

The variability in the TAMSAT daily rainfall estimates is derived entirely from the satellite imagery—with the calibration carried out on 10-daily (v2.0) or 5-daily (v3.0) accumulated rainfall/CCD over regions encompassing hundreds of gauge-CCD pairs[8,18]. The evaluation of the TAMSAT rainfall estimates described here can thus arguably be considered to be against independent data, even though some of the gauge records may have been included in the historical calibration. This is not the case for some of the comparison satellite datasets used in this study, which incorporate contemporaneous gauge records. The Niger gauge dataset however, is not included either in the TAMSAT version 2.0 dekadal calibration, or—to our knowledge—in the comparison satellite datasets.

To ensure a consistent comparison between the satellite estimates and ground-based data, all rain gauge records were interpolated onto a regular 0.25° by 0.25° grid using block kriging. Kriging was chosen as it has been shown to be superior compared to other forms of spatial interpolation[42-44]. Since the uncertainty in the interpolated rainfall amount increases significantly away from a rain gauge, only those grid squares containing at least one gauge were used. For simplicity, it was assumed that all 0.25° grid squares containing only dry gauges were set to zero rainfall. In the event of a grid square containing dry and wet gauges, the kriged rainfall amount was used.

It should be noted that given the high density of the Niger gauge network, the interpolated area-average values will, in general, be much more accurate than the equivalent interpolated grid values over the four other regions whose gauge networks are considerably less dense. Moreover, since the availability of the satellite estimates and gauge data for each region do not cover the same time periods, it is not possible to directly compare the results from one region to another. This is particularly the case for Niger whose gauge data is only available for one year. However, the results presented provide a useful indicator of the expected skill of the TAMSAT daily rainfall estimates, in comparison to the other satellite datasets.

The TAMSAT-2 and TAMSAT-3 rainfall estimates were evaluated alongside six other satellite precipitation datasets providing daily estimates. These datasets are CHIRPS, CHIRP (CHIRPS without stations), ARC, NOAA's African Rainfall Estimates version 2 (hereinafter RFE), the Tropical Rainfall Measuring Mission (TRMM) Multi-satellite Precipitation Analysis (TMPA)-3B42 and NOAA's Climate Prediction Center (CPC) morphing technique (CMORPH) (see Table 2). The latter three datasets include passive microwave (PMW) imagery, and hence are expected to be capable of providing more realistic information on rainfall intensity. A brief description of these datasets is as follows:

Table 2: Summary of the daily satellite rainfall datasets used in this study.

Product	Spatial resolution	Starting year	Inputs*	Latency	Reference
TAMSAT v2.0	0.0375°	1983	TIR, gauge	Up to 12 days	—
TAMSAT v3.0	0.0375°	1983	TIR, gauge	Up to 7 days	—
CHIRP	0.05°	1981	TIR, TMPA-3B42	Up to 7 days	Funk *et al.*[20,45]
CHIRPS v2.0	0.05°	1981	TIR, TMPA-3B42, gauge	Up to 7 weeks	Funk *et al.*[20,45]
ARC v2.0	0.1°	1983	TIR, gauge	1 day	Novella and Thiaw[19]
RFE v2.0	0.1°	2001	TIR, PMW, gauge	1 day	http://www.cpc.ncep.noaa.gov/products/fews/RFE2.0_tech.pdf
TMPA-3B42 v7.0	0.25°	1997	TIR, PMW, radar, gauge	No longer operational	Huffman *et al.*[49]
CMORPH	0.25°	2002	TIR, PMW	18 h	Joyce *et al.*[50]

CHIRPS provides 30+ years of high resolution (0.05° lat-lon grid) quasi-global (50°S-50°N and 180°W–180°E) rainfall estimates at daily, pentadal, and monthly time-steps. CHIRPS depends on several data sources to produce estimates of rainfall. First, TIR imagery are used to produce maps of pentadal CCD. Unlike TAMSAT, which implements a temporally and spatially varying threshold to compute the CCD, a constant rain/no-rain threshold of 235 K is used. Calibration regression coefficients are then derived by comparing TMPA-3B42 rainfall estimates (2000–2013) and CCD. These calibration parameters are then applied to the complete CCD record to produce a time-series of rainfall estimates. Next, these pentadal rainfall estimates are expressed as a fraction of their long-term mean (1981–2013) and then multiplied by the Climate Hazards group Precipitation climatology (CHPclim). This step produces what is known as CHIRP, i.e., the satellite-based estimates with no merging of rain gauge records, and is also evaluated in this study. CHPclim is an attempt to create accurate pentadal and monthly climatologies based on rain gauge records and multiple satellite-based products[45]. Finally, station rain gauge records are merged with CHIRP using a modified form of the inverse distance weighting algorithm to create the CHIRPS product. A preliminary version, CHIRPS-prelim, is created with a 2-day latency based on GTS data, while the final version (evaluated in this paper) makes use of public monthly gauge summaries and additional data from meteorological agencies. Daily estimates of precipitation are created by disaggregating the pentadal estimates using daily CCD observations (analogous to the method described in this paper).

Both ARC and RFE produce daily rainfall estimates solely for Africa and were created to aid drought monitoring across sub-Saharan Africa. RFE uses satellite imagery from two streams, namely (1) TIR imagery to create rainfall estimates based on the GOES Precipitation Index (GPI) algorithm[46] and (2) PMW imagery from the AMSU and SSM/I satellite instruments are used to create rainfall estimates using the method described by Ferraro and Marks[47]. The TIR and PMW rainfall estimates are then merged, before being adjusted to available GTS station data. ARC is a long term (30+ years) dataset and employs a similar method to RFE in that satellite estimates are merged with GTS gauge data, however PMW data are not considered.

The primary objective of the TRMM satellite and the derived products was aimed at improving observations of tropical precipitation[48,49]. The TRMM satellite, equipped with a precipitation radar, as well as microwave imager and a visible-infrared scanner, was used to better estimate precipitation features such as intensity, distribution, and type. 3-hourly TMPA-3B42 (evaluated in this study) estimates are derived from merged-TIR imagery from geostationary and polar-orbiting platforms, adjusted by information derived from the TRMM instruments. The final step used the monthly Global Precipitation Climatology Centre (GPCC) gauge analysis to scale the monthly TMPA estimates to the gauge values. Sub-monthly products, including the 3-hourly TMPA-3B42 estimates, take account of this gauge scaling. TMPA data were issued to provide near-global coverage at a spatial resolution of 0.25°.

CMORPH[50] produces global rainfall estimates from various PMW sensors. Motion vectors are calculated using half-hourly geostationary TIR imagery, which are used to propagate the PMW precipitation fields forward and back in time where no direct PMW data are available. A time-weighted interpolation is applied to the available PMW estimates to provide an estimate of the rainfall distribution and intensity for the intervening missing half-hour periods. This process is referred to as 'morphing' of the observations. For this study, the 3-hourly estimates at a spatial resolution of 0.25° were used.

In the case of CHIRPS, the operational product CHIRPS-Prelim was not considered because at the time this study was conducted, the data were not available prior to 2015. All of the other datasets can be considered fully operational, except TMPA-3B42 which was replaced by the Integrated Multi-satellitE Retrievals for Global Precipitation Measurement (GPM) (IMERG) estimates in 2014. For consistency, all satellite datasets (except TMPA-3B42 and CMORPH) were bilinearly interpolated to a regular grid spacing of 0.25° by 0.25°—same as the kriged gauge grid, and grid squares with coincident gauge measurements were then extracted. When summing the CMORPH and TMPA-3B42 3-hourly estimates to daily totals, the 3-hourly slots corresponding to the TAMSAT day (i.e., 06:00–06:00 the following day) were chosen. Evaluations were then carried out for the period of the gauge data, which differs from region to region (see Table 1).

2.6 Statistical comparison of TAMSAT daily rainfall estimates with rain gauge data and other satellite-based rainfall datasets

The TAMSAT version 2.0 dekadal and monthly estimates and their representation of the Africa-wide climatology and seasonal cycle have been evaluated elsewhere[18] and hence these features are not assessed here. Similar analyses for TAMSAT version 3.0 are documented on the TAMSAT website. Instead, the paper focuses on the ability of TAMSAT to capture daily rainfall characteristics, i.e., occurrence and amount.

2.6.1 Rainfall occurrence

Rainfall occurrence was evaluated using a suite of binary skill scores that encapsulate information on rainy/dry days in a contingency table (see Table 3).

Table 3: Contingency table used for the statistical analysis of rainfall occurrence for the daily satellite rainfall estimates.

		Gauge Observations	
		Rain	**No Rain**
Satellite Rainfall Estimate	**Rain**	Hits 'A'	False Alarms 'B'
	No Rain	Misses 'C'	Correct Negatives 'D'

A contingency table has been constructed for each region using all available data and is used to compute the following statistics:

- Accuracy; defined as the fraction of rainfall estimates that were estimated correctly: $(A+D)/(A+B+C+D)$
- Frequency bias (bias); defined as the rainfall estimate frequency of rainy days compared to the gauge observed frequency of rainy days: $(A+B)/(A+C)$
- Probability of detection (POD); defined as the fraction of rainy days correctly estimated: $A/(A+C)$
- False alarm ratio (FAR): defined as the proportion of estimated rainy days that did not actually occur: $B/(A+B)$
- Probability of false detection (POFD); defined as the fraction of gauge observed dry days incorrectly estimated as a rainy day: $B/(B+D)$
- Equitable threat score (ETS); defined as the fraction of gauge observed rainy days that were correctly estimated allowing for hits due to chance: $(A-A_{random})/(A+B+C-A_{random})$ where $A_{random}=(A+C)(A+B)/(A+B+C+D)$
- Peirce's Skill Score (PSS, also known as Hanssen and Kuipers discriminant); defined as the ability of the satellite estimate to differentiate between a rainy day and a dry day (as given by the gauge observation): $(A/(A+C))-(B/(B+D))$

Figure 3 displays barplots for each binary skill score over each of the study regions and over all regions for TAMSAT-2, TAMSAT-3 and the other six satellite datasets (values are also given in Table 4). In general, the TAMSAT skill scores (both versions) are similar to most of the other satellite products on all skill measures. Across all regions (leftmost column in Fig. 3), the accuracy skill measure indicates around 70% of the estimates were correct (i.e., in estimating dry and rainy days) and that around 70–80% of the observed rainy days were captured (POD). However, around 35–45% of estimated rainy days were falsely estimated (FAR) resulting in all products overestimating the occurrence of rainy days (bias), with the errors most severe in CHIRP, RFE and CMORPH. Similarly, around 20–40% of the gauge observed dry days, were estimated as rainy days (POFD). Of the eight datasets, the TIR-based products show more commonality than the PMW-based products. The similarity between skill scores of the former suggests that this is a result of the use of TIR imagery being used to define those regions which are rainy. The exception to these findings is CHIRP, which, across all countries, grossly overestimates the frequency of rainfall events, leading to a high frequency bias (1.98) and POFD (0.67). However, CHIRPS demonstrates marked improvement on all statistical measures compared to CHIRP.

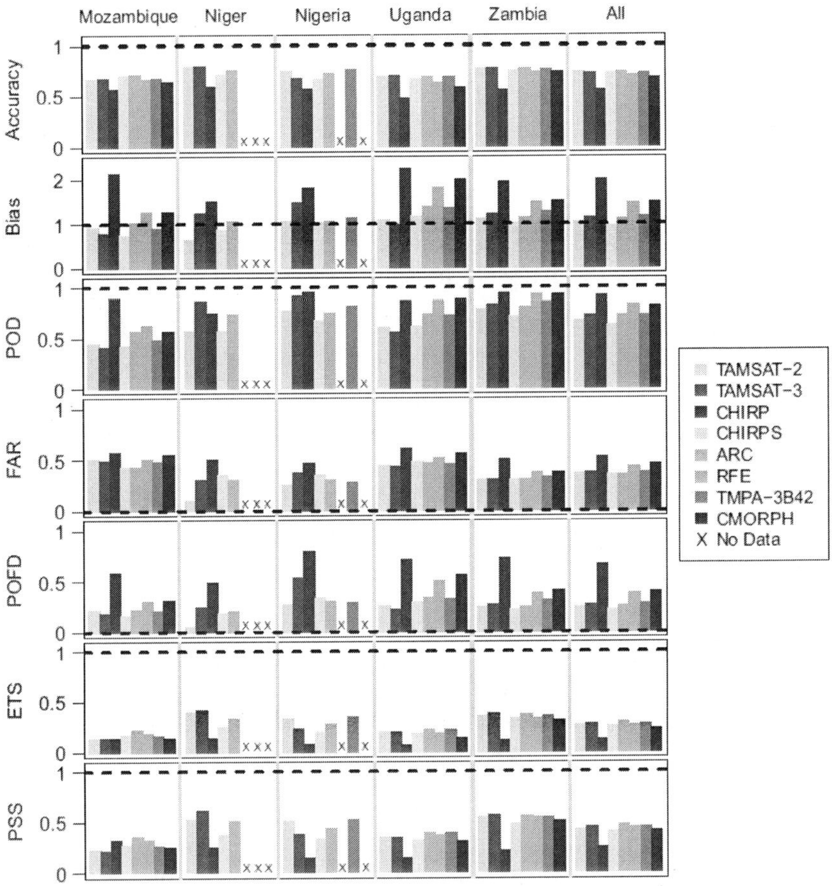

Figure 3: Skill scores for each product and country computed using a threshold of 0 mm (rain gauge and satellite estimate) using all available gauge observations (see Table 1**).**Horizontal black dashed line denotes the perfect skill score. Blue-shaded colour bars denote those datasets that are considered TIR-only, yellow/orange-shaded colour bars denote those that are merged TIR-gauge and pink/red-shaded colour bars denote PMW-based datasets.

Table 4: Skill scores computed using a threshold of 0 mm for each country (the threshold is defined as the rainfall amount that distinguishes a dry day from a rainy day). N is the number of coincident gauge-satellite grid observations used. Values in bold denote the most favourable comparison

Country	Dataset	N	Accuracy	Bias	POD	FAR	POFD	ETS	PSS
Mozambique	TAMSAT-2	147,059	0.67	0.92	0.45	0.51	0.22	0.13	0.23
	TAMSAT-3	123,586	0.68	0.80	0.41	0.49	0.18	0.13	0.22
	CHIRP	153,078	0.57	2.12	**0.90**	0.57	0.59	0.14	0.32
	CHIRPS	114,204	0.70	0.75	0.42	**0.43**	**0.16**	0.17	0.27
	ARC	145,600	**0.71**	**1.03**	0.57	0.44	0.22	**0.22**	**0.36**
	RFE	49,935	0.67	1.27	0.63	0.51	0.30	0.18	0.32
	TMPA-3B42	65,719	0.68	0.91	0.48	0.48	0.21	0.16	0.26
	CMORPH	37,813	0.64	1.26	0.57	0.55	0.32	0.13	0.25
Niger	TAMSAT-2	3,671	**0.80**	0.64	0.57	**0.11**	**0.05**	0.39	0.52
	TAMSAT-3	3,671	**0.80**	1.24	**0.86**	0.31	0.24	**0.42**	**0.62**
	CHIRP	3,671	0.60	1.51	0.74	0.51	0.49	0.13	0.25
	CHIRPS	3,671	0.71	0.87	0.57	0.35	0.19	0.24	0.37
	ARC	3,623	0.76	**1.06**	0.73	0.31	0.21	0.34	0.51
	RFE	—	—	—	—	—	—	—	—
	TMPA-3B42	—	—	—	—	—	—	—	—
	CMORPH	—	—	—	—	—	—	—	—

Nigeria	TAMSAT-2	64,527	0.75	1.05	0.77	**0.26**	**0.27**	0.34	0.51
	TAMSAT-3	63,257	0.68	1.48	0.92	0.38	0.54	0.23	0.38
	CHIRP	65,290	0.57	1.80	**0.95**	0.47	0.80	0.08	0.15
	CHIRPS	65,290	0.67	**1.03**	0.67	0.35	0.34	0.02	0.33
	ARC	64,381	0.72	1.06	0.74	0.30	0.30	0.28	0.44
	RFE	—	—	—	—	—	—	—	—
	TMPA-3B42	17,815	**0.76**	1.13	0.81	0.28	0.29	**0.35**	**0.52**
	CMORPH	—	—	—	—	—	—	—	—
Uganda	TAMSAT-2	31,084	0.69	1.09	0.60	0.45	0.26	0.20	0.34
	TAMSAT-3	30,949	**0.70**	**0.97**	0.55	**0.43**	**0.22**	0.20	0.34
	CHIRP	31,084	0.48	2.23	0.86	0.61	0.71	0.06	0.15
	CHIRPS	31,084	0.67	1.17	0.61	0.48	0.29	0.18	0.32
	ARC	31,020	0.69	1.37	0.73	0.47	0.34	**0.22**	**0.39**
	RFE	25,207	0.63	1.80	0.87	0.52	0.50	0.18	0.37
	TMPA-3B42	30,975	0.69	1.34	0.72	0.46	0.33	**0.22**	**0.39**
	CMORPH	14,082	0.58	1.99	**0.88**	0.56	0.56	0.14	0.31

Zambia	TAMSAT-2	169,021	**0.77**	1.12	0.78	**0.30**	0.24	0.36	0.54
	TAMSAT-3	161,074	**0.77**	1.21	0.83	0.31	0.27	0.38	0.57
	CHIRP	175,080	0.56	1.93	**0.94**	0.51	0.72	**0.11**	**0.22**
	CHIRPS	175,080	0.75	**1.01**	0.71	**0.30**	**0.22**	0.33	0.49
	ARC	167,530	**0.77**	1.14	0.80	**0.30**	0.24	0.37	0.56
	RFE	65,538	0.74	1.48	0.93	0.37	0.39	0.34	0.54
	TMPA-3B42	83,560	0.76	1.28	0.85	0.33	0.31	0.36	0.55
	CMORPH	53,458	0.73	1.50	0.93	0.38	0.41	0.31	0.51
All	TAMSAT-2	415,362	0.73	**1.04**	0.67	0.36	0.24	0.27	0.43
	TAMSAT-3	382,537	0.72	1.14	0.72	0.37	0.27	0.28	0.45
	CHIRP	428,203	0.56	1.98	**0.92**	0.53	0.67	0.12	0.25
	CHIRPS	389,329	0.72	**0.96**	0.62	**0.35**	**0.22**	0.25	0.40
	ARC	412,154	**0.74**	1.11	0.72	**0.35**	0.25	**0.30**	**0.47**
	RFE	140,680	0.70	1.47	0.83	0.44	0.38	0.26	0.45
	TMPA-3B42	198,069	0.72	1.17	0.72	0.38	0.28	0.28	0.45
	CMORPH	105,353	0.68	1.49	0.81	0.46	0.40	0.23	0.41

Regionally however, there are some differences. Across most of the satellite products, scores are generally better for Niger, particularly for TAMSAT-2, which has the best scores for accuracy, FAR and POFD, and TAMSAT-3, which has the best scores for accuracy (same value as TAMSAT-2), POD, ETS and PSS. Scores are generally worst for Mozambique and Uganda. This is consistent with the expectation that satellite rainfall estimation algorithms, even those that incorporate PMW imagery, generally perform worse when the rainfall climate is strongly modulated by large water bodies, and for regions in close proximity to the ocean and complex topography17. Conversely, such algorithms perform well in the Sahel and over Zambia, where rainfall is primarily convective and the rainfall climate is less variable spatially. The high skill across both Niger and Zambia reflects this.

The skill scores were also assessed as a function of rainfall threshold (i.e., changing the satellite rainfall estimate threshold at which the contingency table is constructed). However, for all datasets the skill scores exhibited no improvement in skill as the threshold was increased from 0 mm up to 40 mm (not shown).

2.6.2 Rainfall amount

Figure 4 shows a density scatterplot of TAMSAT-2, TAMSAT-3 and the other satellite rainfall estimates against kriged rain gauge amounts for all regions included in this study. Quantitative assessment of rainfall amount was based on the calculation of bias, coefficient of determination (R^2), root mean square error (RMSE), and mean absolute error (MAE). The kriging process also generates an estimate of the uncertainty of the interpolated gauge grid value. Using this, the fraction of satellite estimates within one and two standard errors of the gauge value was also computed. A summary of the aforementioned statistics is given in Fig. 5 for each dataset and for each region (values are also given in Table 5).

There is some correlation between rainfall estimates and gauge measured rainfall amount for all of the satellite-rainfall estimation datasets, but there are also significant discrepancies (see Fig. 4). For example, TAMSAT-2 systematically underestimates rainfall amount, and does not distinguish between moderate and high rainfall. Figure 5 confirms that there is a negative TAMSAT-2 bias for all countries, with the largest bias being for Mozambique and Nigeria. The correlations (i.e., R^2) between gauge and TAMSAT-2 rainfall amounts range from 0.05 (Mozambique) to 0.61 (Niger). Niger also has the lowest errors (RMSE and MAE) out of the five regions whereas Mozambique has the largest RMSE. TAMSAT-3 however demonstrates improvement on some on the statistics considered when compared to TAMSAT-2, most notably, a reduction in the dry bias. There is also slightly better distinction between moderate and high rainfall (c.f. Figs 4 and 6).

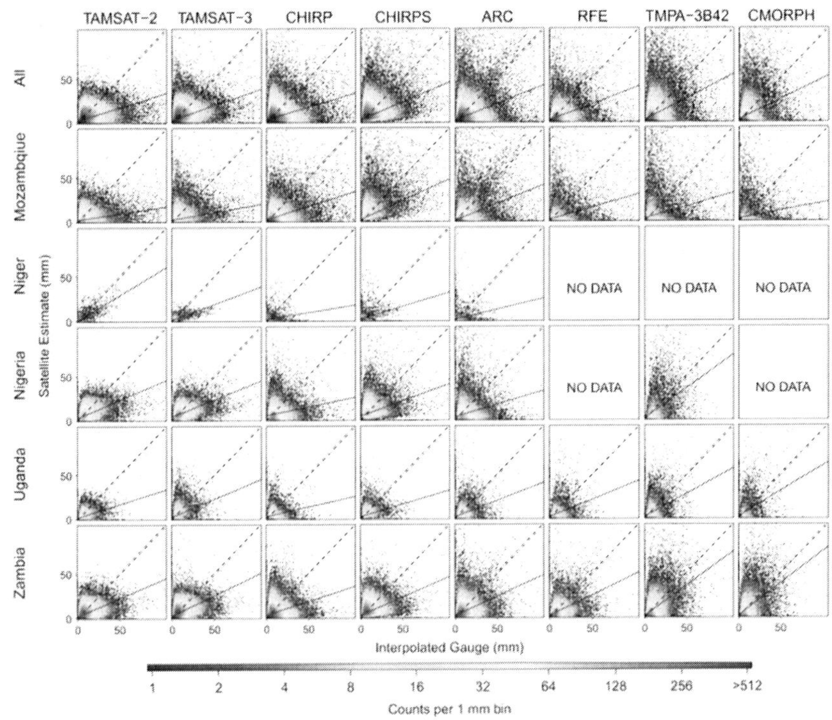

Figure 4: Scatterplot of daily satellite rainfall estimates against kriged gauge estimates for 0.25° grid cells that contain at least one rain gauge. Scale gives the counts per 1 mm bin. Dashed line indicates the one-to-one correspondence; solid line gives the linear regression best fit.

When the TAMSAT estimates are contrasted with the other rainfall datasets, it can be seen that over all countries, TAMSAT is in general, comparable in all skill measures, except for bias. CHIRPS has the smallest bias, which can be attributed to the bias removal procedure implemented in the rainfall estimation approach. TAMSAT-2 and TAMSAT-3 estimates typically have smaller errors, as given by lower RMSE and MAE values. The smallest errors are for Niger. Low R^2 values (with the exception of TAMSAT over Niger) indicate limited skill in representing variability for all datasets. Given the high density gauge network over Niger and the contiguous 0.25° grid squares used here, measures of variability (i.e., R^2) are associated with both spatial and temporal variability.

All datasets typically perform worse over Mozambique as evident by the large spread of data points in Fig. 4. Despite including PMW data in their estimation approaches, neither RFE, TMPA-3B42 and CMOPRH demonstrate substantial improvements in skill over the TIR-based methods, particularly for rainfall amount variability. This indicates that at such fine scales (daily and 0.25°), no dataset considered here can provide robust estimates of daily rainfall amount. This is in agreement with other studies at such scales[31,33,51].

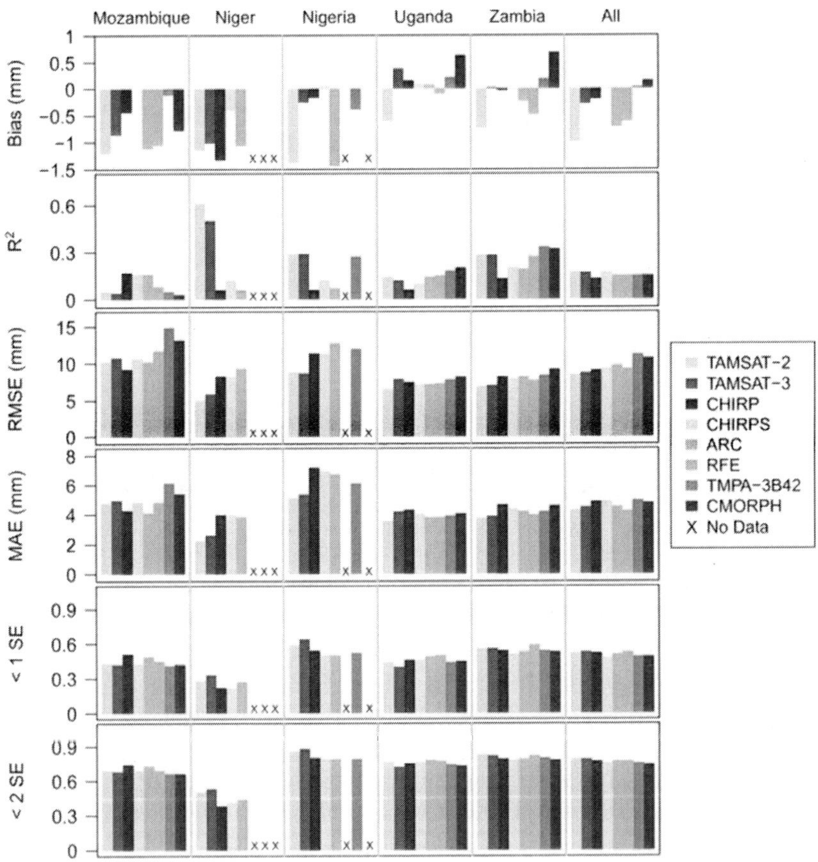

Figure 5. Same as Fig. 3, but for rainfall amount skill measures.

Table 5: Comparison of satellite rainfall estimates and collocated kriged gauge records for each country and for each satellite product.

Country	Dataset	*N*	Bias (mm)	R^2	RMSE (mm)	MAE (mm)	< 1 s.e.	< 2 s.e.
Mozambique	TAMSAT-2	147,059	−1.21	0.05	10.20	4.76	0.43	0.69
	TAMSAT-3	123,586	−0.86	0.04	10.75	4.96	0.42	0.68
	CHIRP	153,078	−0.44	**0.17**	**9.21**	4.27	**0.51**	**0.74**
	CHIRPS	114,204	**0.00**	0.16	10.61	4.82	0.43	0.69
	ARC	145,600	−1.12	0.16	10.19	**4.10**	0.49	0.73
	RFE	49,935	−1.06	0.08	11.67	4.83	0.45	0.69
	TMPA-3B42	65,719	−0.12	0.05	14.85	6.14	0.41	0.66
	CMORPH	37,813	−0.78	0.03	13.17	5.38	0.42	0.66
Niger	TAMSAT-2	3,671	−1.15	**0.61**	**5.04**	**2.22**	0.28	0.50
	TAMSAT-3	3,671	−1.02	0.50	5.85	2.60	**0.33**	**0.53**
	CHIRP	3,671	−1.33	0.06	8.27	3.97	0.22	0.38
	CHIRPS	3,671	**−0.40**	0.12	8.24	3.98	0.22	0.41
	ARC	3,623	−1.07	0.06	9.31	3.83	0.27	0.44
	RFE	—	—	—	—	—	—	—
	TMPA-3B42	—	—	—	—	—	—	—
	CMORPH	—	—	—	—	—	—	—

Nigeria	TAMSAT-2	64,527	−1.38	**0.29**	8.77	**5.12**	0.59	0.86
	TAMSAT-3	63,257	−0.26	**0.29**	**8.65**	5.36	**0.64**	**0.88**
	CHIRP	65,290	−0.17	0.06	11.33	7.19	0.54	0.80
	CHIRPS	65,290	**0.04**	0.12	11.28	6.93	0.50	0.79
	ARC	64,381	−1.44	0.07	12.73	6.72	0.50	0.79
	RFE	—	—	—	—	—	—	—
	TMPA-3B42	17,815	−0.39	0.27	11.93	6.10	0.52	0.79
	CMORPH	—	—	—	—	—	—	—
Uganda	TAMSAT-2	31,084	−0.61	0.14	**6.53**	**3.55**	0.44	0.76
	TAMSAT-3	30,949	0.38	0.12	7.88	4.20	0.40	0.72
	CHIRP	31,084	0.15	0.06	7.43	4.31	0.46	0.75
	CHIRPS	31,084	0.08	0.10	7.13	4.01	0.46	0.76
	ARC	31,020	**0.07**	0.14	7.14	3.80	0.49	**0.78**
	RFE	25,207	−0.10	0.15	7.29	3.81	**0.50**	0.77
	TMPA-3B42	30,975	0.21	0.18	7.78	3.90	0.44	0.74
	CMORPH	14,082	0.63	**0.20**	8.18	4.06	0.45	0.73
Zambia	TAMSAT-2	169,021	−0.74	0.28	**6.89**	**3.74**	0.56	**0.83**

	TAMSAT-3	161,074	0.02	0.28	7.02	3.88	0.56	0.82
	CHIRP	175,080	−0.04	0.13	8.20	4.66	0.54	0.79
	CHIRPS	175,080	**0.01**	0.20	7.85	4.35	0.51	0.78
	ARC	167,530	−0.24	0.19	8.20	4.20	0.53	0.79
	RFE	65,538	−0.49	0.27	7.68	3.97	**0.59**	0.82
	TMPA-3B42	83,560	0.18	**0.33**	8.35	4.19	0.54	0.80
	CMORPH	53,458	0.68	0.32	9.18	4.59	0.53	0.78
All	TAMSAT-2	415,362	−1.00	**0.17**	**8.46**	**4.29**	0.52	**0.79**
	TAMSAT-3	382,537	−0.29	**0.17**	8.71	4.49	**0.53**	**0.79**
	CHIRP	428,203	−0.20	0.13	9.05	4.87	0.52	0.77
	CHIRPS	389,329	**0.01**	**0.17**	9.31	4.89	0.48	0.75
	ARC	412,154	−0.72	0.15	9.69	4.53	0.51	0.77
	RFE	140,680	−0.62	0.15	9.24	4.25	**0.53**	0.77
	TMPA-3B42	198,069	0.03	0.15	11.17	4.96	0.49	0.75
	CMORPH	105,353	0.15	0.15	10.68	4.80	0.49	0.74

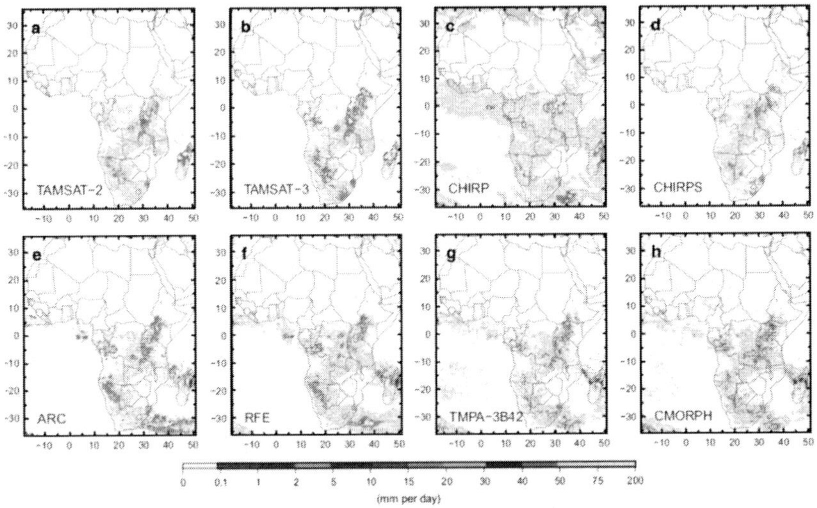

Figure 6: Africa-wide rainfall fields (at 0.25° by 0.25° resolution) for January 1st 2010 from each daily satellite-based dataset.(a) TAMSAT-2, (b) TAMSAT-3, (c) CHIRP, (d) CHIRPS, (e) ARC, (f) RFE, (g) TMPA-3B42 and (h) CMORPH.

For all of the regions other than Niger, it is likely that at least some of the validation gauge records have been ingested into the rainfall estimation process for ARC, RFE and CHIRPS. While the high gauge density may be a factor, it is notable that TAMSAT has significantly more skill than the other datasets for Niger, in particular, the relatively high R^2 values for both TAMSAT-2 and TAMSAT-3. As TAMSAT is the only dataset considered here that is locally calibrated for both rainfall occurrence and rainfall amount, the skill of the TAMSAT data is noteworthy given it does not include contemporaneous information from gauges or PMW imagery. This illustrates the importance and the utility of a local and historical calibration approach.

Figure 6 gives an example of rainfall estimates for January 1st 2010. It can be seen that while the rainfall fields have similar spatial structures, there are fewer intensely rainy pixels in TAMSAT-2 (compared to the other datasets), although this is ameliorated somewhat in TAMSAT-3. While the rainy areas are similar for all of the datasets, the intensities vary considerably. This is consistent with the quantitative analysis described in this study, which showed that for all of the datasets, occurrence is more reliably estimated than amount across the five countries considered.

3. USAGE NOTES

The TAMSAT system was originally designed for seasonal early warning of drought. Until the initial release of daily TAMSAT-2 in 2012, ARC in 2013, and CHIRPS in 2014, long-term satellite-based rainfall data for drought early warning have typically been released at the dekadal time scale. This paper has presented the daily version of the TAMSAT data (versions 2.0 and 3.0). TAMSAT has previously been demonstrated to have good skill for 10-day cumulative rainfall estimates14 and we have shown here that the daily data reliably represents the occurrence of rainfall, capturing, on average, around 70 % of observed rainy days (POD) and falsely estimating less than 40 % of rainy days (FAR) across the case study countries. Regionally, TAMSAT captured rainy and non-rainy days better across Niger and Zambia—regions whose rainfall climates are not significantly modulated by large water bodies and complex topography. Variability in rainfall amount is, however, not well captured. Whilst the ability to differentiate between low and high rainfall amounts is important, it can be argued that across Africa, long dry spells (which, to be detected, require satellite estimation algorithms to skilfully differentiate a rainy day from a non-rainy day) is more damaging to crops than extremes of rainfall1. Many aspects of the skill of the TAMSAT daily data are however similar or better (depending on the skill measure) than other, widely used African operational daily datasets. Since CHIRPS, ARC, and RFE make use of contemporaneous gauges which are likely included in the validation datasets, this complicates the interpretation of the results.

An obvious application for the daily data is the production of rainfall estimates for periods other than 5-day or 10-day accumulations starting on fixed days of the calendar month. The availability of a daily version of the TAMSAT dataset gives a choice of products based on the optimal length and starting point of cumulative rainfall estimates required. This facilitates comparison with other datasets, which are issued at weekly resolutions for example, and allows for greater flexibility for agricultural and hydrological applications.

Many crop and hydrological models require daily input[52-55]. In the case of crop modelling, yield generally depends on cumulative rainfall for key parts of the growing cycle. Daily data are therefore useful because the data can pick out key development phases of crops and is an example of the value of being able to cumulate rainfall over bespoke periods. Although TAMSAT data may be too coarse for analysis of small catchments, hydrological models for medium and large catchments may be able to utilise data at 4-km resolution[56]. The TAMSAT data have most skill when spatially aggregated[4,14], and this is especially the case for rainfall that is not aggregated in time. In this context, the suitability of TAMSAT daily rainfall estimates depends on the hydrological features of the catchment and the purpose of the monitoring or modelling. TAMSAT's poor skill for rainfall amount means that it is most suitable for monitoring large catchments where river discharge is determined by gradual accumulation of rainfall over a period of days. It can be argued that the TAMSAT data is not suitable for providing information on pluvial flood risk.

Unlike the other daily rainfall datasets considered, TAMSAT does not incorporate gauge data in real time. Recent studies have shown that inconsistencies in the gauge record can lead to spurious trends in rainfall, especially in the tropics, where the station network is patchy[21,57]. The TAMSAT cumulative rainfall datasets and the derived daily estimates can therefore be considered temporally consistent, which is important in both assessing climatic risks and for seasonal rainfall monitoring. As such, TAMSAT daily data are well suited to the study of long term changes in daily metrics, relating primarily to occurrence, such as the length of dry spells and the length of the growing season[58]. Since it cannot capture the intensity of high rainfall events well, TAMSAT daily data is less suited for studies of long term changes in rainfall amount.

In conclusion, we present the TAMSAT high-resolution daily rainfall dataset for Africa. The data are back calculated to January 1983 and updated in near-real time (v2.0 is updated every ten days and v3.0 is updated every five days). The recent development of TAMSAT version 3.0 pentadal estimates and derived daily estimates removes spatial artefacts and greatly reduces the dry bias associated with the previous version. A formal statistical assessment indicates that both TAMSAT daily datasets have comparable skill to other remotely sensed rainfall datasets, and can therefore be used for similar applications. Furthermore, TAMSAT's historical calibration suits it well for risk assessment and the investigation of long-term changes in the rainfall climate.

ACKNOWLEDGEMENTS

This paper is dedicated to our friend and mentor, George Dugdale, who sadly passed away on 1 March 2016. George was a founding member of the TAMSAT Group who pioneered the use of satellite imagery for rainfall estimation over Africa. His outstanding contribution to the field has left a lasting legacy that is strongly reflected in the research and other activities carried out by the TAMSAT Group today. He is sorely missed by all that who knew him.

This work was implemented as part of the CGIAR Research Program on Climate Change, Agriculture and Food Security (CCAFS), which is a strategic partnership of CGIAR and Future Earth. It was carried out with funding by CGIAR Fund Donors, the Danish International Development Agency (DANIDA), Australian Government (ACIAR), Irish Aid, Environment Canada, Ministry of Foreign Affairs for the Netherlands, Swiss Agency for Development and Cooperation (SDC), Instituto de Investigação Científica Tropical (IICT), UK Aid, Government of Russia, the European Union (EU), New Zealand Ministry of Foreign Affairs and Trade, with technical support from the International Fund for Agricultural Development (IFAD). The views expressed in this document cannot be taken to reflect the official opinions of CGIAR or Future Earth. The authors would like to thank Keith Shine for his support and input into the initial CCAFS study that led to this paper. The development of the TAMSAT version 3.0 dataset was supported by the European Commission MARSOP4 programme. R.I.M. is supported by the HyCRISTAL project (NE/M020371/1). E.B. is supported by the NCAS-Climate core programme, and the following NERC grants: ERADACS NE/P015352/1, BRAVE2 NE/M008983/1 and HyCRISTAL NE/M020371/1. M.Y. is supported by the NCAS-Climate core programme and the European Commission MARSOP4 programme.

COMPETING INTERESTS

The authors declare no competing financial interests

REFERENCES

1. Lesk, C., Rowhani, P. & Ramankutty, N. Influence of extreme weather disasters on global crop production. Nature 529, 84–87 (2016).
2. United Nations Office for the Coordination of Humanitarian Affairs. Horn of Africa Drought Crisis Situation Report No. 5 (21 July 2011) (2011).
3. Boyd, E., Cornforth, R. & Lamb, P. Building resilience to face recurring environmental crisis in African Sahel. Nat. Clim. Chang 3, 631–637 (2013).
4. Black, E. et al. The Use of Remotely Sensed Rainfall for Managing Drought Risk: a Case Study of Weather Index Insurance in Zambia. Remote Sens 8, 342 (2016).
5. Greatrex, H., Grimes, D. & Wheeler, T. Advances in the Stochastic Modeling of Satellite-Derived Rainfall Estimates Using a Sparse Calibration Dataset. J. Hydrometeorol. 15, 1810–1831 (2014).
6. Washington, R. et al. African Climate Change: Taking the Shorter Route. Bull. Am. Meteorol. Soc 87, 1355–1366 (2006).
7. Flitcroft, I. D., Milford, J. R. & Dugdale, G. Relating Point to Area Average Rainfall in Semiarid West Africa and the Implications for Rainfall Estimates Derived from Satellite Data. J. Appl. Meteorol. 28, 252–266 (1989).
8. Tarnavsky, E. et al. Extension of the TAMSAT Satellite-Based Rainfall Monitoring over Africa and from 1983 to Present. J. Appl. Meteorol. Climatol. 53, 2805–2822 (2014).

9. Laurent, H., Jobard, I. & Toma, A. Validation of satellite and ground-based estimates of precipitation over the Sahel. Atmos. Res. 47, 651–670 (1998).
10. Grimes, D. I. F., Pardo-Igúzquiza, E. & Bonifacio, R. Optimal areal rainfall estimation using raingauges and satellite data. J. Hydrol. 222, 93–108 (1999).
11. Dinku, T., Chidzambwa, S., Ceccato, P., Connor, S. J. & Ropelewski, C. F. Validation of high-resolution satellite rainfall products over complex terrain. Int. J. Remote Sens. 29, 4097–4110 (2008).
12. Jobard, I., Chopin, F., Berges, J. C. & Roca, R. An intercomparison of 10-day satellite precipitation products during West African monsoon. Int. J. Remote Sens. 32, 2353–2376 (2011).
13. Thiemig, V., Rojas, R., Zambrano-Bigiarini, M., Levizzani, V. & De Roo, A. Validation of Satellite-Based Precipitation Products over Sparsely Gauged African River Basins. J. Hydrometeorol. 13, 1760–1783 (2012).
14. Maidment, R. I. et al. Evaluation of satellite-based and model re-analysis rainfall estimates for Uganda. Meteorol. Appl. 20, 308–317 (2013).
15. Roca, R. et al. Comparing Satellite and Surface Rainfall Products over West Africa at Meteorologically Relevant Scales during the AMMA Campaign Using Error Estimates. J. Appl. Meteorol. Climatol. 49, 715–731 (2010).
16. Pfeifroth, U., Trentmann, J., Fink, A. H. & Ahrens, B. Evaluating satellite-based diurnal cycles of precipitation in the African tropics. J. Appl. Meteorol. Climatol. 55, 23–39 (2015).
17. Toté, C. et al. Evaluation of Satellite Rainfall Estimates for Drought and Flood Monitoring in Mozambique. Remote Sens 7, 1758–1776 (2015).
18. Maidment, R. I. et al. The 30 year TAMSAT African Rainfall Climatology And Time series (TARCAT) data set. J. Geophys. Res. Atmos. 119, 2014JD021927 (2014).
19. Novella, N. S. & Thiaw, W. M. African Rainfall Climatology Version 2 for Famine Early Warning Systems. J. Appl. Meteorol. Climatol. 52, 588–606 (2013).
20. Funk, C. et al. The climate hazards infrared precipitation with stations—a new environmental record for monitoring extremes. Sci. Data 2, 150066 (2015).
21. Maidment, R. I., Allan, R. P. & Black, E. Recent observed and simulated changes in precipitation over Africa. Geophys. Res. Lett. 42, 2015GL065765 (2015).
22. Awange, J. L. et al. Uncertainties in remotely sensed precipitation data over Africa. Int. J. Climatol. 36, 303–323 (2015).
23. Sylla, M. B., Giorgi, F., Coppola, E. & Mariotti, L. Uncertainties in daily rainfall over Africa: assessment of gridded observation products and evaluation of a regional climate model simulation. Int. J. Climatol. 33, 1805–1817 (2013).
24. Gerland, P. et al. World population stabilization unlikely this century. Science 346, 234–237 (2014).
25. Dinku, T., Cousin, R., Corral, J. & Vadillo, A. The ENACTS Approach: Transforming climate services in Africa one country at a time. J. World's Policy 1–24 (2016).
26. Tarhule, A., Saley-Bana, Z. & Lamb, P. J. Rainwatch. Bull. Am. Meteorol. Soc 90, 1607–1614 (2009).
27. Milford, J. R., McDougall, V. D. & Dugdale, G. in Colloques et Seminaires. Validation problems of rainfall estimation methods by satellite in intertropical Africa. Proceedings of the Niamey Workshop (1994) (ORSTROM, 1996).
28. Dugdale, G., McDougall, V. D. & Milford, J. R. Rainfall estimates in the Sahel from cold cloud statistics: accuracy and limitations of operational systems (Proceedings of the Niamey Workshop, February 1991). Soil Water Balanc. Sudano-Sahelian Zo. IAHS 199, 65–74 (1991).
29. Thorne, V., Coakeley, P., Grimes, D. & Dugdale, G. Comparison of TAMSAT and CPC rainfall estimates with raingauges, for southern Africa. Int. J. Remote Sens. 22, 1951–1974 (2001).
30. Tucker, M. R. & Sear, C. B. A comparison of Meteosat rainfall estimation techniques in Kenya. Meteorol. Appl. 8, 107–117 (2001).
31. Dinku, T. et al. Validation of satellite rainfall products over East Africa's complex topography. Int. J. Remote Sens. 28, 1503–1526 (2007).

32. Chadwick, R. S., Grimes, D. I. F., Saunders, R. W., Francis, P. N. & Blackmore, T. A. The TAMORA algorithm: satellite rainfall estimates over West Africa using multi-spectral SEVIRI data. Adv. Geosci 25, 3–9 (2010).
33. Young, M. P., Williams, C. J. R., Chiu, J. C., Maidment, R. I. & Chen, S.-H. Investigation of discrepancies in satellite rainfall estimates over Ethiopia. J. Hydrometeorol. 15, 2347–2369 (2014).
34. Schumacher, C. & Houze, R. A. Stratiform precipitation production over sub-Saharan Africa and the tropical East Atlantic as observed by TRMM. Q. J. R. Meteorol. Soc 132, 2235–2255 (2006).
35. Laing, A. G., Fritsch, J. M. & Negri, A. J. Contribution of Mesoscale Convective Complexes to Rainfall in Sahelian Africa: estimates from Geostationary Infrared and Passive Microwave Data. J. Appl. Meteorol. 38, 957–964 (1999).
36. Lebel, T., Diedhiou, A. & Laurent, H. Seasonal cycle and interannual variability of the Sahelian rainfall at hydrological scales. J. Geophys. Res. 108, 8389 (2003).
37. Le Barbé, L. & Lebel, T. Rainfall climatology of the HAPEX-Sahel region during the years 1950–1990. J. Hydrol. 189, 43–73 (1997).
38. Lebel, T. & Le Barbé, L. Rainfall monitoring during HAPEX-Sahel. 2. Point and areal estimation at the event and seasonal scales. J. Hydrol. 189, 97–122 (1997).
39. Lebel, T., Taupin, J. & D'Amato, N. Rainfall monitoring during HAPEX-Sahel. 1. General rainfall conditions and climatology. J. Hydrol. 189, 74–96 (1997).
40. Lebel, T. et al. Rainfall estimation in the Sahel: the EPSAT-NIGER experiment. Hydrol. Sci. J 37, 201–215 (1992).
41. Uribe Alcántara, E. M. Gridded Analysis of Meteorological Variables in Mozambique (The World Bank, Washington, D.C., USA: 2011).
42. Creutin, J. D. & Obled, C. Objective analyses and mapping techniques for rainfall fields: An objective comparison. Water Resour. Res. 18, 413–431 (1982).
43. Tabios, G. & Salas, J. A comparative analysis of techniques for spatial interpolation of precipitation. JAWRA J. Am. Water Resour. Assoc 21, 365–380 (1985).
44. Goovaerts, P. Geostatistical approaches for incorporating elevation into the spatial interpolation of rainfall. J. Hydrol. 228, 113–129 (2000).
45. Funk, C. et al. A global satellite-assisted precipitation climatology. Earth Syst. Sci. Data 7, 275–287 (2015).
46. Arkin, P. A. & Meisner, B. N. The Relationship between Large-Scale Convective Rainfall and Cold Cloud over the Western Hemisphere during 1982–84. Mon. Weather Rev. 115, 51–74 (1987).
47. Ferraro, R. R. & Marks, G. F. The Development of SSM/I Rain-Rate Retrieval Algorithms Using Ground-Based Radar Measurements. J. Atmos. Ocean. Technol. 12, 755–770 (1995).
48. Kummerow, C. et al. The Status of the Tropical Rainfall Measuring Mission (TRMM) after Two Years in Orbit. J. Appl. Meteorol. 39, 1965–1982 (2000).
49. Huffman, G. J. et al. The TRMM Multisatellite Precipitation Analysis (TMPA): Quasi-Global, Multiyear, Combined-Sensor Precipitation Estimates at Fine Scales. J. Hydrometeorol. 8, 38–55 (2007).
50. Joyce, R. J., Janowiak, J. E., Arkin, P. A. & Xie, P. CMORPH: a method that produces global precipitation estimates from passive microwave and infrared data at high spatial and temporal resolution. J. Hydrometeorol. 5, 487–503 (2004).
51. Diem, J. E., Hartter, J., Ryan, S. J. & Palace, M. W. Validation of Satellite Rainfall Products for Western Uganda. J. Hydrometeorol. 15, 2030–2038 (2014).
52. Best, M. J., Pryor, M., Clark, D. B., Rooney, G. G. & Essery, R. L. H. Model Development The Joint UK Land Environment Simulator (JULES), model description—Part 1 : energy and water fluxes. Geosci. Model Dev 4, 677–699 (2011).
53. Clark, D. B. et al. Model Development The Joint UK Land Environment Simulator (JULES), model description—Part 2 : carbon fluxes and vegetation dynamics. Geosci. Model Dev 4, 701–722 (2011).
54. Osborne, T. et al. JULES-crop : a parametrisation of crops in the Joint UK Land Environment Simulator. Geosci. Model Dev 8, 1139–1155 (2015).

55. Challinor, A. J., Wheeler, T. R., Craufurd, P. Q., Slingo, J. M. & Grimes, D. I. F. Design and optimisation of a large-area processbased model for annual crops. Agric. For. Meteorol 124, 99–120 (2004).
56. Seyyedi, H., Anagnostou, E. N., Beighley, E. & McCollum, J. Satellite-driven downscaling of global reanalysis precipitation products for hydrological applications. Hydrol. Earth Syst. Sci. Discuss 11, 9067–9112 (2014).
57. Balan Sarojini, B., Stott, P. A., Black, E. & Polson, D. Fingerprints of changes in annual and seasonal precipitation from CMIP5 models over land and ocean. Geophys. Res. Lett. 39, L21706 (2012).
58. Kniveton, D., Layberry, R., Williams, C. J. R. & Peck, M. Trends in the start of the wet season over Africa. Int. J. Climatol. 1225, 1216–1225 (2009).

CHAPTER 14

The Guanacaste Volcanic Arc Sliver of Northwestern Costa Rica

Walter Montero, Jonathan C. Lewis & Maria Cristina Araya

CONTRIBUTIONS

W.M. completed all of the aerial photo and DEM analyses of the fault traces. W.M. and J.C.L. completed the field investigations of active faulting. M.C.A. revised the seismic velocity model, relocated crustal earthquakes and generated new focal mechanism solutions. J.C.L. and W.M. wrote the manuscript. All authors contributed to the final text and figures.

ABSTRACT

Recent studies have shown that the Nicoya Peninsula of northwestern Costa Rica is moving northwestward ~11 mm a^{-1} as part of a tectonic sliver. Toward the northwest in El Salvador the northern sliver boundary is marked by a dextral strike-slip fault system active since Late Pleistocene time. To the southeast there is no consensus on what constitutes the northern boundary of the sliver, although a system of active crustal faults has been described in central Costa Rica. Here we propose that the Haciendas-Chiripa fault system serves as the northeastern boundary for the sliver and that the sliver includes most of the Guanacaste volcanic arc, herein the Guanacaste Volcanic Arc Sliver. In this paper we provide constraints on the geometry and kinematics of the boundary of the Guanacaste Volcanic Arc Sliver that are timely and essential to any models aimed at resolving the driving mechanism for sliver motion. Our results are also critical for assessing geological hazards in northwestern Costa Rica.

1. INTRODUCTION

Ever since forearc slivers were first identified[1] considerable attention has been paid to the impact that sliver motion has on plate motion budgets[2] and forearc deformation[3, 4], but less attention has focused on the geology of their trailing edges[5,6,7]. Geodetic work has demonstrated that in northwestern Costa Rica, the Nicoya Peninsula is moving northwestward ~11 mm a^{-1} relative to eastern Costa Rica[8,9,10,11]. Forearc motion in El Salvador, to the northwest, is accommodated by a well characterized dextral strike-slip fault system[9, 12, 13]. The configuration of the sliver to the southeast in the more populous parts of Costa Rica has to date remained unresolved. As a consequence, the scale of the tectonic sliver and the potential roles that it might play in the tectonic, seismologic and volcanic history of Costa Rica are poorly understood. Here we document a fault system that we suggest serves as the northeastern margin of the sliver, and that traverses the Guanacaste volcanic arc. The configuration of this tectonic boundary, herein referred to as the Haciendas-Chiripa fault system, requires that what has previously been described as a forearc sliver[10] includes most of the Guanacaste volcanic arc. By documenting the geometry and kinematics of this fault system on the margin of the Guanacaste Volcanic Arc Sliver (GVAS) we provide constraints that are essential for models of sliver motion that aim to disentangle the relative contributions of ridge collision[5, 10], oblique subduction2 or trailing edge plate forces[14].

2. Tectonic Setting

Central Costa Rica occupies the southern boundary of the Caribbean plate (CA) where it is in contact with the Panama microplate, and is bounded to the southwest by the Middle America Trench (MAT, Fig. 1). The broad, concave-to-the-SE zone of diffuse deformation that separates CA and Panama microplate is the Central Costa Rica Deformed Belt (CCRDB)[15, 16]. Widespread mesoscale faulting in the CCRDB has been attributed to the subduction of rough seafloor at the MAT[16]. This interpretation is in agreement with the observation that the western edge of the CCRDB meets the Pacific coast at the northeast-trending boundary on the downgoing Cocos plate that separates smooth seafloor to the northwest from rough seafloor to the southeast[17, 18]. In western Costa Rica the Cocos plate is subducting toward the northeast at ~81 mm a^{-1} relative to CA, and at a high angle to the MAT[2]. Recent mapping suggests that locally the CCRDB is being crosscut by the dextral Atirro-Río Sucio fault system that accommodates lateral extrusion of the upper plate in response to Cocos Ridge collision[6].

Northwestern Costa Rica is dominated by the northwest-trending, extinct (~1.0–2.2 Ma) Monteverde volcanic arc, and the active Guanacaste volcanic arc (Fig. 1a) reflecting subduction of the Cocos plate[19]. The volcanic-arc rocks are locally covered by epiclastic and alluvial deposits on both the southern and northern flanks of the arc[19]. At the Pacific coast, the Nicoya Peninsula and coastal areas to the east (across the Gulf of Nicoya) expose oceanic basement and cover sediments of fluvial and marine origin[15, 20]. The Costa Rican forearc appears to reflect the influence of lower plate architecture. For example, immediately inboard of the Quepos Plateau (Fig. 1) the rocks record differential tilting, associated mostly with NE-striking faults, and this has produced asymmetric watersheds[21]. This influence ceases toward the northwest at approximately the Gulf of Nicoya[21], with the exception of the southernmost tip of the Nicoya Peninsula where the incoming Fisher Seamount Chain drives high Holocene uplift rates (~6 m ka^{-1}) and northward tilting[22].

(a)

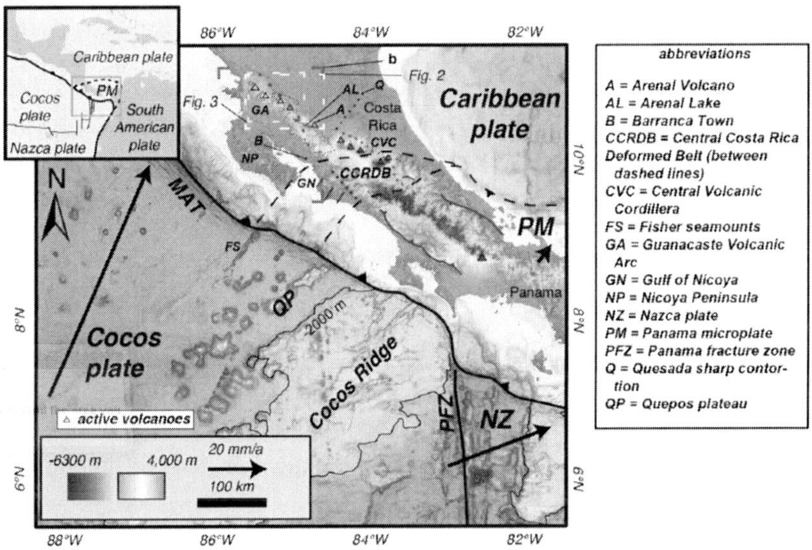

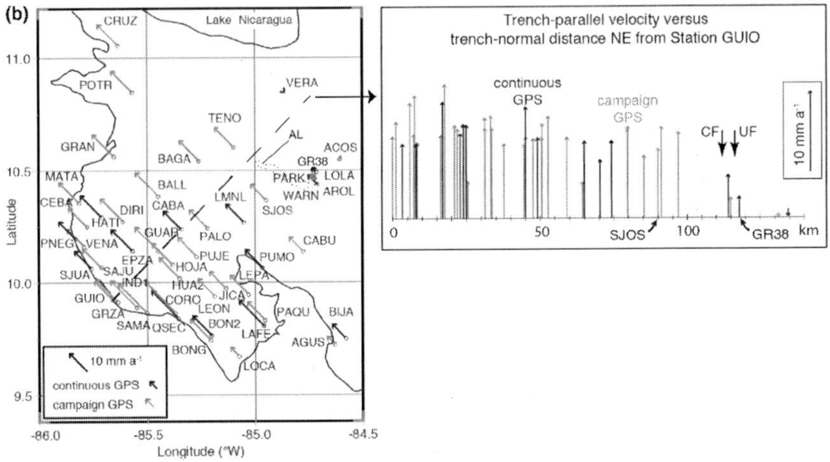

Figure 1 (**a**) Tectonic setting of Costa Rica. Topographic and bathymetric base map simplified from Morell[31]. Dotted outlines are Monteverde arc of Carr *et al.*[18]. Location of the Quesada Sharp Contortion from Protti *et al.*[29]. Red dashed line outlines the Haciendas-Chiripa fault system. The dashed magenta and blue lines show the positions of the San Miguel-Atirro-Rio Sucio dextral fault system[6] and the Candelaria fault[6], respectively. (**b**) Trench-parallel component of GPS velocities from Feng *et al.*[8]. Magenta vectors are from campaign GPS and black vectors are for permanent GPS stations. The profile at the upper right of (**b**) shows the same velocity plotted versus distance perpendicular to the trench. UF = Upala Fault and CF = Chiquero Fault. The location of Fig. 1b is indicated by red corners. The location of Figs 2 and 3 are shown by the dashed white and yellow boxes, respectively. Map at (**a**) and inset were provided with permission to modify by K. Morell and modified using Adobe Illustrator CC 2015 release. Map and plot at (**b**) were generated using Adobe Illustrator CC 2015 release.

3. RESULTS

On the basis of existing GPS velocities, well-located earthquake hypocenters, new focal mechanism solutions, and field observations we document the northeastern boundary of the GVAS from the area south of Lake Arenal Well-located crustal earthquakes and focal mechanism solutions provide robust constraints on the northeastern GVAS boundary (Figs 1a and 2).

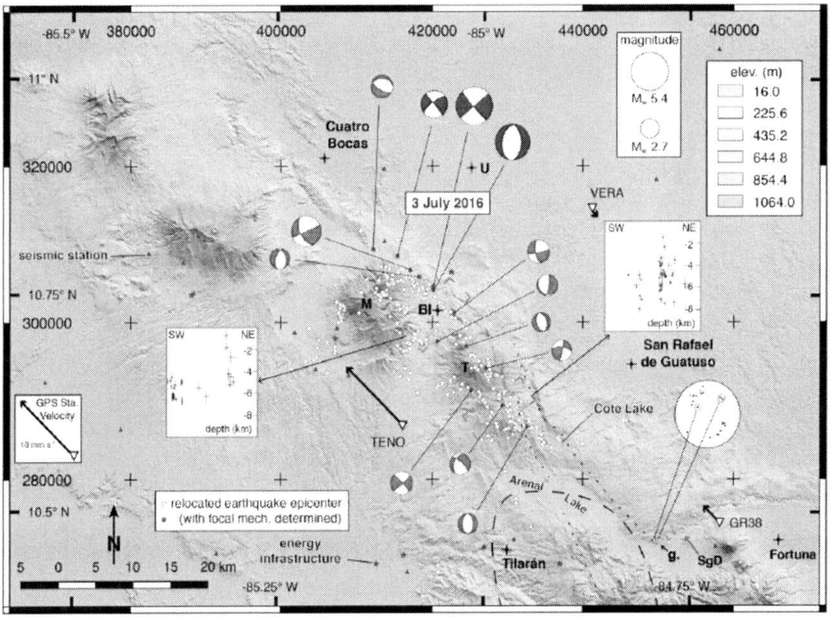

Figure 2 Seismotectonic setting. Geographic reference system here and in Fig. 3 is that found on most topographic maps of Costa Rica, the Ocotepeque 1935/Costa Rica Sur. Latitude and longitude graticules using the WGS 84 reference frame also shown here and in Fig. 3. Digital elevation model and shaded relief map derived from enhanced Shuttle Radar Topography Mission (SRTM) data with an effective resolution of 30 m. Lighting from 020°. Lower-case bold letter g. with view direction for photograph at Fig. 4g. The stereonet shows poles to faults at the outcrop at location g. The red poles show the geometry of the bounding strike-slip fault, the principal displacement zone[27] at the eastern end of the exposure (right end of white symbol), whereas the black poles show normal faults that record dip separations that average ~0.6 m. The circled black pole is the westernmost fault. Portions of the Cote-Arenal fault and Chiripa faults from Fig. 3 are included as a dotted line. The dashed line outlines the aftershock area for the 13 April 1973 M_S6.5 Tilarán earthquake[29]. The dashed magenta boxes show the locations of the two inset seismicity profiles in which the uncertainty in the earthquake locations is shown by vertical and horizontal red lines. Towns are shown with a dot-plus symbol. Town abbreviations: BI = Bijagua and U = Upala. Volcanoes: M = Miravalles and T = Tenorio. SgD with red arrow = Sangregado Dam. Map generated using QGIS version 2.14 with SRTM 1 Arc-Second Void Filled digital elevation data courtesy of the U.S. Geological Survey from https://earthexplorer.usgs.gov/. Keys, labels and symbols were added using Adobe Illustrator CC 2015 release.

We have taken advantage of an improved crustal 1D velocity model for western Costa Rica[23] (±0.147 km in latitude, ±0.072 km in longitude and ±0.074 km in depth) and used HypoDD[24] to locate earthquakes precisely and these define a seismic source zone where field observations confirm evidence of active or recent faulting. In Fig. 2 we show locations for earthquakes recorded by a local network from December 14, 2006 until July 29, 2016. These include three recent moderate magnitude events on July 3, 2016 that reveal strike slip motion and normal motion (Fig. 2). Our new focal mechanisms are consistent with historical events included in the Centroid Moment Tensor (CMT) catalog.

Field observations coupled with analyses of aerial photos and Digital Elevation Models (DEMs) reveal evidence of recent or active faulting along the Haciendas-Chiripa fault system. The northern flank of the Guanacaste volcanic arc in northwestern Costa Rica is marked by a clear topographic lineament (Fig. 3) that includes fault exposures and well-preserved fault scarps (Fig. 4). Fresh outcrops can be found at quarries and roadcuts, but owing to the humid tropical climate they tend to be ephemeral. This is in fact one of the major challenges to mapping active or recently active faults in Central America. Along streams most bedrock exposures are deeply weathered, however, lithologies and primary structures are sometimes preserved. At location **a** (Fig. 3) there is a recent, irregular roadcut that exposes the fault trace at two locations separated by several meters (Fig. 4a). The steeply dipping fault juxtaposes relatively fresh ignimbrite to the northeast with a clayey unit to the southwest that we interpret as fault gouge. The trend of the fault trace where it is cut by the excavation is ~295° and based on the topographic lineament we infer a fault strike of ~300° in this area. We refer to this as the Haciendas fault segment. The presence of the fault is highlighted here by the lateral change in soil character. Above the ignimbrite is a dm-scale reddish brown soil horizon whereas above the gouge is a 0.5 m thick, bright red lateritic soil. The ignimbrite includes recognizable, fresh volcanic lithic fragments of varying size whereas the clay gouge unit consists of a deeply weathered matrix of highly plastic clay containing small relict volcanic clasts.

Along strike to the southeast at location **b** (Fig. 3) there is a very similar roadcut through an elongate E-W ignimbrite unit that is truncated on its eastern end by the topographic lineament that we map as the Haciendas fault. The main body of the ignimbrite is cut by numerous conjugate, m-scale normal faults that record minor offsets (dm-scale). Many of these faults cut the entire thickness of the unit, and although kinematic indicators are rarely preserved, overall they suggest ~E-W stretching (see station b stereonet at Fig. 3).

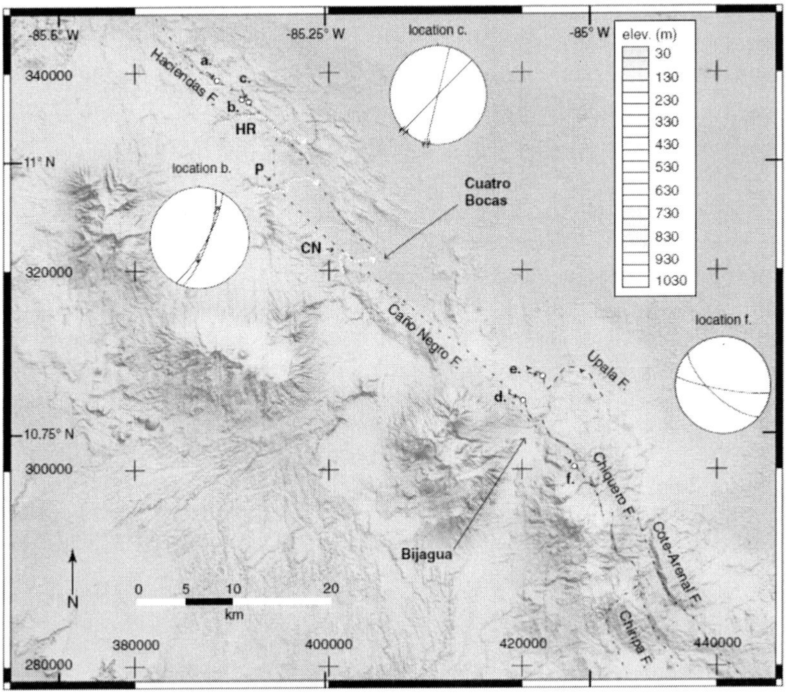

Figure 3 Haciendas-Chiripa fault system. Digital elevation model and shaded relief map derived from enhanced SRTM data with an effective resolution of 30 m. Lighting is directed from an azimuth of 020°. Mapped fault segments are named and the locations of the major fault traces are shown with dashed lines. Lower-case bold letters a.–f. with short arrows showing view direction of lettered photos at Fig. 4. Lower-hemisphere equal-area projections show structural data described in text. CN = Caño Negro River, HR = Haciendas River and P = Pizote River. Map generated using QGIS version 2.14 with SRTM 1 Arc-Second Void Filled digital elevation data courtesy of the U.S. Geological Survey from https://earthexplorer.usgs.gov/. Keys, labels and symbols were added using Adobe Illustrator CC 2015 release.

We interpret these to be minor shear zones that reflect damage-zone processes adjacent to the mature Haciendas fault. Downslope and to the east at location **c** is a stream cutbank exposure of deeply weathered clayey ignimbrite that is cut by anastomosing sub-vertical, lineated antithetic subsidiary (Riedel) shear planes (Fig. 4c). This exposure is along strike from location **a** described above and we interpret it to lie within the fault zone. Immediately to south is the Haciendas River which shows a strong right deflection (see HR in Fig. 3).

The eastern end of the ignimbrite, within ~2 meters of where it is truncated, is cut by numerous, cm-scale, sub-vertical anastomosing clay seams that trend west northwest (Fig. 4b).

Southeast from the Haciendas fault we trace additional fault segments based on topographic expression, offset drainages, exposures of highly sheared rock, and seismicity. These fault segments are from west to east the Caño Negro, Upala, Chiquero, Cote-Arenal and Chiripa. In the Caño Negro segment the topographic lineament bifurcates where the fault displays a right (releasing) step (Fig. 3). Near the town of Cuatro Bocas, there are additional abundant deflected streams and local exposures of highly sheared Quaternary volcanic strata that mark the fault trace. Terrace risers where the Pizote and Caño Negro Rivers cross the fault trace display right-deflections on the scale of hundreds of meters (see P and N in Fig. 3).

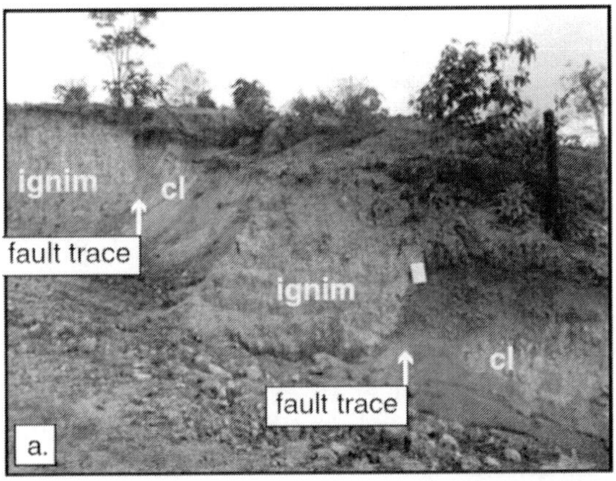

c.
↑ relict volcanic clast
person
d.
↘ break in slope
person from
knees up
e.

Figure 4 Field photographs. Locations as lower-case bold letters in Figs 2 and 3. (**a**) Fault trace where it juxtaposes fresh ignimbrite (ignim) against clay gouge (cl). White fieldbook is 14 cm in width. (**b**) Looking down on ignimbrite with thin, ~vertical clay seams that strike ~305°. White arrows show three seams. Pocket transit 7 cm wide. (**c**) Cutbank exposure of fault gouge containing anastomosing vertical subsidiary shear planes, slip shown by magenta half arrow. (**d**) Drainage deflected ~5.5 m dextrally within a small pull-apart basin. Dashed white lines

highlight the curvature of the margins of the stream. (**e**) Thrust scarp in the Upala fault segment. (**f**) Planar, steeply dipping fault trace in deeply weathered exposure on the road to Tenorio Volcano. Detail at bottom shows steeply raking, subtle lineations (~parallel to pencil) on a polished fault plane preserved in deeply weathered, clay rich fault rock. (**g**) Fault exposed in an ephemeral outcrop that extended >120 m west from this location along the shores of Lake Arenal in 2008. The poles to fault planes observed in this outcrop are shown in the stereonet in Fig. 2. Photographs annotated using Adobe Illustrator CC 2015 release.

In the eastern sector, the Caño Negro fault is marked by geomorphic expressions of active or recent slip. North of the town of Bijagua (location **d** in Fig. 3) two fault strands define a narrow, km-scale pull-apart basin that hosts parallel, right-deflected headwater drainages (Fig. 4d). The deflections are sharp and aligned, forming a linear array on the north margin of the pull-apart basin that can be traced for more than 440 meters. We measured a dextral offset of ~5.5 m at location **d** (Fig. 4d). To the east southeast the fault is apparent as a strong topographic lineament on the north flank of Tenorio Volcano (T in Fig. 2). We have generated new focal mechanisms for three moderate magnitude earthquakes that occurred near Bijagua on 3 July 2016 (Fig. 2), two revealing strike-slip faulting and one showing normal faulting. All three events indicate approximately E-W extension. Importantly, <10 km to the northeast a M_w 5.4 thrust earthquake occurred on 27 January 2002[25] and we have identified the probable scarp at location **e** (Fig. 4e). We map this structure as the Upala fault. The scarp occurs on the northern edge of a landscape that: (1) includes surface slopes that are tilted southward, toward the edifices of Tenorio and Miravalles volcanoes; and (2) hosts alluvial sediments and small meandering streams in an area otherwise dominated by higher stream gradients. Exposures of bedrock in a stream that has incised the hangingwall of the Upala fault includes an ignimbrite unit with a dip angle >70°, attesting to significant tilting (for example near location **e**). A local resident reported to us in June of 2016 that he noted changes to the landscape in the immediate aftermath of the 2002 earthquake that are consistent with overall northward motion of the hangingwall – specifically open cracks parallel to the fault trace. The occurrence of thrusting north of the main trace of the Caño Negro segment suggests local partitioning of slip within the fault system.

Southeast from Bijagua town the fault system curves slightly taking on a more south southeast trend in the subparallel Chiquero, Cote-Arenal and Chiripa segments. Clear fault scarps mark the fault traces and a recent roadcut on the road from Bijagua to the Tenorio Volcano National Park (location **f** in Fig. 3) exposes a zone of localized shearing (Fig. 4f). This outcrop also showcases the deeply weathered nature of the rocks that we suggest mark the locus of deformation in the fault system. The rock here is highly altered to clay minerals and the preservation of fault-kinematic indicators is poor (Fig. 4f bottom image).

Well-located earthquakes (Fig. 2) also provide constraints on the fault system in this area, in particular illuminating the eastern end of the Caño Negro fault from northwest of Bijagua town to the east side of Tenorio Volcano (Figs 2 and 3). Our earthquake catalog covers a relatively short period of time so it is not surprising that some fault scarps do not have co-located epicenters. Northeast-trending seismicity profiles (Fig. 2) show that the earthquakes define steep planes, consistent with our new focal mechanisms, the tectonic setting and the linear surface expression of the fault traces. Most of the hypocenters are deeper than ~2 km, and in spite of small errors in locations, it is not possible to discern details of the fault system architecture at this point. Locally we note offsets between epicenters and fault scarps that highlight the need for more work. Such offsets might reflect the weak mechanical link between fault ruptures at depth and the deeply weathered cover strata, thick soils and relatively steep slopes at the surface. In such a case the overburden may serve to mask or mute the relation between faults and their surface expression. Alternatively, the mismatch is a reflection of an evolving zone of deformation in which slip is not strongly localized and for which the surface traces of additional faults have yet to be identified.

To the southeast, ephemeral outcrops of high strain zones attest to the trace of the Chiquero, Chiripa, and Cote-Arenal segments as they cross Lake Arenal near the Sangregado dam site (location **g** in Fig. 2). In this region the earthquake epicenters tend to be distributed along the mapped fault traces, including a small cluster of events at Cote Lake along strike from where the Cote-Arenal fault traverses the north shore of Lake Arenal. A now-covered exposure west of the Sangregado dam was documented previously[26], and in 2008 fault data were collected along this >120 m long outcrop. The eastern end of the outcrop is interpreted to be the principal displacement zone (PDZ)[27] and is marked by the west-facing scarp seen at Fig. 4g and a similarly oriented fault just to the west. These two faults are shown as red poles in the stereonet at Fig. 2. The easternmost of these two faults displaces a paleosol dated at 1370 ± 215 B.C. but with a minimum age estimated at 900 BC[26]. To the west, the outcrop is cut mostly by normal faults oriented at high angles to the PDZ and that record an average of ~0.6 m of dip separation (black poles in stereonet at Fig. 2). The westernmost of these shows 6.35 m of vertical separation and is marked by an east-facing scarp (circled black pole in stereonet in Fig. 2). The geometry and location of the faults are consistent with a local releasing step in the fault system. Importantly, however, ~30 m west of the eastern PDZ a broad antiform occurs in the lower part of the stratigraphic section that is overlain by tephra that is not folded. These observations suggest transient horizontal shortening akin to that observed in other transtensional systems[28].

South of Lake Arenal the Chiripa segment is marked by linear ridges we interpret to reflect the fault trace (see dotted line south of Lake Arenal in Fig. 2). This region hosted the 13 April 1973 M_s 6.5 Tilarán strike slip earthquake[29] and the aftershocks clustered near the Chiripa fault (Figs 2 and 3). This area also marks a substantial break in GPS velocities (Fig. 1b, see also GPS station trench-parallel velocities for GR38, TENO, and VERA in Fig. 2). The nature of the linkage between these faults and other active structures in eastern Costa Rica remains an important open question. If the faults connect to the San Miguel-Rio Sucio-Atirro fault system north of the Central Volcanic Cordillera (magenta dashed line in Fig. 1a), the geometry implies a regional-scale transpressional (restraining) step. Alternatively, if the faults continue toward the southeast they may link to the Candelaria fault (dark blue dashed line in Fig. 1a) along the Pacific coast.

4. DISCUSSION

Our results have immediate implications for hazards mitigation, specifically in the wake of three damaging crustal earthquakes on 3 July 2016 that occurred near Bijagua (Fig. 2). The area that is transected by the Haciendas-Chiripa fault system hosts several important population centers, including Fortuna, Bijagua, Tilarán, San Rafael de Guatuso and Upala (see locations in Figs 2 and 3), as well as numerous hydroelectric projects (see red stars in Fig. 2). Moreover, the volcanic rock that dominates the steep slopes is subject to deep tropical weathering that promotes formation of clay minerals30. Many of the ephemeral outcrops in the region expose dm-scale clay horizons that reflect the alteration of volcanic ash to highly plastic bentonite. Mixed with the ignimbrites, tuffs and lava flows that are abundant here are occasional chaotic, block-in-matrix lahar units. All of these lithologies are susceptible to clay alteration rendering them inherently unstable.

Our findings also highlight important targets for future geodetic and geologic work essential to solving fundamental geodynamics questions. Existing GPS data8 are inadequate for resolving whether sliver motion is accommodated entirely on the Haciendas-Chiripa fault system. Answering this will require dense GPS transects across the fault, and these would provide constraints on fault locking depths that are essential for evaluating seismic hazards (e.g., estimating maximum moment release). The geodetic results would also highlight targets for complementary geomorphology and paleoseismology investigations. More fundamentally, expanded geodetic work spanning northwestern Costa Rica is essential for addressing the mechanisms that drive sliver motion. Outstanding questions include: (1) When did the sliver form? (2) What is the distribution of slip partitioning on the margins of the sliver? and (3) What is the relationship between the sliver and the segmentation of the volcanic arc? If the sliver reflects only Cocos-ridge collision then it should be no older than the first arrival of the ridge at the MAT ~2–3 Ma31. Moreover, the faults accommodating lateral escape might be expected to show an age-progression that reflects northward and radial expansion of the collision zone. This might be reflected by older slip on the Candelaria fault and more recent slip on the San Miguel-Rio Sucio-Atirro fault system (Fig. 1). In contrast, if the sliver is driven by oblique subduction and/or trailing edge plate forces, then the sliver may pre-date Cocos ridge collision. It has been shown through numerical modeling that MAT megathrust coupling is weak32, and forearc sliver motion is likely driven by the eastward motion of the Caribbean plate relative to the North American Plate14. Such an interpretation is in accord with geodetic modeling results that suggest that oblique convergence is inadequate to explain trench-parallel motion of the Nicaraguan forearc33. Understanding the relative contributions of strike slip motion and interplate thrusting will be critical to unraveling the slip budget, and thus important regardless of what drives sliver motion.

With regard to question 3 above, establishing the history of slip on the sliver boundary will also help resolve questions regarding the evolution of volcanism in Central America, including the contributions of upper-plate and lower-plate architecture. Geochronology data (Carr *et al.*19) suggest that the current volcanic front of Costa Rica initiated at ~600 Ka, a time of near complete cessation of volcanism in the gap between the Guanacaste Volcanic arc and the Central Volcanic Cordillera (Fig. 1). This gap also marks a poorly understood change from a 100% Ocean Island Basalt (OIB) derived mantle source in the Central Volcanic Cordillera to 20% OIB derived mantle source in Guanacaste arc volcanoes other than Arenal19. It is also an area known for the northeast striking Quesada Sharp Contortion (Fig. 1a), which marks a steepening of the Wadati-Benioff zone to the northwest34. Recognition that the through-going Haciendas-Chiripa fault system serves as the GVAS boundary in this region provides new constraints with which to explore arc evolution mechanistically.

Lastly, our findings suggest a mechanism for the genesis of allochthonous volcanic terranes. The western cordillera of North America hosts many such terranes that have been attributed to both collisional35, 36 and strike-slip tectonics36,37,38,39. The GVAS therefore provides opportunity to document the lithofacies and structure of a nascent strike-slip allochthonous volcanic arc.

5. METHODS

All the crustal earthquakes were registered by the Observatorio Sismológico y Vulcanológico Arenal-Miravalles (OSIVAM) from March 2006 through August 2016 (Supplementary Table S1). The earthquakes occurred near or at the volcanic arc and were selected for manual relocation from the seismic network database depending of the following characteristics: a minimum of 8 P wave arrivals at different stations, and with all the evident S wave arrivals at the closest stations from the epicenter, a maximum azimuthal gap of 180° and an initial rms of 1.0 second. The final rms values were consistently much lower, typically ~0.2 or less for the hand-picked events (Supplementary Table S1) and <0.1 for the relocated events (Supplementary Table S2). All the resulting earthquakes are between 1 and 30 km of depth below sea level. The OSIVAM network is an arrangement of stations with Lenartz 3D Lite sensors and Reftek's 130 s digitizers. The digitizers use a unity gain, and a sample rate of 100 samples per second.

The velocity model used for the manual locations is one-dimensional and consists of 6 layers made for the area and with the station corrections obtained with Velest[23]. After finding the manual locations the events were relocated with a double difference algorithm using HypoDD[24] (Supplementary Table S2).

The configuration for the double difference relocation was for a medium size dataset, with high certainty of the manual picks. The data used consisted of a catalog of P and S arrival manual picks with a maximum distance of 200 km, with a maximum hypocentral separation of 10 km between events, 10 as maximum number of neighbors per event and a minimum of 5 links to define a neighbor. The initial location is from the catalog and the relocation was resolved in two sets of 5 iterations. The weight for P arrivals was set at one for both sets and a weight of −9 and 0.75 for the S arrivals. The residual threshold in seconds for the first set of iterations was not taken into account and a 5 second threshold for the second set of iterations, and a maximum distance between linked pairs was not restricted for the first iteration and a maximum of 3 km for the second set of iterations. The damping factor used for the first and second iterations was set at 50 and 60, respectively.

Lastly, the new earthquake locations were used to generate focal mechanism solutions. For this we used events with magnitudes greater than M_w 3.5 using the program Focmec[40] and searching every 5 degrees for best solutions (Supplementary Table S2).

ACKNOWLEDGEMENTS

We are grateful to W. Taylor, G. Alvarado and the Instituto Costarricense de Electricidad (ICE) for their willingness to share ideas and data from the ICE earthquake catalog. We also thank L. Linkimer and the Red Sismológica Nacional from the University of Costa Rica for their earthquake catalog. This work is a product from the research projects 830-A8-133 "Investigación de fallas activas en áreas claves de Costa Rica" (W.M.P.) and 113-B5-A02, "Estudio del catálogo de sismos de la Red Sismológica Nacional de Costa Rica" (C.A.) both from the University of Costa Rica. J.C.L. appreciates support from several IUP University Senate Research Committee small grants program awards, one of which helped get former IUP undergraduate student K. McDannell to Costa Rica to ably help us with field work. We offer our gratitude for incisive comments and suggestions by two anonymous reviewers.

REFERENCES

1. Fitch, T. J. Plate Convergence, Transcurrent Faults, and Internal Deformation Adjacent to Southeast Asia and the Western Pacific. *Journal of Geophysical Research* **77**, 4432–4460, doi:10.1029/JB077i023p04432 (1972).
2. DeMets, C. A new estimate for present-day Cocos-Caribbean plate motion: Implications for slip along the Central American volcanic arc. *Geophysical Research Letters* **28**, 4043–4046, doi:10.1029/2001GL013518 (2001).
3. McCaffrey, R. Slip vectors and stretching of the Sumatran fore-arc. *Geology* **19**, 881–884, doi:10.1130/0091-7613 (1991).
4. McCaffrey, R. Estimates of modern arc-parallel strain rates in fore arcs. *Geology* **24**, 27–30, doi:10.1130/0091-7613 (1996).
5. Lewis, J. C., Boozer, A. C., López, A. & Montero, W. Collision versus sliver transport in the hanging wall at the Middle America subduction zone: Constraints from background seismicity in central Costa Rica. *Geochemistry Geophysics Geosystems* **9**, n/a–n/a, doi:10.1029/2007GC001711 (2008).
6. Montero, W., Lewis, J. C., Marshall, J. S., Kruse, S. & Wetmore, P. Neotectonic faulting and forearc sliver motion along the Atirro–Río Sucio fault system, Costa Rica, Central America. *Geological Society of America Bulletin* **125**, 857–876, doi:10.1130/B30471.1 (2013).
7. Wells, R. E., Weaver, C. S. & Blakely, R. J. Fore-arc migration in Cascadia and its neotectonic significance. *Geology* **26**, 759–762, doi:10.1130/0091-7613 (1998).
8. Feng, L. *et al.* Active deformation near the Nicoya Peninsula, northwestern Costa Rica, between 1996 and 2010: Interseismic megathrust coupling. *Journal of Geophysical Research: Solid Earth* **117**, n/a–n/a, doi:10.1029/2012JB009230 (2012).
9. Funk, J., Mann, P., McIntosh, K. & Stephens, J. Cenozoic tectonics of the Nicaraguan depression, Nicaragua, and Median Trough, El Salvador, based on seismic-reflection profiling and remote-sensing data. *Geological Society of America Bulletin* **121**, 1491–1521, doi:10.1130/B26428.1 (2009).
10. LaFemina, P. *et al.* Fore-arc motion and Cocos Ridge collision in Central America. *Geochemistry Geophysics Geosystems* **10**, doi:10.1029/2008GC002181 (2009).

11. Norabuena, E. *et al.* Geodetic and seismic constraints on some seismogenic zone processes in Costa Rica. *Journal of Geophysical Research* **109**, doi:10.1029/2003JB002931 (2004).
12. Corti, G., Carminati, E., Mazzarini, F. & Garcia, M. O. Active strike-slip faulting in El Salvador, Central America. *Geology* **33**, 989–992, doi:10.1130/G21992.1 (2005).
13. Martínez-Díaz, J. J., Álvarez-Gómez, J. A., Benito, B. & Hernández, D. Triggering of destructive earthquakes in El Salvador. *Geology* **32**, 65–68, doi:10.1130/g20089.1 (2004).
14. Álvarez-Gómez, J. A., Meijer, P. T., Martínez-Díaz, J. J. & Capote, R. Constraints from finite element modeling on the active tectonics of northern Central America and the Middle America Trench. *Tectonics* **27**, doi:10.1029/2007TC002162 (2008).
15. Fisher, D. M., Gardner, T. W., Marshall, J. S. & Montero, W. Kinematics associated with late Cenozoic deformation in central Costa Rica: Western boundary of the Panama microplate. *Geology* **22**, 263–266, doi:10.1130/0091-7613 (1994).
16. Marshall, J. S., Fisher, D. M. & Gardner, T. W. Central Costa Rica deformed belt: Kinematics of diffuse faulting across the western Panama block. *Tectonics* **19**, 468–492, doi:10.1029/1999TC001136 (2000).
17. Hey, R. Tectonic evolution of the Cocos-Nazca spreading center. *Geological Society of America Bulletin* **88**, 1404–1420, doi:10.1130/0016-7606 (1977).
18. Lonsdale, P. & Klitgord, K. D. Structure and tectonic history of the eastern Panama Basin. *Geological Society of America Bulletin* **89**, 981–999, doi:10.1130/0016-7606 (1978).
19. Carr, M. J. *et al.* Element fluxes from the volcanic front of Nicaragua and Costa Rica. *Geochemistry Geophysics Geosystems* **8**, doi:10.1029/2006GC001396 (2007).
20. Gardner, T. W. *et al.* Quaternary uplift astride the aseismic Cocos Ridge, Pacific coast, Costa Rica. *Geological Society of America Bulletin* **104**, 219–232, doi:10.1130/0016-7606 (1992).
21. Fisher, D. M., Gardner, T. W., Marshall, J. S., Sak, P. B. & Protti, M. Effect of subducting sea-floor roughness on fore-arc kinematics, Pacific coast, Costa Rica. *Geology* **26**, 467–470, doi:10.1130/0091-7613 (1998).
22. Gardner, T. *et al.* Holocene forearc block rotation in response to seamount subduction, southeastern Peninsula de Nicoya, Costa Rica. *Geology* **29**, 151–154, doi:10.1130/0091-7613 (2001).
23. Araya, C., Linkimer, L. & Taylor, W. Modelo mínimo unidimensional de velocidades de la onda P para la cordillera volcánica de Guanacaste. *Rev. Geol. Amér. Central* **54**, 179–191, doi:10.15517/rgac.v54i0.23283 (2016).
24. Waldhauser, F. & Ellsworth, W. L. A double-difference earthquake location algorithm: Method and application to the northern Hayward fault. *Bulletin of the Seismological Society of America* **90**, 1353–1368, doi:10.1785/0120000006 (2000).
25. Taylor, W., Barquero, R., Rojas, W., Montero, W. & Linkimer, L. La crisis sísmica de Bijagua de Upala (enero a marzo del 2002), Costa Rica. *Boletín OSIVAM* Año **12**, 48–66 (2002).
26. Alvarado, G. Consideraciones neotectónicas recientes en los alrededores de la laguna de Arenal, Costa Rica. *Boletin OSIVAM* Año **2**, 6–21 (1989).
27. McClay, K. & Dooley, T. Analogue models of pull-apart basins. *Geology* **23**, 711–714, doi:10.1130/0091-7613 (1995).
28. Mann, P., Hempton, M. R., Bradley, D. C. & Burke, K. Development of Pull-Apart Basins. *The Journal of Geology* **91**, 529–554, doi:10.1086/628803 (1983).
29. Plafker, G. Field reconnaissance of the effects of the earthquake of April 13, 1973, near Laguna de Arenal, Costa Rica. *Bulletin of the Seismological Society of America* **63**, 1847–1856 (1973).
30. Driese, S. G., Orvis, K. H., Horn, S. P., Li, Z.-H. & Jennings, D. S. Paleosol evidence for Quaternary uplift and for climate and ecosystem changes in the Cordillera de Talamanca, Costa Rica. *Palaeogeography, Palaeoclimatology, Palaeoecology* (2007).
31. Morell, K. D. Late Miocene to recent plate tectonic history of the southern Central America convergent margin. *Geochemistry Geophysics Geosystems* **16**, doi:10.1002/2015GC005971 (2015).

32. Lyon-Caen, H. *et al.* Kinematics of the North American–Caribbean-Cocos plates in Central America from new GPS measurements across the Polochic-Motagua fault system. *Geophysical Research Letters* **33**, n/a–n/a, doi:10.1029/2006GL027694 (2006).
33. Turner, H. L. *et al.* Kinematics of the Nicaraguan forearc from GPS geodesy. *Geophysical Research Letters* **34**, L02302, doi:10.1029/2006GL027586 (2007).
34. Protti, M., Gïendel, F. & McNally, K. Correlation between the age of the subducting Cocos plate and the geometry of the Wadati- Benioff zone under Nicaragua and Costa Rica. *Geological Society of America Special Papers* **295**, 309–326, doi:10.1130/SPE295-p309 (1995).
35. Edelman, S. H. & Sharp, W. D. Terranes, early faults, and pre-Late Jurassic amalgamation of the western Sierra Nevada metamorphic belt, California. *Geological Society of America Bulletin* **101**, 1420–1433, doi:10.1130/0016-7606 (1989).
36. Cowan, D. S., Brandon, M. T. & Garver, J. I. Geological tests of hypotheses for large coastwise displacements; a critique illustrated by the Baja British Columbia controversy. *American Journal of Science* **297**, 117–173, doi:10.2475/ajs.297.2.117 (1997).
37. Housen, B. A. & Beck, M. E. Testing terrane transport: An inclusive approach to the Baja B.C. controversy. *Geology* **27**, 1143–1146, doi:10.1130/0091-7613 (1999).
38. Umhoefer, P. J. Northward translation of "BAJA British Columbia" along the Late Cretaceous to Paleocene margin of western North America. *Tectonics* **6**, 377–394, doi:10.1029/TC006i004p00377 (1987).
39. Hildebrand, R. S. Dismemberment and northward migration of the Cordilleran orogen: Baja-BC resolved. *GSA Today* **25**, 4–11, doi:10.1130/GSATG255A.1 (2015).
40. Snoke, J. A., Munsey, J. W., Teague, A. G. & Bollinger, G. A. A program for focal mechanism determination by combined use of polarity and SV-P amplitude ratio data. *Earthquake Notes* **55**, 15–20 (1984).

CHAPTER 15

Flood hazard assessment for extreme flood events

Davor Kvočka[1] Roger A. Falconer[1] Michaela Bray[1]

1 Hydro-Environmental Research Centre, School of Engineering, Cardiff University, Cardiff CF24 3AA, UK

ABSTRACT

Climate change is expected to result in an increase in the frequency and intensity of extreme weather events. In turn, this will result in more frequent occurrences of extreme flood events, such as flash flooding and large-scale river flooding. This being the case, there is a need for more accurate flood risk assessment schemes, particularly in areas prone to extreme flooding. This study investigates what type of flood hazard assessment methods should be used for assessing the flood hazard to people caused by extreme flooding. Two flood hazard assessment criteria were tested, namely: a widely used, empirically derived method, and recently introduced, physically based and experimentally calibrated method. The two selected flood hazard assessment methods were: (1) validated against experimental data, and (2) used to assess flood hazard indices for two different extreme flood events, namely: the 2010 Kostanjevica na Krki extreme river flood and the 2007 Železniki flash flood. The results obtained in this study suggest that in the areas prone to extreme flooding, the flood hazard indices should be based on using the formulae derived for a mechanics-based analysis, as these formulations consider all of the physical forces acting on a human body in floodwaters, take into account the rapid changes in the flow regime, which often occur for extreme flood events, and enable a rapid assessment of the degree of flood hazard risk in a short time period, a feature particularly important when assessing flood hazard indices for high Froude numbers flows.

Keywords: Extreme floods, Large floods, Flash floods, Flood hazard, Flood risk, DIVAST-TVD

1. INTRODUCTION

Flooding is the most frequently occurring natural disaster in the world (Jonkman 2005). However, it is expected that flooding is going to occur even more frequently in the future as a result of climate change (Lenderink and Van Meijgaard 2008; Min et al. 2011; Rojas et al. 2013; Bruwier et al. 2015). As well as the expected increase in the magnitude and frequency of future flooding, there will also be an increase in the world's population from the current level of 7.3–9.7 billion by 2050 (UN 2015). Furthermore, it is projected that 66 % of the world's population will live in urban areas by 2050 (UN 2014), while in the near future 40 % of the global urban land will be located in areas highly prone to flooding (Güneralp et al. 2015). In addition, 90 % of the projected urban expansion is going to take place in Asia and Africa (UN 2014), e.g. in countries where millions of people already live in high-frequency flood zones (Muis et al. 2015). Therefore, the number of people affected annually by flooding will almost certainly increase significantly in the future.

As a result of climate change, there will also be an increase in the frequency and intensity of extreme rainfall events (Beniston 2009), which in turn is expected to result in more frequent occurrences of extreme flood events (Alfieri et al. 2015). Categorizing past flood events as normal or extreme can be a matter of debate. However, there are some orientation guides that can be used for categorizing the scale of past flood events. Brakenridge (2012) proposed two indices for characterizing flood events: flood severity and flood magnitude. The flood severity is not an exact descriptive statistic, but more of an orientation method that allows the use of expert judgment to estimate how unusual was the flood or discharge (Kundzewicz et al. 2013). Three flood severity classes were defined, which are based on the flood recurrence interval: Class 1 includes large flood events with a return period of the order 10–20 years, Class 2 includes very large flood events with a return period of between 20 and 100 years, and Class 3 includes extreme flood events with a return period equal to, or greater than, 100 years (Brakenridge 2012; Kundzewicz et al. 2013). Severity is an important flood characteristic, but it does not provide information on other critical aspects of flooding, such as flood duration or the extent of flooding (Kundzewicz et al. 2013). Therefore, a second statistic is needed, e.g. flood magnitude. Flood magnitude is defined as a function of flood severity, flood duration and flood inundation area (Brakenridge 2012). Flood magnitude is designed to mimic the Richter scale for earthquakes and thus provides a continuous metric, instead of just artificially classifying floods into different flood classes (Kundzewicz et al. 2013).

Determining the scale of past flood events should therefore be based on statistical descriptors (i.e. flood severity), spatiotemporal descriptors (i.e. flood magnitude) and also on socio-economical descriptors (i.e. the extent of flood damage, human casualties, psychological impact). This being the case, an extreme flood event can be simply characterized as a flood event with a small probability of occurrence, but with a significant impact on human society in terms of general damage, human casualties and overall social disruption. Taking into account all of these considerations, four types of flood events can be generally characterized as extreme flood events, including: (1) dam-break floods (Duffaut 2013; Bergman et al. 2014; Raška and Emmer 2014), (2) storm surges (Chau et al. 2013; Breilh et al. 2014; Androulidakis et al. 2015), (3) flash floods (Moussa and Bocquillon 2009; Martínez Ibarra 2012; Foulds et al. 2014; Amengual et al. 2015; Kvočka et al. 2015), and (4) extreme/large river floods (Zhi-Yong et al. 2013; Bruwier et al. 2015; Herget et al. 2015; Schröter et al. 2015; Antico et al. 2016). Among these types of extreme flood events, flash floods and large river floods are the most common and generally the most serious extreme events (Ashley and Ashley 2008; Di Baldassarre et al. 2010), which pose the greatest flood risk to the general population.

Flood risk assessment methods mainly consist of four steps, including: hazard assessment, exposure assessment, vulnerability assessment and risk assessment (Penning-Rowsell et al. 2005b; Foudi et al. 2015). Even though all of these steps are important in the flood risk assessment process, flood hazard assessment usually receives most attention as flood hazard maps are used for estimating the danger to people due to flooding (Koks et al. 2015). According to the Department for the Environment, Food and Rural Affairs (Defra) at the UK Environment Agency, flood hazard *"describes the flood conditions in which people are likely to be swept over or drown in a flood"*, with these conditions being as a result of the combined effect of: (1) the depth of flow, (2) the velocity of flow, (3) the presence of debris in the flow, and (4) the spatial and temporal dynamics of these parameters (Ramsbottom et al. 2006). In other words, flood hazard assessment defines stability threshold for people in floodwaters. In floodwaters, people predominately lose stability due to two hydrodynamic or stability mechanisms: toppling or moment instability and sliding or friction instability (Jonkman and Penning-Rowsell 2008a, b; Cox et al. 2010; Xia et al. 2014).

There are two different types of criteria for assessing the human stability in floodwaters, i.e. the degree of flood hazard (Xia et al. 2011), including: (1) formulae based on mechanical analysis and calibrated using laboratory experiments with models and/or real human subjects (Foster and Cox 1973; Abt et al. 1989; Takahashi et al. 1992; Karvonen et al. 2000; Yee 2003; Jonkman and Penning-Rowsell 2008a, b; Russo et al. 2013), and (2) formulae based on empirical or quasi-theoretical studies (Keller and Mitsch 1993; Lind et al. 2004; Penning-Rowsell et al. 2005a; Ramsbottom et al. 2003, 2006; Ishigaki et al. 2005, 2009). However, flood hazard assessments methods based only on laboratory experiments with models and/or real human subjects are usually too dependent on the physical characteristics of the model or the human subject, and cognitive characteristics of the tested human subjects, whereas flood hazard assessment methods based on empirical or quasi-theoretical work often excessively over-simplify the anatomy of human body and the hydraulic characteristics of the flow (Jonkman and Penning-Rowsell 2008a, b; Xia et al. 2014). These shortcomings could particularly be relevant when it comes to designing safety criteria in areas prone to extreme flood events, since such events are generally associated with deep floodwaters, high velocity flows and the occurrences of abrupt changes in the flow regime (e.g. hydraulic jumps). Such floods require full physical representation in order to be adequately integrated in the prediction of flood hazard indices. Therefore, there is a need to improve on the current flood hazard assessment methods used, or to develop new methods, that would be based more on representing the physics accurately rather than on simple empirical relations.

The main aim of this study is to raise awareness about limitations of flood hazard assessment schemes currently in use, and in particular for areas susceptible to the occurrence of extreme flood events, such as flash flooding or large river flooding. Therefore, the key objective is to investigate and determine what type of flood hazard assessment methods should be used for assessing the hazard risk to people caused by extreme flooding. In this study, two different types of flood hazard criteria were tested, namely a classical and widely used empirically based approach, and a recently proposed physically based and experimentally calibrated approach. The two selected flood hazard assessment methods were: (1) validated against different experimental datasets, including datasets based on experiments using real humans as test subjects and datasets based on experiments using a scaled model human body, and (2) used to assess flood hazard indices for two different extreme flood events, including one extreme river flood event and one flash flood event. The results obtained suggest that the flood hazard assessment in areas prone to extreme flooding should be conducted with flood hazard assessment methods based on a mechanics-based analysis, i.e. the recently introduced physically based and experimentally calibrated methods, such as the method considered in this study. These methods are highly adaptable to complex hydrodynamic processes associated with extreme flood events and are generic. The findings in this paper should be of particular interest to government departments (national and regional) and regulatory and planning authorities, as the findings presented herein should help to improve on the general flood hazard assessment data, and procedures used for emergency services and planning assessment.

2. STUDY AREAS

2.1 Kostanjevica na Krki

Kostanjevica na Krki is one of the oldest and smallest towns in Slovenia, EU. It is located in the south-eastern part of Slovenia, near the border with Croatia. The centre of the town lies on a small artificial island, surrounded by the River Krka (see Fig. 1), and is protected as a cultural town with historical monuments. Due to rich historical and cultural heritage, and frequent flooding, the town is also promoted as the "Venice of Lower Carniola".

Figure. 1 Town of Kostanjevica na Krki

During the period between 17th and 19th September 2010, Slovenia was hit by heavy and extensive precipitation, which led to a sharp rise in the water levels in watercourses and flooding occurred all over the country

(Slovenian Environment Agency 2010a). In the territory of Slovenia, an average of 170–180 mm of rainfall fell over 48 h period, which was the highest amount of rainfall accumulated in two-day period over the past 60 years (Slovenian Environment Agency 2010b). The River Krka started flooding on the evening of 18th September, and on the 19th September the town of Kostanjevica na Krki was flooded (Kobold 2011). The data from the nearby stream gage station at Podbočje revealed that the peak discharge of 468 m^3/s occurred on 20th of September (Slovenian Environment Agency 2010a). This was the highest measured discharge for the River Krka since the beginning of measurements in 1926 and statistically corresponds to nearly a 1 in 1000 year flood event (Kobold 2011; Slovenian Environment Agency 2013).

Topographic data for the Kostanjevica na Krki study domain were collected during a national project of LiDAR (Light Detection and Ranging) scanning of the Slovenian surface (Ministry of the Environment and Spatial Planning of the Republic of Slovenia 2015). These LiDAR data were then used to set up a hydraulic model for the 2010 Kostanjevica na Krki extreme river flood event. The study domain (see Fig. 2) was 705 m long, 641 m wide and divided into square cells, with each cell having an area of 1 m^2. The western boundary of the domain was set as an inflow boundary for the River Krka, with the peak discharge being specified as the upstream boundary condition. The area where the River Krka leaves the study domain was set as the downstream boundary, with a prescribed water level being specified as the downstream boundary condition.

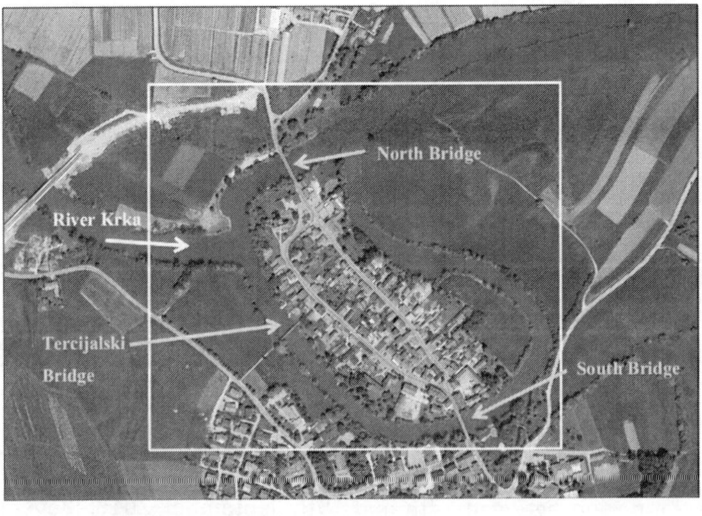

Figure. 2 Kostanjevica na Krki study domain

Three wooden bridges connect the old town centre with the surrounding area, namely the North Bridge, the South Bridge and the Tercijalski Bridge. There were no official reports or eye-witness accounts suggesting that any of the three bridges were blocked during the time of the flood event. This being the case, there was no need to specifically model bridges in the hydraulic model for the Kostanjevica na Krki study domain.

2.2 Železniki

Železniki is a small town in the north-west part of Slovenia, EU. The town is situated in the long and narrow Selca Valley, which stretches along the River Selška Sora. The Selca Valley, and in particular the Selška Sora catchment area upstream of Železniki, is characterized by a steep topographic terrain and relatively thin soils on the hillslopes (Zanon et al. 2010). The town of Železniki is the administrative centre and the largest settlement in the Selca Valley. Železniki is also the economic hub of the area, with relatively large metalworking and furniture factories being based in the town (see Fig. 3).

Figure. 3 Town of Železniki

On the 18th of September 2007, there was a region of low air pressure over Northern Europe, which was moving over western and middle Europe towards the Alps. In addition to a low-pressure weather system, a high valley of cold air was moving over Western Europe towards the east, while the south-western wind was getting stronger and stronger over Slovenia. As a result of these complex weather conditions, an extensive convective system formed over the north-western part of Slovenia, which led to the occurrence of an extreme rainfall event that affected approximately one-third of the country (Slovenian Environment Agency 2008). More than 300 mm of rainfall was recorded on some rain-gauge stations within six hours after the start of the storm, with the return period of the highest precipitation being more than 100 years (Kobold et al. 2008). In Železniki, the observed maximum daily amount of rainfall was nearly 200 mm, which was the highest recorded amount of rainfall since the beginning of the measurements in 1930 (Grillakis et al. 2010). The accumulated rainfall resulted in a flood event that by far exceeded the 100-year return period of maximal floods, with the estimated peak discharge being around 300 m^3/s (Rusjan et al. 2009; Marchi et al. 2009; Zanon et al. 2010). The flash flood triggered by the intense precipitation devastated the town of Železniki: three people lost their lives, while it was estimated that the flood caused nearly 100 million Euros of damage (Bouilloud et al. 2009).

Topographic data for the Železniki study domain were collected during the national project of LiDAR scanning of the Slovenian surface (Ministry of the Environment and Spatial Planning of the Republic of Slovenia 2015). These LiDAR data were then used to set up a hydraulic model for the 2007 Železniki flash flood event. The study domain (see Fig. 4) was 1966 m long, 1285 m wide and divided into square cells, with each cell having an area of 1 m^2. The western boundary of the domain was set as an inflow boundary for the River Selška Sora, with the peak discharge being specified as the upstream boundary condition. The eastern boundary of the study domain was set as the downstream boundary, with a prescribed water level being specified as the downstream boundary condition.

3. NUMERICAL MODEL

The research presented in this paper was conducted using the DIVAST-TVD numerical model. The DIVAST-TVD model is a shock-capturing flood simulation model, which was developed by Liang et al. (2007a). The DIVAST-TVD model was specifically developed to simulate the hydrodynamic processes in rivers and coastal waters by solving the shallow water equations, with the model being predominately used for simulating rapidly varying flood events and high Froude number flows (Liang et al. 2007b, 2014; Kvočka et al. 2015).

Figure. 4 Železniki study domain

This numerical model combines the standard MacCormack scheme with a symmetric five-point total variation diminishing (TVD) term. The MacCormack scheme is a numerical method ideally suited for solving the time-dependent compressible Navier–Stokes equations (MacCormack 1976). The MacCormack scheme is a predictor–corrector scheme and is a variation of the Lax-Wendroff scheme (Lax and Wendroff 1960). In the MacCormack scheme, the predictor step calculates a rough approximation of the desired variable, whereas the corrector step refines the initial approximation. The standard MacCormack scheme has second-ordered accuracy (MacCormack 1976). It is well known that all schemes of accuracy greater than one will generate spurious numerical oscillations in the regions where the gradient is high (Godunov 1959). Therefore, a symmetric five-point TVD term is appended to the corrector step of the MacCormack scheme to prevent non-physical oscillations. Total variation diminishing schemes prevent the emergence of numerical oscillations through added nonlinear artificial dissipation terms (Mingham et al. 2001). The concept of TVD was first introduced by Harten (1983), whereas the TVD term implemented in the DIVAST-TVD model was first presented by Davis (1984). Davis (1984) proposed a total variation diminishing scheme where a symmetric five-point TVD term is added to the Lax-Wendroff scheme. The symmetric five-point TVD term is used to adjust the introduced numerical diffusion: a second-order accurate MacCormack scheme is deployed where the solution is smooth, whereas a first-order accurate upwind scheme is deployed to avoid spurious numerical oscillations (Liang et al. 2007a).

The DIVAST-TVD model is an effective tool for analysing storm surges, dam-break scenarios, flash floods, etc., i.e. any flow scenario that could involve rapid changes in the flow regime. The information about model development and extensive model verifications can be found in the literature (Liang et al. 2006, 2007b; Hunter et al. 2008; Neelz and Pender 2009; Liang et al. 2010; Kvočka et al. 2015).

4. FLOOD HAZARD ASSESSMENT METHODS

Two different flood hazard assessment methodologies were considered in this paper: (1) an empirically derived method (Ramsbottom et al. 2003, 2006), and (2) a physically based and experimentally calibrated method (Xia et al. 2014).

4.1 Empirically derived method

Ramsbottom et al. (2003) developed a methodology for assessing and mapping the risk to people caused by flooding, with this methodology being developed for the Department for the Environment, Food and Rural Affairs (DEFRA) and the UK Environment Agency. Based on testing various empirical formulae by comparing the predictions to experimental datasets obtained from laboratory and field-based studies conducted by Abt et al. (1989), Karvonen et al. (2000), (Cox et al. 2010) and Ramsbottom et al. (2003) proposed an empirical formula, which assesses the flood hazard to people as follows:

$$HR = d\,(v + 1.5) + DF \tag{1}$$

where HR is the flood hazard rating (m^2/s), d is the water depth (m), v is the velocity of the flow (m/s) and DF is the debris factor (m^2/s), which can have a value of 0, 1 or 2, depending on the place of the flood and on the features of the flow.

In the review of the original study, Ramsbottom et al. (2006) revised the initially proposed formula, i.e. Eq. (1), wherein: the velocity coefficient was reduced from 1.5 to 0.5, while the values for the debris factor were reduced from the initial values of 0, 1 and 2 to values of 0, 0.5 and 1. This being the case, the revised formula is given as:

$$HR=d(v+0.5)+DF \tag{2}$$

where HR is the flood hazard rating (m^2/s), d is the water depth (m), v is the velocity of the flow (m/s) and DF is the debris factor (m^2/s).

Based on the revised formula, i.e. Eq. (2), Ramsbottom et al. (2006) proposed various flood hazard classifications, which are given in Table 1.

Table 1 Flood hazard to people (Ramsbottom et al. 2006)

HR	Degree of flood hazard	Description
<0.75	Low	Caution
0.75–1.5	Moderate	Dangerous for some (i.e. children)
1.5–2.5	Significant	Dangerous for most people
>2.5	Extreme	Dangerous for all

The empirical expression, i.e. Eq. (2), presented by Ramsbottom et al. (2006) has some shortcomings, including (Cox et al. 2010): (1) the flow regime values from the experimental datasets were averaged, and thus some training (e.g. the ability of the test subject to learn how to manoeuvre in the flow with time) was incorporated in the derived expression; however, this is usually not the case with the general population, as majority of people do not have the experience of standing or manoeuvring in floodwaters, (2) the assigned values for the debris factor were not based on, or supported by, any experimental testing, and (3) the proposed expression has no upper depth limit, and therefore large depth/low velocity flood flows are not necessarily considered as hazardous, i.e. the flow conditions in which a person would be floating, and thus completely dependent upon swimming ability, are not automatically classed as dangerous. Nevertheless, flood hazard maps issued by the regulatory authorities in the UK (e.g. DEFRA) are based on the flood hazard assessment method presented by Ramsbottom et al. (2006). Furthermore, the criterion presented herein is also well established outside the UK (Penning-Rowsell et al. 2005a; Kaźmierczak and Cavan 2011; Purwandari et al. 2011; Foudi et al. 2015). This being the case, the formula presented by Ramsbottom et al. (2006) is generally considered as an accurate criterion for assessing and mapping hazard to people due to flooding.

4.2 Physically based and experimentally calibrated method

Xia et al. (2014) derived a new method for assessing human stability in floodwaters, which is based on extensive theoretical and experimental studies. The formulae proposed by Xia et al. (2014) are based on the mechanisms of toppling and sliding instability, and were derived by considering all forces acting on a human body in floodwater, e.g. drag force, frictional force, gravitational force, buoyancy force and normal reaction force. Furthermore, the new formulae also (1) take into account the effect of a non-uniform upstream velocity profile on the stability of a person standing in a floodwater, and (2) consider the impact of the body buoyancy for rapidly varying water depths (Xia et al. 2014).

Two formulae were proposed. Firstly, the incipient velocity is given for a human body in floodwater experiencing sliding instability:

$$U_c = \alpha\left(\frac{h_f}{h_p}\right)^{\beta}\sqrt{\frac{m_p}{\rho_f h_p h_f} - \left(a_1\frac{h_f}{h_p} + b_1\right)\frac{(a_2 m_p + b_2)}{h_p^2}} \quad (3)$$

where U_c is the incipient velocity, h_f is the water depth (m), h_p is the height of a person (m), $m_p$is the weight of a person (kg), ρ_f is the density of water (kg/m^3), α and β are empirical coefficients and a_1, a_2, b_1 and b_2 are coefficients based on the characteristics of a human body.

Secondly, the incipient velocity for a human body in floodwater experiencing toppling instability is given as:

$$U_c = \alpha\left(\frac{h_f}{h_p}\right)^{\beta}\sqrt{\frac{m_p}{\rho_f h_f^2} - \left(\frac{a_1}{h_p^2} + \frac{b_1}{h_f h_p}\right)(a_2 m_p + b_2)} \quad (4)$$

where U_c is the incipient velocity, h_f is the water depth (m), h_p is the height of a person (m), $m_p$is the weight of a person (kg), ρ_f is the density of water (kg/m^3), α and β are empirical coefficients and a_1, a_2, b_1 and b_2 are coefficients based on the characteristics of a human body.

Finally, the degree of flood hazard for each instability mechanism can be quantified with the following expression:

$$\mathrm{HR} = \mathrm{MIN}\left(1, \frac{U}{U_c}\right) \quad (5)$$

where HR is the flood hazard rating, U is the velocity of the flow and U_c is the incipient velocity, e.g. the velocity at which person loses stability in floodwater.

The main difference between the majority of empirically derived flood hazard formulae and this physically based and experimentally calibrated method is in the way they take into account forces induced by flow conditions. In Eqs. (3) and (4), it can be seen that the overturning force on the body is proportional to the water depth times the velocity squared (i.e. hv^2), whereas for the empirically derived formula [see Eq. (2)] the overturning force on the body is proportional to the water depth times velocity only (i.e. hv). This means that the physically based and experimentally calibrated method can be much more influenced by higher velocities and momentum, with such conditions frequently occurring with extreme flood events. This being the case, this physically based and experimentally calibrated method is highly adaptable to abrupt changes in the flow regime and can rapidly and more accurately assess the degree of flood hazard risk in a short time period, characteristic particularly important for flood hazard assessment of extreme flood events.

A further advantage of the physically based and experimentally calibrated methods is that they can be adjusted to a specific body type (Xia et al. 2014; Milanesi et al. 2015). For example, by applying a biomechanical model, such as the 3D biomechanical model presented by Nikolova and Toshev (2007), Eqs. (3) and (4) can assess the stability of people according to their height, weight and body parameter characteristics, e.g. length, mass and the corresponding volume of a particular body part, such as legs, torso, arms. This means that stability thresholds can be defined specifically for: (1) different sub-population groups of the general population, such as males, females and children, and (2) that stability thresholds can be defined explicitly for different geographic regions or countries, such as Europe, America or the Far East.

5. METHODOLOGY

The initial assumption of this study was that the physically based and experimentally calibrated method would more accurately assess the flood hazard indices for people than the empirically derived method when the flood hazard is assessed for extreme flood events. In order to test this assumption, two different validation procedures were undertaken.

Firstly, both flood hazard assessment criteria were validated against two different types of experimental data, i.e. the data based on the experiments including real human test subjects and the data based on the experiments using model human bodies. The experiments, including real human test subjects, conducted by Abt et al. (1989) and Karvonen et al. (2000), were considered herein. There are several other studies where real humans as test subjects have been used, such as the studies conducted by Foster and Cox (1973), Takahashi et al. (1992) and Jonkman and Penning-Rowsell (2008a, b). However, the focus in this study is on the experimental data by Abt et al. (1989) and Karvonen et al. (2000), as these two datasets were included in the derivation process of the empirically based method considered herein and have the largest amount of experimental observations among the available experimental datasets. It should be noted that Russo et al. (2013) have collected the largest amount of experimental observations to date (more than 800 tests were performed in this study). Unfortunately, this dataset was not available and thus was not considered in this study.

In the experiments conducted by Abt et al. (1989) and Karvonen et al. (2000), the water depth and the corresponding velocity were recorded when the human test subject lost manoeuvrability or stability in the flume due to the flow conditions. The authors of these two studies did not specifically record which instability mechanism led to the loss of stability of the test subject during a particular test. Nonetheless, it can be established from the recorded data that only in a few tests the stability was lost due to sliding, whereas in the vast majority of the tests the stability was lost due to toppling (Jonkman and Penning-Rowsell 2008a, b; Xia et al. 2014). In contrast to experiments with real human test subjects, Xia et al. (2014) used a scaled human body (i.e. a dummy) for their experiments. The scaled human body strictly followed the principles of geometric, kinematic and dynamic scaling, which meant that the flow conditions were ideally similar to those in the prototype (Chanson 2004), e.g. density, drag and friction coefficient of the selected human body model were approximately equal to those of the prototype (Xia et al. 2014). This being the case, the prototype, at 1.70 m in height and 60 kg in weight, was scaled down to a model human body of 0.3 m in height and 0.334 kg in weight, with the model human body strictly following the geometric similarity in each dimension. For each test, the water depth and the corresponding velocity were recorded at the moment when the model human body started to become unstable. In addition, the instability mechanism (e.g. toppling or sliding), which led to instability of the model human body, was identified for each test.

As mentioned previously, the physically based and experimentally calibrated method considered herein can be adjusted to a specific body type. The characteristics of a specific body type in Eqs. (3) and (4) are represented with coefficients a_1, b_1, a_2 and b_2. More precisely, the coefficients a_1, b_1, a_2 and b_2 form part of an expression, which represents the effect of the buoyancy force as a function of the human height and mass for a given water depth (Xia et al. 2014). These coefficients can be determined from the characteristics of the human body, e.g. the human height and mass, and the volume and mass of the body segment parameters, such as legs, arms, torso. For example, based on the body segment characteristics of a typical Bulgarian body (Nikolova and Toshev 2007), the values of a_1 and b_1 were assumed to be 0.612 and 0.388, while according to the body structure of an American body (Drillis et al. 1964), the value of the coefficients a_1 and b_1 was 0.737 and 0.263, respectively (Xia et al. 2014). For the coefficients $a_2$and b_2, these can be obtained from the relationship between the total body volume and the corresponding mass of a human body. Hence, the values of the coefficients a_2 and b_2 for a typical Bulgarian body, for example, were assumed to be 0.9748×10^{-3} m^3/kg and -0.7111×10^{-3} m^3. In addition, the coefficients a_1, b_1, a_2 and b_2 are constant, which means that the same values of these coefficients apply for both instability mechanisms, as given by Eqs. (3) and (4). The value of the coefficients a_1, b_1, a_2 and b_2 for the model human body used in this study was: $a_1 = 0.633$, $b_1 = 0.367$, $a_2 = 1.015 \times 10^{-3}$ m^3/kg and $b_2 = -4.927 \times 10^{-3}$ m^3 (Xia et al. 2014).

Two parameters are needed to calibrate the physically based and experimentally calibrated method considered herein, i.e. parameters α and β [see Eqs. (3) and (4)]. These two parameters can be evaluated from the relevant experimental data and are influenced by the shape of the test subject, the friction coefficient between the soles and the ground surface, the drag coefficient, the effect of a non-uniform velocity distribution along the vertical direction, and the effect of the ability of a test subject to adjust its position in the flow (Xia et al. 2014). Hence, the values of the parameters α and β are different for each body type and for each instability mechanism. However, as the height, mass and the values of the aforementioned coefficients are constant for a particular human body, the values of parameters α and β can be evaluated by statistical analysis from the relevant experimental data (Xia et al. 2014). As we were modelling sites in Slovenia, it would be reasonable to calibrate Eqs. (3) and (4) for the body characteristics of people living in these two countries. Unfortunately, such calibrations are not possible at present due to the lack of body segment parameter data needed to undertake the calibration process. Alternatively, Eqs. (3) and (4) could be calibrated for a typical Bulgarian body, since the necessary body segment parameter data exists (Nikolova and Toshev 2007). These data can be used to represent a model human body based on the typical Bulgarian body prototype, with the model human body being used for the experimental tests and thus for calibration of Eqs. (3) and (4). However, as the difference in size and body characteristics between a typical Bulgarian body and the body considered in the study of Xia et al. (2014) is relatively small, it was decided to use the values of the parameters α and β obtained in the study by Xia et al. (2014)

for the calibration of Eqs. (3) and (4) in this study. This saved valuable resources and time that would otherwise be spent on experimental procedures, yet still enabled satisfactory quality of the final results as the potential computational error is in an acceptable range. The values of the parameters α and β used in this study for the calibration of Eqs. (3) and (4), for both the model human body considered by Xia et al. (2014) and the real human body (i.e. data collected in the experiments considering real human test subjects), are presented in Table 2.

Table 2 Values of the parameters α and β used to calibrate the physically based and experimentally calibrated method for the model human body and real human body (Xia et al. 2014)

Parameter	Model human body		Real human body	
	Toppling instability	Sliding instability	Toppling instability	Sliding instability
α	3.472	7.975	7.867	10.253
β	0.188	0.018	0.462	0.139

Secondly, the assessed flood hazard indices from both methods were compared at different stages of three extreme flood events, including one extreme river flood event and two flash flood events. In order to focus solely on the hydrodynamic conditions of the flow that can lead to loss of stability in floodwaters, any external factors that present a risk to people in floodwaters (such as floating debris) were omitted from the flood hazard assessment process. As mentioned, the physically based and experimentally calibrated method considered in this study can assess the hazard to people by considering both the sliding and toppling instability mechanisms. This being the case, the limiting stability threshold for the physically based and experimentally calibrated method was defined as the minimum of the toppling and sliding incipient velocities:

$$\mathrm{HR} = \mathrm{MIN}\left(1, \frac{U}{\mathrm{MIN}(U_{\mathrm{toppling}}, U_{\mathrm{sliding}})}\right) \quad (6)$$

where HR is the flood hazard rating, U is the velocity of the flow, U_{toppling} is the toppling incipient velocity and U_{sliding} is the sliding incipient velocity.

The empirically derived method categorizes flood hazard into four flood hazard classifications, namely: low, moderate, significant and extreme. On the other hand, the physically based and experimentally calibrated method considered herein quantifies flood hazard by mimicking the principle of bivalence, i.e. there is only one threshold that defines whether the stability of a person in floodwater is lost or not. This means that there is only one flood hazard class, i.e. extreme. However, in order to allow a more detailed comparison to be undertaken between the two flood hazard assessment methods, the assessed degree of flood hazard according to the physically based and experimentally calibrated method was divided into three additional flood hazard classifications that correspond to the flood hazard classifications of the empirically derived method, i.e. low, moderate and significant. The subdivision of the quantifying flood hazard criteria of the physically based and experimentally calibrated method, which ranges between 0 and 1, was undertaken in such a manner that the ratio of the threshold values that separate the subdivided flood hazard classes was identical to the ratio of the threshold values that separate the flood hazard classes of the empirically based method. Therefore, if the threshold values in the empirical method are 0.75, 1.5 and 2.5, then the corresponding values for the physically based and experimentally calibrated method would be 0.3, 0.6 and 1. The subdivision of the physically based and experimentally calibrated method into three additional flood hazard classifications, corresponding to the flood hazard classifications of the empirically derived method, should enable better evaluation of how the two flood hazard assessment methods adapt to the violent nature of extreme flood events and thus provide more meaningful results.

In this study, extreme flood events were simulated with rapid changes occurring in the flow regime and the Froude numbers being relatively high. Therefore, there was a need to determine what type of flood simulation model should be used for predicting the main flood parameters (i.e. flood depths and velocities), which were needed for an accurate flood hazard assessment process. The shock-capturing flood simulation DIVAST-TVD model was used for predicting the main flood parameters for flash flood scenario, as only shock-capturing flood simulation model produces numerically accurate predictions of flood depths and velocities when simulating flash flood events (Kvočka et al. 2015). In order to be consistent, the DIVAST-TVD model was also used for predicting the main parameters in the extreme river flood scenario.

6. RESULTS AND DISCUSSION

6.1 Comparison with the experimental datasets

Figure 5 shows a comparison of the flood hazard prediction ability of the two methods with the experimental observations of Abt et al. (1989), while Fig. 6 shows a comparison of the flood hazard prediction ability of the two methods with the experimental observations of Karvonen et al. (2000). As mentioned, in the experiments conducted by Abt et al. (1989) and Karvonen et al. (2000), the dominant mode of instability was due to the toppling instability mechanism (Jonkman and Penning-Rowsell 2008a, b; Xia et al. 2014), and therefore Eq. (4) was used for the comparisons with the empirically derived method. Table 3 shows the values of the parameters α and β used for the calibration of Eq. (4) when the two datasets are considered separately.

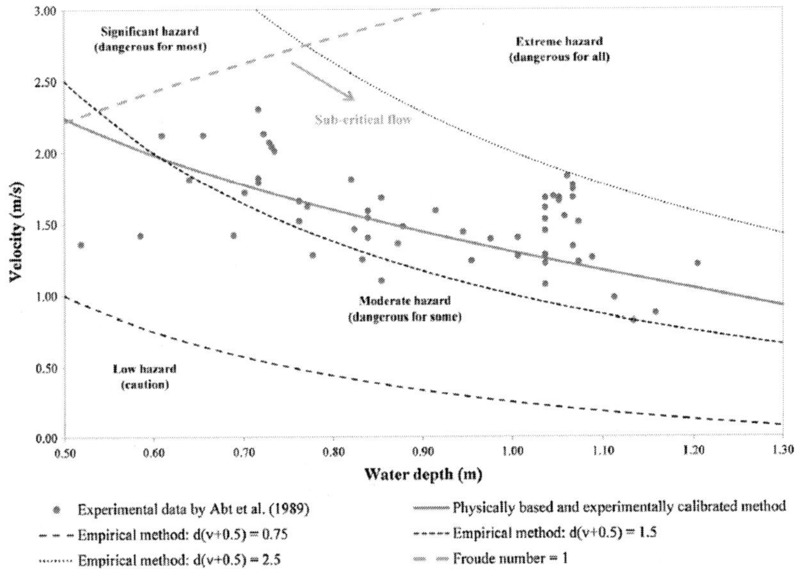

Figure. 5 Comparison of the flood hazard prediction ability of the two methods with the experimental observations of Abt et al. (1989)

In both Figs. 5 and 8, the blue circles represent the depth and velocity when stability was lost, i.e. the flow conditions which led to instability of the test subject. In Fig. 5, it can be seen that the empirically derived method classified all experimental observations into two flood hazard classes, i.e. moderate hazard and significant hazard. The majority of the experimental observations (i.e. the point in the experiments when the real human test subject was losing, or completely lost, stability) were classified as significant hazard, which indicated that the empirical method generally accurately assessed the flood hazard indices. Figure 5 also shows that the physically based and experimentally calibrated method agreed well with the experimental observations, as 38 out of 58 experimental observations can be found above the stability threshold (red line), which indicated that stability was lost, and with at least 15 experimental observations being in the relative proximity of the stability threshold, which indicated that subject was close to losing stability, i.e. was greatly endangered by the flow conditions. In Fig. 6, it can be seen that the empirically derived method classified the majority of the experimental observations in the moderate flood hazard class, which was regarded as the stability threshold for children. This indicated that the empirically derived method underestimated the flood hazard indices for the dataset of Karvonen et al. (2000). On the other hand, Fig. 6 also shows that the physically based and experimentally calibrated method agreed, almost for all of the experimental observations, as 21 out of 29 of the experimental observations fall above the stability threshold (red line), which indicated that stability was lost. In addition, Figs. 5 and 6 also show that in practically all test runs the stability of the real human test subjects was lost in the sub-critical flow regime. Therefore, it is hard to evaluate the prediction ability of the empirically derived method for high Froude number flows, which often occur with extreme flood events. Nonetheless, it can be seen from Figs. 5 and 6 that the physically based and the experimentally calibrated method performed better than the empirically derived method.

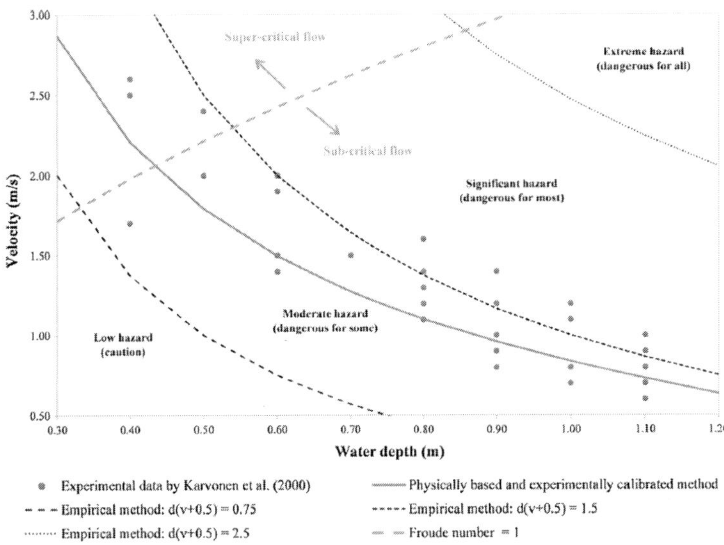

Figure. 6 Comparison of the flood hazard prediction ability of the two methods with the experimental observations of Karvonen et al. (2000)

Table 3 Values of the parameters α and β used to calibrate Eq. (4) separately for each of the two datasets (Xia et al. 2014)

Parameter	Data by Abt et al. (1989) Equation (4)	Data by Karvonen et al. (2000) Equation (4)
α	8.855	4.825
β	0.473	0.160

In contrast, the results are somewhat different when both datasets are merged. For example, Fig. 7 shows a comparison of the flood hazard prediction ability of the two methods, with the experimental data from both aforementioned experimental studies. The blue circles and orange squares represent the depth and velocity when stability was lost, i.e. the flow conditions which led to instability of the test subject. In Fig. 7, it can be seen that the empirically derived method classified all data into two flood hazard classes, i.e. moderate hazard and significant hazard. This means that nothing changed from the perspective of the empirically derived method when compared to the results presented in Figs. 5 and 8. On the other hand, Fig. 7 also shows that there was some change in the predictive ability of the physically based method. It can be seen in Fig. 7 that the physically based and experimentally calibrated method classified 57 out of 87 experimental observations as extreme hazard. However, there are now 48 experimental observations (compared to 38 in Fig. 5) from the dataset by Abt et al. (1989), and only nine experimental observations (compared to 20 in Fig. 6) from the dataset by Karvonen et al. (2000) above the stability threshold (red line) of the physically based and experimentally calibrated method.

The results in Fig. 7 show that the predictive ability of the physically based and experimentally calibrated method is closely related to the body characteristics of the test subjects, which agrees with the observations of other authors (Xia et al. 2014; Milanesi et al. 2015). As the two datasets were combined, the body characteristics (and thus the effect of buoyancy) of an average test subject changed, which consequently changed the values of the calibrating parameters α and β (see Table 2), and thus stability threshold of the physically based and experimentally calibrated method. For example, average height and weight in the study of Abt et al. (1989) were 178 cm and 76 kg, while for a test subject in the study of Karvonen et al. (2000) they were 174.6 cm and 73.6 kg, respectively. However, when both datasets are merged the average test subject is 177.1 cm in height and weighs 75.3 kg, which explains why there are more experimental observations of Abt et al. (1989) above the stability threshold (red line) of the physically based and experimentally calibrated method when compared to Fig. 5, and why there are less experimental observations for Karvonen et al. (2000) above the stability threshold (red line) when compared to Fig. 6.

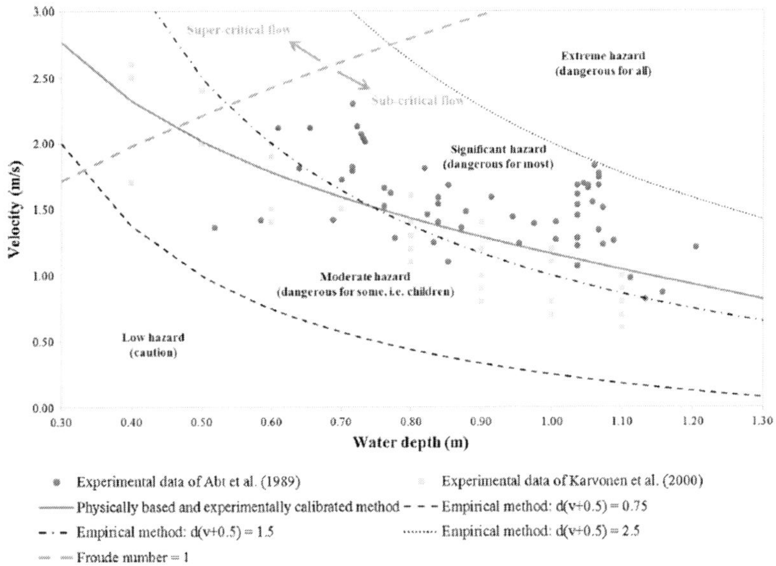

Figure. 7 Comparison of the flood hazard prediction ability of the two methods with the experimental observations of Abt et al. (1989) and Karvonen et al. (2000)

On the other hand, the results in Fig. 7 also indicate that there are some reservations regarding the experimental data based on real human test subjects. For example, Lind et al. (2004) noted that these two datasets should not be aggregated in the application process, because these datasets are too small, not random and consequently not representative enough. In addition, Abt et al. (1989) reported that their experiments had some constraints, such as optimal experiment conditions, the presence of the safety equipment, healthy test subjects and the ability of the test subject to learn how to manoeuvre in the flow with time. Similar findings were reported by Karvonen et al. (2000), who noted that their experiments also had some shortcomings, such as excessive safety features, the use of survival suits, which increased the buoyancy and cross-sectional area of a person, and the use of a slippery surface. It is therefore reasonable to assume that these experiments did not represent the actual response of a person in the event of real-life flooding.

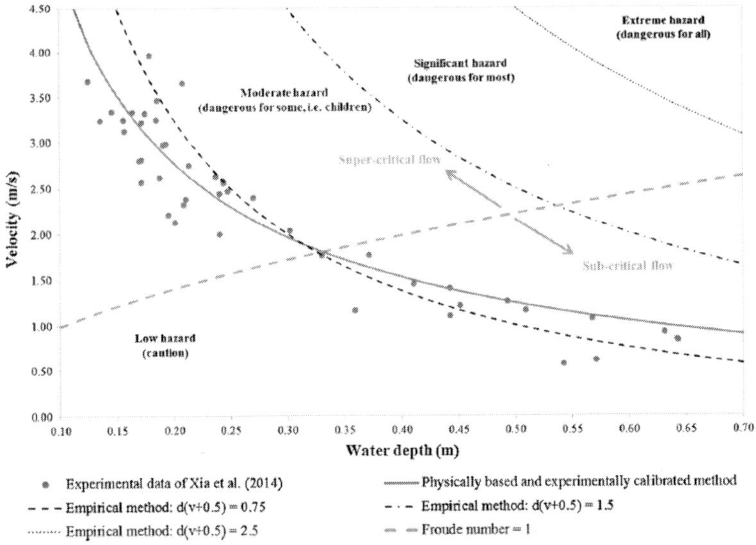

Figure. 8 Comparison of the flood hazard prediction ability of the two methods with the experimental observations of Xia et al. (2014)

In general, an average person has probably never been directly faced with rising floodwaters, which means that such a person does not know how to readjust the body position in order to maintain stability for longer, as a real human test subject would after a few test runs (i.e. training). Furthermore, it is important to point out that not all people in the general population are healthy or physically fit and are not therefore able to adequately face the dangers due to flooding. This being the case, including only healthy and relatively young test subjects in the experimental studies leads to safety criteria that exclude some of the most vulnerable groups of the general population, such as frail and/or elderly citizens, and people with physical disabilities. Finally, one of the key factors that have a major effect on human stability in floodwaters is human psychology. If a human test person is attached with safety ropes, and therefore feels completely safe, then the psychological factors, such as shock or fear, do not play a significant role in maintaining or losing stability in floodwaters. However, the psychological factors are sometimes even more influential than physical ability. For example, in the study presented by Cox et al. (2004) two child test subjects were similar in size and thus should have had similar safety characteristics. However, as these two child test subjects were of different age and physiological development, they had differing safety characteristics. This highlights the importance of demographic and psychological characteristics, as they can have a significant impact on a person's response time and consequently on a person's ability to adequately react to the dangers due to flooding. Therefore, it is reasonable to conclude that any stability thresholds based on experimental data collected using real human test subjects could be misleading and could lead to hazardous and unsafe situations in the case of a real-life flooding (Chanson et al. 2014). This being the case, the predictive ability of both methods was also validated against the experimental data of Xia et al. (2014).

As mentioned, Xia et al. (2014) used a scaled human body (i.e. a dummy) for their experiments. This meant that the scaled human body could not readjust its position or learn how to manoeuvre itself in the flow with time, and it could not be influenced by any sort of safety feature. This being the case, the experimental results obtained in this study tend to be more conservative from the flood risk point of view when compared to the experimental data obtained in the studies using real human test subjects (Xia et al. 2014). Figure 8 shows a comparison of the flood hazard prediction ability of the two methods using the experimental data of Xia et al. (2014). The blue circles identify the conditions when stability was lost due to the toppling instability mechanism, i.e. the flow conditions which led to instability of the human model due to the momentum. In Fig. 8, it can be seen that the empirically derived method classified the majority of the data as of low hazard, which indicates that the empirically derived method generally under-predicts the flood hazard indices. This is probably due to the fact that the experimental data of Abt et al. (1989) and Karvonen et al. (2000) were used in the derivation process for the empirical method, and therefore the ability of the test subjects to learn how to manoeuvre in the flow is incorporated in the empirically derived method (Cox et al. 2010). However, as mentioned earlier this means that the flood hazard assessment with the empirically derived method tends to be too optimistic regarding safety. Furthermore, it can also be seen that the empirically derived method showed a higher flood hazard index (i.e. moderate hazard) for the tests where the stability was lost in the sub-critical flow regime, whereas almost all tests in which the stability was lost in the super-critical flow regime were classified as low hazard. This indicates that the empirically derived method, despite the potential shortcomings, still better assesses flood hazard indices for relatively slowly changing flow conditions than for the high Froude number flows, i.e. for the rapidly varying velocities that often occur with extreme flood events.

Figure 8 also shows that the physically based and experimentally calibrated method agreed well with the experimental observations, as 20 out of 46 experimental observations can be found above the stability threshold (red line), which indicated that stability was lost, and with the vast majority of the remaining experimental observations being in the relative proximity of the stability threshold, which indicated that the model was close to losing stability, i.e. was greatly endangered by the flow conditions. In Fig. 8, it can also be seen that the physically based and experimentally calibrated method adapts well to high Froude number flows, which are generally characterized with abrupt changes in the flow regime and often occur with extreme flood events. From the experimental observations 17 out of 31 in the super-critical flow regime can be found above the stability threshold (red line), while the remaining experimental observations in the super-critical flow regime are in the low hazard stability regime. Thus, the physically based and experimentally calibrated method assesses the flood hazard indices much better for super-critical flows when compared to the empirically derived method.

In summary, the comparisons of the two methods with the experimental data have shown that the physically based and experimentally calibrated method is highly adaptable to the characteristics of a particular human body, which allows more accurate assessment of flood hazard indices to be made when compared to the empirically based method. Furthermore, the physically based and experimentally calibrated method also generally agreed better with the observations which recorded the loss of stability in the super-critical flow regime when compared to the empirically based method. Therefore, the results obtained in the comparisons with the experimental observations suggest that the methods derived from a mechanics-based analysis, such as the physically based and experimentally calibrated method considered in this study, would be more appropriate for flood hazard assessment of extreme flood events than the classically used methods, such as the empirically based method considered herein.

6.2 Comparisons of the flood hazard indices assessed using the two methods in the numerical simulations of extreme flood events

Two extreme flood events were considered herein: the 2010 Kostanjevica na Krki large river flood and the 2007 Železniki flash flood. As the empirical method considered herein is based on the experimental data of Abt et al. (1989) and Karvonen et al. (2000) (e.g. real human body), the calibrating parameters α and β used to calibrate the physically based and experimentally calibrated method for the numerical simulations of extreme flood events were also based on the same datasets, i.e. a real human body (see Table 2). This being the case, neither of the considered methods were ideal in terms of being able to predict accurately to the calibration dataset.

Figures 9 and 11 show a comparison between the empirically derived and the physically based and experimentally calibrated flood hazard assessment method for the 2010 Kostanjevica na Krki extreme river flood event.

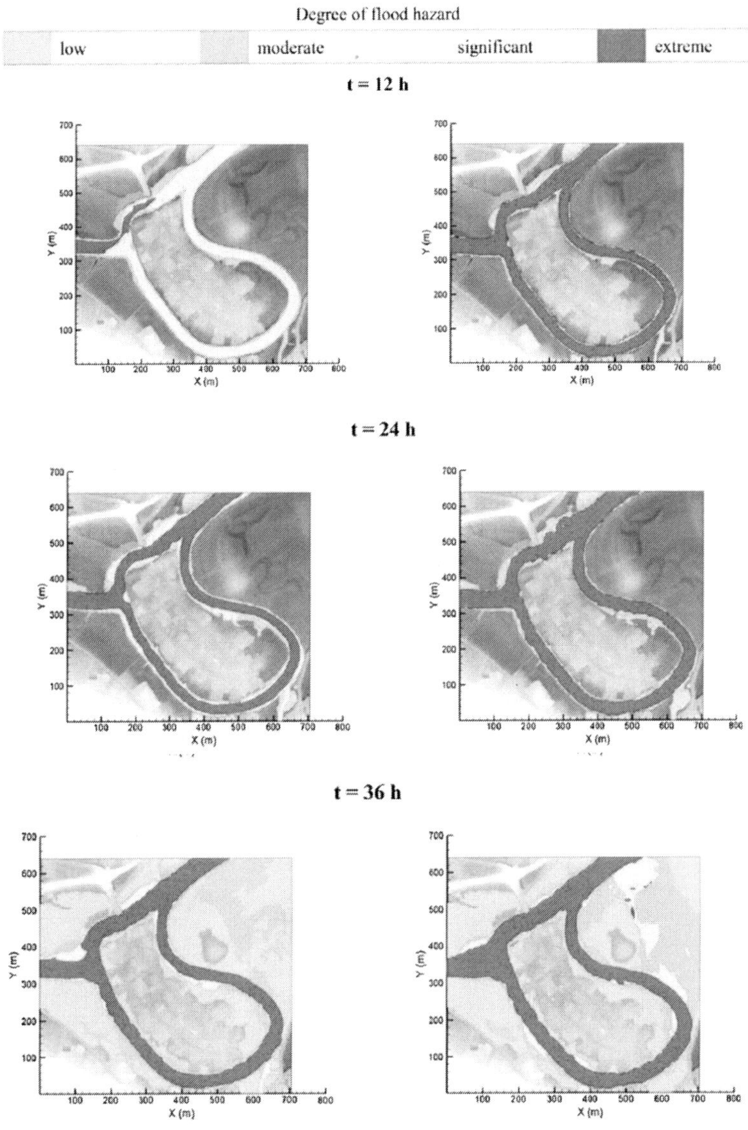

Figure. 9 Flood hazard rating at 12, 24 and 36 h after the start of the 2010 Kostanjevica na Krki flood simulation according to the empirically derived method (*left*) and the physically based experimentally calibrated method (*right*)

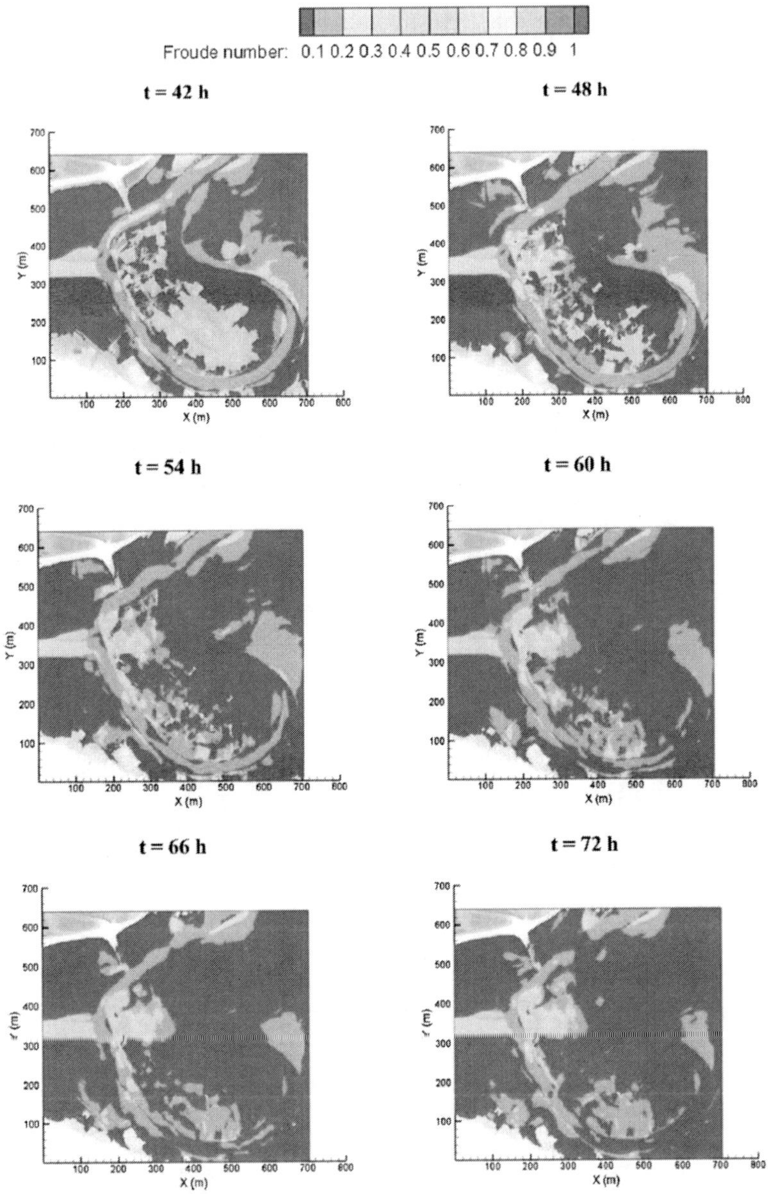

Figure. 10 Froude number for the last 36 h of the 2010 Kostanjevica na Krki flood simulation

A step-by-step presentation of the assessed flood hazard shows that both methods were assessing a relatively similar degree of flood hazard for the first half of the simulation, i.e. for the first 36 h (see Fig. 9). This is not surprising, as the flood was gradually increasing in intensity during the first half of the simulation period and therefore both methods could easily replicate the relatively gradual changes in the flow regime. However, the flooding becomes much more intense during the second part of the simulation, i.e. over the last 36 h (see Fig. 10). This clearly reflects on the flood hazard assessment predictions, as it can be seen in Fig. 11. In the last 36 h of the 2010 Kostanjevica na Krki flood simulation scenario, the physically based and experimentally calibrated method assessed a higher degree of flood hazard at every time mark when compared to the flood hazard assessment obtained with the empirically derived method. These results were expected, as the physically based and experimentally calibrated method is based on being linked to the momentum of the flow, which is proportional to the square of the velocity, as compared to the empirically derived formulation, which is far less sensitive to the velocity of the flood flow. These results suggest that the empirically derived flood hazard assessment method cannot accurately assess the degree of flood hazard for high Froude number flows, or violent flood events, and that flood hazard assessment methods based on a mechanics-based analysis, such as the physically based and experimentally calibrated method considered herein, should be used for flood hazard assessment of extreme flood events and where the flood velocity of flow is relatively high (Fig. 12).

Figures 13 and 14 show a comparison between the empirically derived, and the physically based and experimentally calibrated flood hazard assessment method for the 2007 Železniki flash flood event. Step-by-step presentation of the assessed flood hazard shows that the results from the two selected methods differ from the beginning of the flood simulation. In Figs. 13 and 14, it can be seen that the physically based and experimentally calibrated method predicted higher flood hazard indices at every stage of the simulation when compared to the results obtained with the empirically derived method. These results were expected, as the physically based and experimentally calibrated method is much more influenced by higher velocities and momentum, associated with higher Froude number flows and which occur more frequently with flash floods (see Fig. 12). On the other hand, the empirically derived method is rather a simple approach for flash flood events, or other extreme flood events where the Froude number is relatively high, with the method being a function of the velocity only, vis-à-vis the square of the velocity for the physics based method.

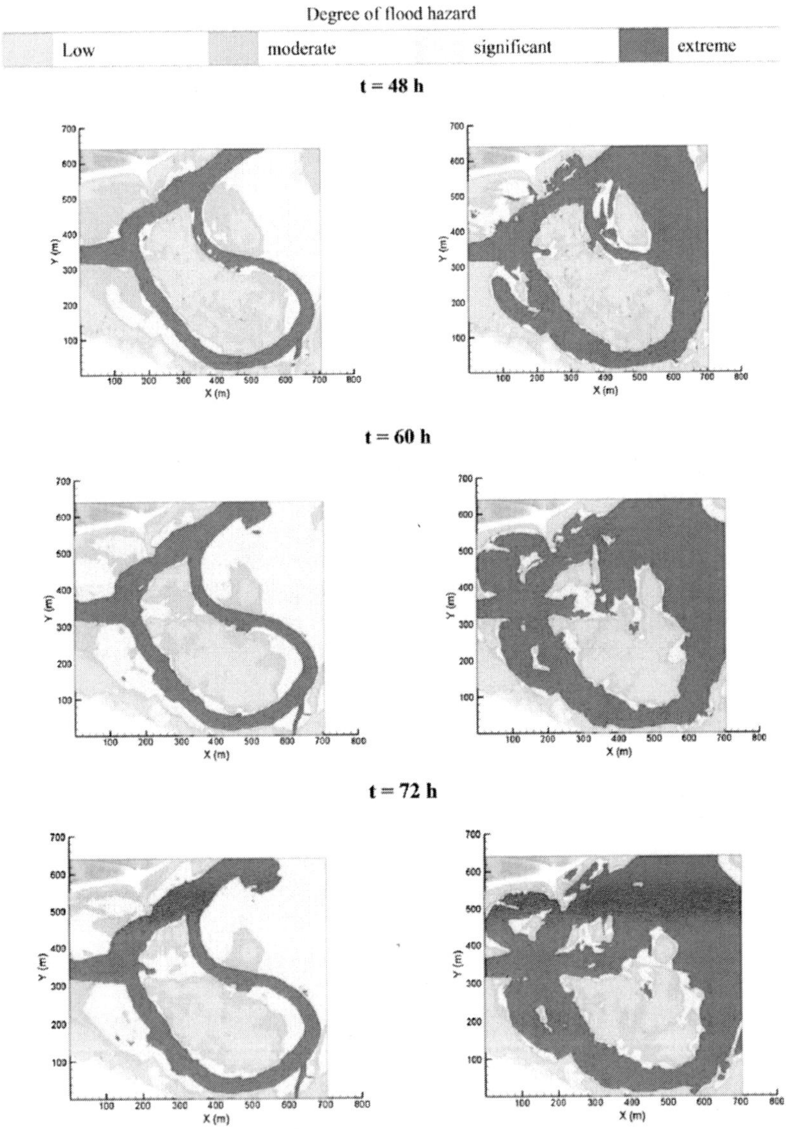

Figure. 11 Flood hazard rating at 48, 60 and 72 h after the start of the 2010 Kostanjevica na Krki flood simulation according to the empirically derived method (*left*) and the physically based experimentally calibrated method (*right*)

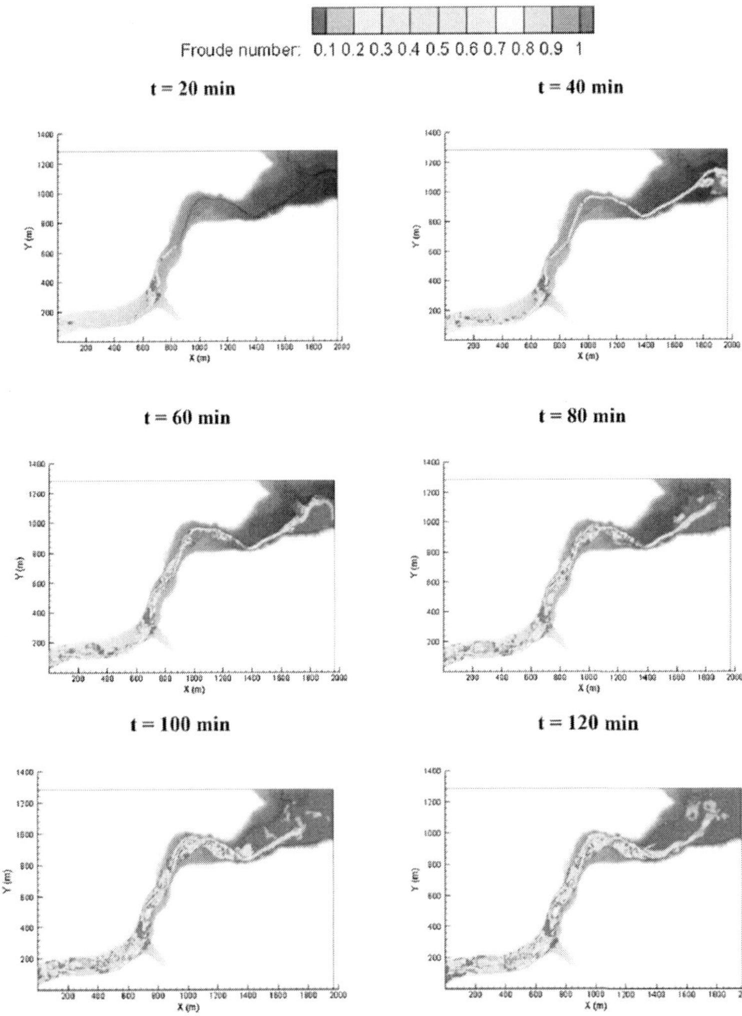

Figure. 12 Froude number for the 2007 Železniki flood simulation

All in all, these results agree well with the results obtained for the Kostanjevica na Krki extreme river flood event and further indicate that flood hazard assessment methods based on a mechanics-based analysis, such as the physically based and experimentally calibrated method considered herein, should be used for flood hazard assessment of extreme flood events.

It should be again noted that the empirically derived flood hazard assessment method presented herein is well established in the UK and that flood hazard maps issued by the governing authorities in the UK, e.g. DEFRA, are based on this criterion. The formulae based on empirical or quasi-theoretical studies, such as the empirically derived method considered in this study, are suitable and accurate for low land floods, with a low Froude number.

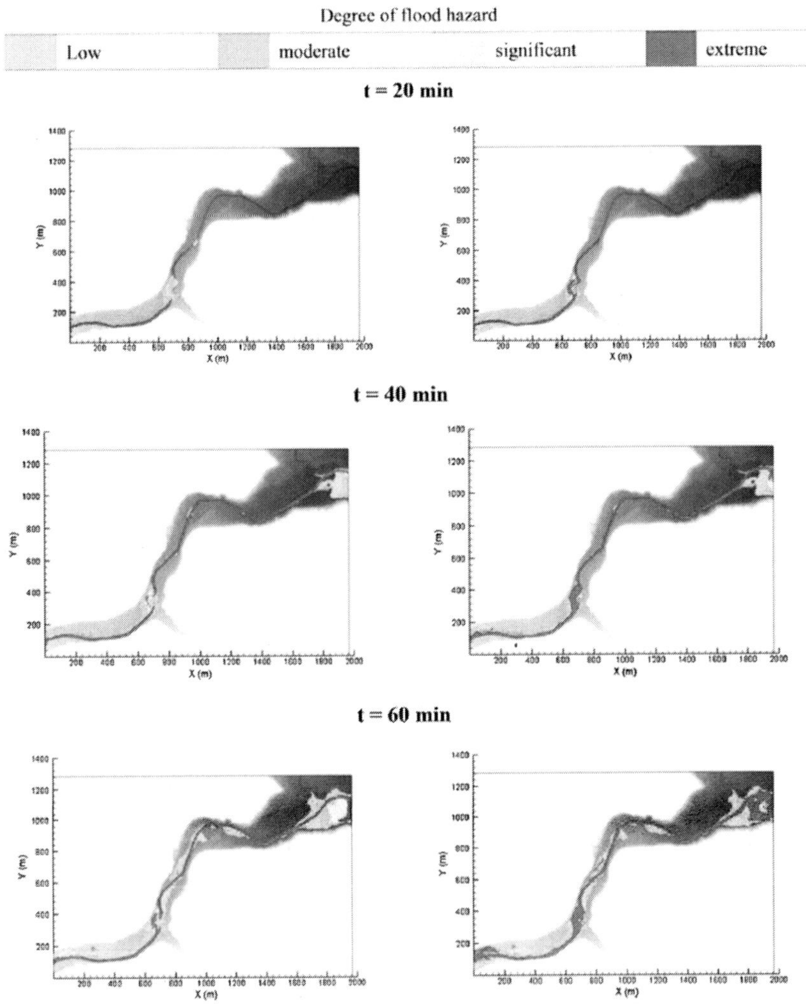

Figure. 13 Flood hazard rating at 20, 40 and 60 min after the start of the 2007 Železniki flash flood simulation according to the empirically derived method (*left*) and the physically based method and experimentally calibrated method (*right*)

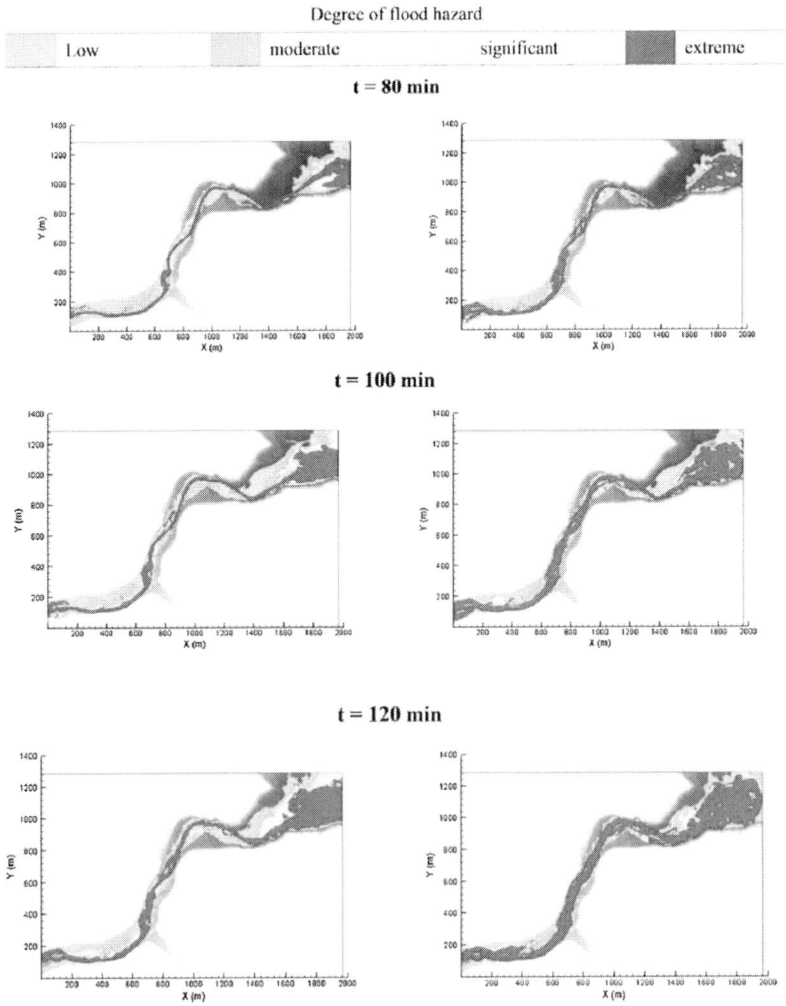

Figure. 14 Flood hazard rating at 80, 100 and 120 min after the start of the 2007 Železniki flash flood simulation according to the empirically derived method (*left*) and the physically based experimentally calibrated method (*right*)

However, for flood events where the velocity conditions change rapidly and the Froude numbers are relatively large, such as during flash floods or extreme river floods (see Figs. 10 and 12), these methods fail to accurately predict flood hazard indices. Instead, a flood hazard assessment in areas prone to flash flooding, or in areas known for large-scale and violent river flooding, should be undertaken using flood hazard assessment methods based on a mechanics-based analysis, because these methods are able to efficiently take into account rapid changes in the flow regime and enable a rapid assessment of the degree of flood hazard risk in a short time period.

Although the differences between the tested two flood hazard assessment methods might seem insignificant for some cases, the difference in the accuracy could be a crucial factor when it comes to real-life rescue actions and the need to decide on the priority areas for the emergency services, etc. The rescue services can acquire much more accurate and meaningful information from the step-by-step presentation of the development of a potential flood event, or from a video simulation of flood propagation, using the physically based approach for flood hazard prediction, as compared to standard flood hazard maps. This is particularly important in the case of violent flood events, such as flash flooding, as it allows such flood hazards to be more accurately determined and for the emergency services, etc., to determine how much time they have for a rescue operation, as well as determining the optimum rescue routes from a flood prone region. Therefore, flood hazard assessment methods based on a mechanics-based analysis, such as the physically based and experimentally calibrated method considered herein, could provide an additional response time and more efficient deployment of the rescue services, particularly during the most critical stages of flooding.

Even though the use of more sophisticated flood hazard assessment methods can improve the prediction of flood hazard indices, the precondition for an adequate flood hazard assessment is still an accurate flood inundation model. Firstly, there is a need to use appropriate flood inundation models and flood risk assessment techniques when modelling specific flood events, such as flash flooding (Kvočka et al. 2015). Secondly, an important factor in flood inundation modelling, and thus flood hazard assessment processes, is also the selected grid size. In the study conducted by Smith and Wasko (2012), it was shown that model resolutions of up to 10 m were adequate for representing peak flood levels, whereas model resolutions of 2 m or less were required to represent the complex flow patterns in urban areas. This being the case, a change in the grid resolution can have a significant effect on the predicted flow velocities, flow directions, flow discharge distributions and ultimately on the prediction of flood hazard indices (Smith and Wasko 2012). Therefore, any flood hazard assessment should not be focused exclusively on the selection of an appropriate flood hazard assessment method, but it should also take into account the complexity of the modelling area and the nature of the considered flood flow (e.g. the expected hydraulic characteristics of the considered flood flow).

7. CONCLUSIONS

In this paper, we have investigated two very different flood hazard assessment methods and analysed the different predictions for flood hazard assessment in the areas where extreme flood events are likely to occur, i.e. in areas prone to flash flooding or large river flooding. Two flood hazard assessment methods were tested, including: (1) a widely used, empirically derived method and (2) a recently introduced, physically based and experimentally calibrated method. The two flood hazard assessment methods considered herein were first evaluated against three different experimental datasets, including two datasets based on testing real human subjects and one dataset based on experiments using model human bodies, and later used to assess the flood hazard rating for two extreme flood events, i.e. the 2010 Kostanjevica na Krki extreme river flood event and the 2007 Železniki flash flood.

The results obtained in this study show that in areas prone to extreme flooding, the flood hazard indices should be predicted based on methods derived from a mechanics-based analysis, such as the physically based and experimentally calibrated method considered in this study. Such methods have a number of benefits since they: (1) consider all of the physical forces acting on a human body in floodwaters, (2) are able to efficiently take into account rapid changes in the flow (or velocity) regime, something that usually occurs during flash flooding or large river flooding, and (3) are very sensitive to significant changes locally in the hydrodynamic conditions of the flow. Furthermore, the recently introduced flood hazard assessment methodologies derived from a mechanics-based analysis, such as the method by Xia et al. (2014) considered herein, or the methodology proposed by Milanesi et al. (2015), can be adjusted to: (1) a specific body type, such as Asian, Caucasian or African-American, (2) a specific region, such as Europe, USA or China and (3) a specific sub-population group, such as men, women or children. This being the case, the criterion derived from a mechanics-based analysis can provide valuable information on the response time and enable more efficient rescue operations, as they are highly adaptable to extreme flood conditions and human body characteristics, and can therefore provide an accurate real-time assessment of the risk to people caused by real-time extreme flooding.

Even though extreme flood events rarely occur, they are the most devastating and life-threatening form of all natural hazards. This being the case, there is a need to improve our knowledge of predicting extreme weather events, extreme flooding processes and flood hazard assessment methodologies, in particular for areas prone to extreme flooding. The development of recently introduced physically based methods, such as the method considered herein or the methodology presented by Milanesi et al.

(2015), has shown the importance of the physical interpretation and the need for a realistic description of the processes that influence human stability in floodwaters. However, future research in the field of flood hazard assessment should not focus solely on the hydraulic parameters that influence the human stability in floodwaters, but should also: (1) include the demographic and psychological characteristics of the general population in improved, or newly proposed, flood hazard assessment criteria, and (2) consider other indirect parameters (such as grid resolution) in the flood hazard assessment process. As indicated herein, and also by other authors, these characteristics are sometimes even more important than the hydraulic characteristics of the flood flow and therefore should be considered in more detail in future studies. Although extreme flooding cannot be prevented, more can be done in the future to limit the consequence of these flood events, especially by using appropriate flood inundation prediction models and improved flood hazard assessment methodologies.

ACKNOWLEDGMENTS

This research project was funded by High Performance Computing (HPC) Wales and Fujitsu, part funded by the European Regional Development Fund and European Social Fund. The bathymetric data for the Kostanjevica na Krki and the Železniki study were provided by the Slovenian Environment Agency. The contribution of all of these organizations is gratefully acknowledged. The authors would especially like to thank Professor Junqiang Xia, Wuhan University, for providing the experimental data and all of his additional help. Finally, the authors would also like to thank Gergana Stefanova Nikolova, Žiga Šebenik and Špela Tomažič for their comments and assistance.

REFERENCES

1. Abt SR, Wittier RJ, Taylor A, Love DJ (1989) Human stability in a high flood hazard zone. J Am Water Resour As 25:881–890
2. Alfieri L, Burek P, Feyen L, Forzieri G (2015) Global warming increases the frequency of river floods in Europe. Hydrol Earth Syst Sci 19:2247–2260
3. Amengual A, Homar V, Jaume O (2015) Potential of a probabilistic hydrometeorological forecasting approach for the 28 September 2012 extreme flash flood in Murcia, Spain. Atmos Res 166:10–23
4. Androulidakis YS, Kombiadou KD, Makris CV, Baltikas VN, Krestenitis YN (2015) Storm surges in the Mediterranean Sea: variability and trends under future climatic conditions. Dyn Atmos Oceans 71:56–82

5. Antico A, Torres ME, Diaz HF (2016) Contributions of different time scales to extreme Paraná floods. Clim Dyn 46(11):3785–3792
6. Ashley ST, Ashley WS (2008) Flood fatalities in the United States. J Appl Meteorol Climatol 47:805–818
7. Beniston M (2009) Trends in joint quantiles of temperature and precipitation in Europe since 1901 and projected for 2100. Geophys Res Lett 36:L07707
8. Bergman N, Sholker O, Roskin J, Greenbaum N (2014) The Nahal Oz Reservoir dam-break flood: geomorphic impact on a small ephemeral loess-channel in the semi-arid Negev Desert, Israel. Geomorphology 210:83–97
9. Bouilloud L, Delrieu G, Boudevillain B, Borga M, Zanon F (2009) Radar rainfall estimation for the post-event analysis of a Slovenian flash-flood case: application of the Mountain Reference Technique at C-band frequency. Hydrol Earth Syst Sci 13:1349–1360
10. Brakenridge G (2012) Global Active Archive of Large Flood Events, Dartmouth Flood Observatory, University of Colorado [Online]. http://floodobservatory.colorado.edu/Archives/index.html
11. Breilh J-F, Bertin X, Chaumillon É, Giloy N, Sauzeau T (2014) How frequent is storm-induced flooding in the central part of the Bay of Biscay? Glob Planet Change 122:161–175
12. Bruwier M, Erpicum S, Pirotton M, Archambeau P, Dewals B (2015) Assessing the operation rules of a reservoir system based on a detailed modelling chain. Nat Hazards Earth Syst Sci 15:365–379
13. Chanson H (2004) Hydraulics of open channel flow. Heinemann, Butterworth
14. Chanson H, Brown R, Mcintosh D (2014) Human body stability in floodwaters: the 2011 flood in Brisbane CBD. In: Proceedings of the 5th international symposium on hydraulic structures: engineering challenges and extremes, 2014. The University of Queensland
15. Chau VN, Holland J, Cassells S, Tuohy M (2013) Using GIS to map impacts upon agriculture from extreme floods in Vietnam. Appl Geogr 41:65–74
16. Cox R, Yee M, Ball J (2004) Safety of people in flooded streets and floodways. National conference on hydraulics in water engineering, Engineers Australia
17. Cox R, Shand T, Blacka M (2010) Revision project 10: appropriate safety criteria for people. Stage 1 report. Australian rainfall and runoff. Water Research Laboratory, The University of New South Wales
18. Davis SF (1984) TVD finite difference schemes and artificial viscosity. National Aeronautics and Space Administration, Langley Research Center, Hampton, Virginia
19. Di Baldassarre G, Montanari A, Lins H, Koutsoyiannis D, Brandimarte L, Blöschl G (2010) Flood fatalities in Africa: from diagnosis to mitigation. Geophys Res Lett 37:L22402
20. Drillis R, Contini R, Bluestein M (1964) Body segment parameters. Artif Limbs 8:44–66
21. Duffaut P (2013) The traps behind the failure of Malpasset arch dam, France, in 1959. J Rock Mech Geotech Eng 5:335–341
22. Foster D, Cox R (1973) Stability of children on roads used as floodways. Technical Report No.73/13. Water Research Laboratory of the University of New South Wales, Manly Vale, Australia
23. Foudi S, Osés-Eraso N, Tamayo I (2015) Integrated spatial flood risk assessment: the case of Zaragoza. Land Use Policy 42:278–292
24. Foulds SA, Macklin MG, Brewer PA (2014) The chronology and the hydrometeorology of catastrophic floods on Dartmoor, South West England. Hydrol Process 28:3067–3087
25. Godunov SK (1959) A difference method for numerical calculation of discontinuous solutions of the equations of hydrodynamics. Matematicheskii Sbornik 89:271–306
26. Grillakis M, Tsanis I, Koutroulis A (2010) Application of the HBV hydrological model in a flash flood case in Slovenia. Nat Hazards Earth Syst Sci 10:2713–2725
27. Güneralp B, Güneralp İ, Liu Y (2015) Changing global patterns of urban exposure to flood and drought hazards. Glob Environ Change 31:217–225
28. Harten A (1983) High resolution schemes for hyperbolic conservation laws. J Comput Phys 49:357–393
29. Herget J, Kapala A, Krell M, Rustemeier E, Simmer C, Wyss A (2015) The millennium flood of July 1342 revisited. Catena 130:82–94

30. Hunter N, Bates P, Neelz S, Pender G, Villanueva I, Wright N, Liang D, Falconer RA, Lin B, Waller S (2008) Benchmarking 2D hydraulic models for urban flood simulations. In: Proceedings of the institution of civil engineers: water management, 2008. Thomas Telford (ICE publishing) pp 13–30
31. Ishigaki T, Baba Y, Toda K, Inoue K (2005) Experimental study on evacuation from underground space in urban flood. In: Proceedings of 31st IAHR Congress on CD-ROM, Seoul, 2005
32. Ishigaki T, Kawanaka R, Onishi Y, Shimada H, Toda K, Baba Y (2009) Assessment of safety on evacuating route during underground flooding. In: Advances in water resources and hydraulic engineering. Springer, Berlin, pp 141–146
33. Jonkman SN (2005) Global perspectives on loss of human life caused by floods. Nat Hazards 34:151–175
34. Jonkman S, Penning-Rowsell E (2008a) Human instability in flood flows. J Am Water Resour As 44:1208–1218
35. Jonkman SN, Penning-Rowsell E (2008b) Human instability in flood flows. J Am Water Resour As 44:1208–1218
36. Karvonen R, Hepojoki H, Huhta H, Louhio A (2000) The use of physical models in dam-break flood analysis, Development of Rescue Actions Based on Dam-Break Flood Analysis (RESCDAM). Final report of Helsinki University of Technology. Finnish Environment Institute
37. Kaźmierczak A, Cavan G (2011) Surface water flooding risk to urban communities: analysis of vulnerability, hazard and exposure. Landsc Urban Plan 103:185–197
38. Keller RJ, Mitsch B (1993) Safety aspects of the design of roadways as floodways. Urban Water Research Association of Australia, Australia
39. Kobold M (2011) Comparison of floods in September 2010 with registered historic flood events (in Slovenian). Ujma 25:48–56
40. Kobold M, Sušnik M, Robič M, Ulaga F, Lalić B (2008) Hydrological analysis of high waters and flash floods occurred in September 2007 in Slovenia. In: IOP conference series: earth and environmental science, 2008. IOP Publishing, 012008
41. Koks EE, Jongman B, Husby TG, Botzen WJW (2015) Combining hazard, exposure and social vulnerability to provide lessons for flood risk management. Environ Sci Policy 47:42–52Kundzewicz ZW, Pińskwar I, Brakenridge GR (2013) Large floods in Europe, 1985–2009. Hydrol Sci J 58:1–7
42. Kvočka D, Falconer RA, Bray M (2015) Appropriate model use for predicting elevations and inundation extent for extreme flood events. Nat Hazards 79:1791–1808
43. Lax PD, Wendroff B (1960) Systems of conservation laws. Commun Pure Appl Math 13:217–237
44. Lenderink G, van Meijgaard E (2008) Increase in hourly precipitation extremes beyond expectations from temperature changes. Nat Geosci 1:511–514
45. Liang D, Falconer RA, Lin B (2006) Comparison between TVD-MacCormack and ADI-type solvers of the shallow water equations. Adv Water Resour 29:1833–1845
46. Liang D, Lin B, Falconer RA (2007a) A boundary-fitted numerical model for flood routing with shock-capturing capability. J Hydrol 332:477–486
47. Liang D, Lin B, Falconer RA (2007b) Simulation of rapidly varying flow using an efficient TVD-MacCormack scheme. Int J Numer Meth Fluids 53:811–826
48. Liang D, Wang X, Falconer RA, Bockelmann-Evans BN (2010) Solving the depth-integrated solute transport equation with a TVD-MacCormack scheme. Environ Model Softw 25:1619–1629
49. Liang D, Xia J, Falconer RA, Zhang J (2014) On the refinement of a boundary-fitted shallow water model. Coast Eng J 56:1450001
50. Lind N, Hartford D, Assaf H (2004) Hydrodynamic models of human stability in a flood. J Am Water Resour As 40:89–96
51. Maccormack RW (1976) An efficient numerical method for solving the time-dependent compressible Navier-Stokes equations at high Reynolds number. NASA Ames Research Center, Moffett Field

52. Marchi L, Borga M, Preciso E, Sangati M, Gaume E, Bain V, Delrieu G, Bonnifait L, Pogačnik N (2009) Comprehensive post-event survey of a flash flood in Western Slovenia: observation strategy and lessons learned. Hydrol Process 23:3761–3770
53. Martínez Ibarra E (2012) A geographical approach to post-flood analysis: the extreme flood event of 12 October 2007 in Calpe (Spain). Appl Geogr 32:490–500
54. Milanesi L, Pilotti M, Ranzi R (2015) A conceptual model of people's vulnerability to floods. Water Resour Res 51:182–197
55. Min S-K, Zhang X, Zwiers FW, Hegerl GC (2011) Human contribution to more-intense precipitation extremes. Nature 470:378–381
56. Mingham CG, Causon DM, Ingram D, Mingham CG, Causon DM, Ingram D (2001) A TVD MacCormack scheme for transcritical flow. Proc Inst Civ Eng: Water Marit Eng 148:167–175
57. Ministry of the Environment and Spatial Planning of the Republic of Slovenia (2015) LIDAR [Online]. http://evode.arso.gov.si/indexd022.html?q=node/12
58. Moussa R, Bocquillon C (2009) On the use of the diffusive wave for modelling extreme flood events with overbank flow in the floodplain. J Hydrol 374:116–135
59. Muis S, Güneralp B, Jongman B, Aerts JC, Ward PJ (2015) Flood risk and adaptation strategies under climate change and urban expansion: a probabilistic analysis using global data. Sci Total Environ 538:445–457
60. Neelz S, Pender G (2009) Desktop review of 2D hydraulic modelling packages. Environment Agency, Bristol
61. Nikolova GS, Toshev YE (2007) Estimation of male and female body segment parameters of the Bulgarian population using a 16-segmental mathematical model. J Biomech 40:3700–3707
62. Penning-Rowsell E, Floyd P, Ramsbottom D, Surendran S (2005a) Estimating injury and loss of life in floods: a deterministic framework. Nat Hazards 36:43–64
63. Penning-Rowsell E, Johnson C, Tunstall S, Tapsell S, Morris J, Chatterton J, Green C (2005b) The benefits of flood and coastal risk management: a handbook of assessment techniques. Middlesex University Press, London
64. Purwandari T, Hadi MP, Kingma NC (2011) A GIS modelling approach for flood hazard assessment in part of Surakarta city, Indonesia. Indones J Geogr 43:63–80
65. Ramsbottom D, Floyd P, Penning-Rowsell E (2003) Flood risks to people: Phase 1. R&D Technical Report FD2317, Department for the Environment, Food and Rural Affairs (DEFRA), UK Environment Agency
66. Ramsbottom D, Wade S, Bain V, Hassan M, Penning-Rowsell E, Wilson T, Fernandez A, House M, Floyd P (2006) Flood risks to people: Phase 2. R&D Technical Report FD2321/IR2, Department for the Environment, Food and Rural Affairs (DEFRA), UK Environment Agency
67. Raška P, Emmer A (2014) The 1916 catastrophic flood following the Bílá Desná dam failure: the role of historical data sources in the reconstruction of its geomorphologic and landscape effects. Geomorphology 226:135–147
68. Rojas R, Feyen L, Watkiss P (2013) Climate change and river floods in the European Union: socio-economic consequences and the costs and benefits of adaptation. Glob Environ Change 23:1737–1751
69. Rusjan S, Kobold M, Mikoš M (2009) Characteristics of the extreme rainfall event and consequent flash floods in W Slovenia in September 2007. Nat Hazards Earth Syst Sci 9:947–956
70. Russo B, Gómez M, Macchione F (2013) Pedestrian hazard criteria for flooded urban areas. Nat Hazards 69:251–265
71. Schröter K, Kunz M, Elmer F, Mühr B, Merz B (2015) What made the June 2013 flood in Germany an exceptional event? A hydro-meteorological evaluation. Hydrol Earth Syst Sci 19:309–327
72. Slovenian Environment Agency (2008) Viskoe vode in poplave 18. septembra 2007 (in Slovenian) [Online]. http://www.arso.gov.si/vode/poro%C4%8Dila %20in%20 publikacije/Visoke%20vode%20in%20poplave%2018.%20septembra%202007.pdf

73. Slovenian Environment Agency (2010a) Hidrološko porocilo o povodnji v dneh od 17. do 21. septembra 2010 (in Slovenian) [Online]. http://www.arso.gov.si/vode/ poro%C4%8Dila%20in%20publikacije/Poplave%2017.%20-%2021.%20september%202010.pdf
74. Slovenian Environment Agency (2010b) Porocilo o izjemno obilnih padavinah od 16. do 19. septembra 2010 (in Slovenian) [Online]. http://meteo. arso.gov.si/ uploads/ probase/www/climate/text/sl/weather_events/padavine_16-19sep10.pdf
75. Slovenian Environment Agency (2013) Povratne dobe velikih in malih pretokov za merilna mesta državnega hidrološkega monitoringa površinskih voda (in Slovenian) [Online]. http://www.arso.gov.si/vode/podatki/Povratne%20dobe%20Qvk,Qnp.pdf
76. Smith G, Wasko C (2012) Revision Project 15: two dimensional simulations in urban areas—representation of buildings in 2D numerical flood models. In: Ball J (ed) Australian rainfall and runoff. Engineers Australia, Canberra
77. Takahashi S, Endoh K, Muro Z (1992) Experimental study on people's safety against overtopping waves on breakwaters. Rep Port Harb Inst 34:4–31
78. UN (2014) World Urbanizations prospects: the 2014 revision. Population Division of the Department of Economic and Social Affairs of the United Nations Secretariat, New York
79. UN (2015) World population prospects: the 2015 revision. Population Division of the Department of Economic and Social Affairs of the United Nations Secretariat, New York
80. Xia J, Falconer RA, Lin B, Tan G (2011) Numerical assessment of flood hazard risk to people and vehicles in flash floods. Environ Model Softw 26:987–998
81. Xia J, Falconer RA, Wang Y, Xiao X (2014) New criterion for the stability of a human body in floodwaters. J Hydraul Res 52:93–104
82. Yee M (2003). Human stability in floodways. Undergraduate Honours Thesis, University of New South Wales
83. Zanon F, Borga M, Zoccatelli D, Marchi L, Gaume E, Bonnifait L, Delrieu G (2010) Hydrological analysis of a flash flood across a climatic and geologic gradient: the September 18, 2007 event in Western Slovenia. J Hydrol 394:182–197
84. Zhi-Yong W, Gui-Hua L, Zhi-Yu L, Jin-Xing W, Heng X (2013) Trends of extreme flood events in the Pearl River Basin during 1951–2010. Adv Clim Change Res 4:110–116

INDEX